T0350925

Autonomous Mobile Robots and Multi-Robot Systems

Autonomous Mobile Robots and Multi-Robot Systems

Motion-Planning, Communication, and Swarming

Edited by

Eugene Kagan
Ariel University
Israel

Nir Shvalb
Ariel University
Israel

Irad Ben-Gal
Tel-Aviv University
Israel

This edition first published 2020

Registered Offices
John Wiley & Sons, Inc., 111 River Street, Hoboken, NJ 07030, USA
John Wiley & Sons Ltd, The Atrium, Southern Gate, Chichester, West Sussex, PO19 8SQ, UK

Editorial Office
The Atrium, Southern Gate, Chichester, West Sussex, PO19 8SQ, UK

For details of our global editorial offices, customer services, and more information about Wiley products visit us at www.wiley.com.

Wiley also publishes its books in a variety of electronic formats and by print-on-demand. Some content that appears in standard print versions of this book may not be available in other formats.

Library of Congress Cataloging-in-Publication Data

Names: Kagan, Eugene, editor. | Shvalb, Nir, editor. | Ben-Gal, Irad, editor
Title: Autonomous Mobile Robots and Multi-Robot Systems : Motion-Planning, Communication, and Swarming / [edited by] Eugene Kagan, Ariel University, Nir Shvalb, Ariel University, Irad Ben-Gal, Tel-Aviv University.
Description: First edition. | Hoboken, NJ : John Wiley & Sons, Inc., [2020] | Includes bibliographical references and index. |
Identifiers: LCCN 2019014826 (print) | LCCN 2019017606 (ebook) | ISBN 9781119213178 (Adobe PDF) | ISBN 9781119213161 (ePub) | ISBN 9781119212867 (hardcover)
Subjects: LCSH: Mobile robots–Automatic control. | Multiagent systems. | Swarm intelligence.
Classification: LCC TJ211.415 (ebook) | LCC TJ211.415 .A873 2019 (print) | DDC 629.8/932–dc23
LC record available at https://lccn.loc.gov/2019014826

Cover Design and Image: Luna Romano

Set in 10/12pt Warnock by SPi Global, Pondicherry, India

Printed and bound by CPI Group (UK) Ltd, Croydon, CR0 4YY

10 9 8 7 6 5 4 3 2 1

Contents

List of Contributors

Prof. Irad Ben-Gal
Department of Industrial Engineering and Management
Tel Aviv University
Israel

Dr. Michael Ben Chaim
Department of Mechanical Engineering and Mechatronics
Ariel University
Israel

Prof. Boaz Ben-Moshe
Department of Computer Science
Ariel University
Israel

Dr. Shlomi Hacohen
Department of Mechanical Engineering and Mechatronics
Ariel University
Israel

Dr. Eugene Kagan
Department of Industrial Engineering and Management
Ariel University
Israel

Prof. Eugene Khmelnitsky
Department of Industrial Engineering and Management
Tel Aviv University
Israel

Dr. Simon Lineykin
Department of Mechanical Engineering and Mechatronics
Ariel University
Israel

Dr. Oded Medina
Department of Mechanical Engineering and Mechatronics
Ariel University
Israel

Dr. Alexander Novoselsky
Department of Earth and Planetary Sciences
Weizmann Institute of Science
Israel

Prof. Nir Shvalb
Department of Mechanical Engineering and Mechatronics
Ariel University
Israel

Dr. Shraga Shoval
Department of Mechanical Engineering and Mechatronics
Ariel University
Israel

Dr. Roi Yozevitch
Department of Computer Science
Ariel University
Israel

Preface

In the famous lecture "The question concerning technology,"[1] Martin Heidegger argues that (p. 12):

> Technology is a way of revealing. If we give heed to this, then another whole realm for the essence of technology will open itself up to us. It is the realm of revealing, i.e., of the truth.

And, certainly, robotics is not an exclusion. While the things are the material implementations of mathematical abstractions, the robots, and especially mobile robots and mobile robot systems, imply our imagination of motion.

Attempts at creating tools that can autonomously execute certain tasks can be tracked back to the ancient Greek philosophers and Egyptian inventors. In the Middle Ages and then in the new era, such mechanisms were enriched with mobile devices mimicking animals and humans, and the first efforts of building flying machines were conducted. Certainly, with the invention of steam and gasoline engines and electric motors, these devices became more complicated; however, despite the fact that even the simplest of them demonstrate all the main features of the modern machinery, none of them can be considered as a *robot*[2] in its modern sense.

The modern history of robotics began in the late 1940s, when slave arm manipulators were used in nuclear manufacturing. Following Bernard Roth[3] (p. V):

> The first academic activity was the thesis of H. A. Ernst, in 1961, at MIT. He used a slave arm equipped with touch sensors, and ran it under computer control. The idea in his study was to use the information from the touch sensors to guide the arm.

Probably, the idea of usage of the sensed information formed a basis for robotics, while such an informational feedback distinguishes the robots from the other automatic

1 Heidegger, M. (1954). The question concerning technology. *Technology and Values: Essential Readings.* 99 113; quoted from Heidegger, M. (1977). *The Question Concerning Technology and Other Essays.* New York and London: Garland Publishing, Inc.

2 The term *robot,* which indicates a machine that could perform the human's work like a human, was suggested in 1921 by the Czech writer Karel Capek in his play *R.U.R.* (*Rossum's Universal Robots*).

3 Siciliano, B., Khatib, O. (Eds.) (2008). *Springer Handbook of Robotics.* Springer: Berlin.

tools and machines. In particular, the definition of *robot* was suggested by Vladimir Lumelsky[4] (p. 15):

> A robot is an automatic or semiautomatic machine capable of purposeful motion in response to its surroundings in an unstructured environment.

For practical needs, the book follows this definition.

The book is concentrated on positioning and motion planning of mobile robots with respect to available information about their location in space and communication and sensing abilities. The first part of the book considers the models and algorithms of navigation and motion planning in global coordinates system with complete information about the robot's location and velocity. The second part deals with motion in the potential field, which is defined by the environmental states of the robot's expectations and knowledge. The third part addresses the robot's motion in unknown environments and the corresponding tasks of environment mapping using sensed information. Finally, the fourth part of the book considers two- and three-dimensional multi-robot systems and swarm dynamics.

The book provides theoretical and practical guidance and ready-to-use algorithms for navigation of mobile robots, which can be directly implemented in the laboratory and can be used as a starting point for further research and as a basis for solving engineering tasks. As a textbook, it is aimed to support the course in mobile robotics for undergraduate and graduate students specializing in applied mathematics and engineering, and is structured with respect to the program of a single semester course including complete theoretical material and algorithmic solutions. As a research text, the book is aimed to provide a starting point for the research in different directions of mobile robots navigation and can be used as a guide in the field. As a practical guide, the book is self-contained and includes ready-to-use algorithms, numerical examples, and simulations, which can be directly implemented in both simple and advanced mobile robots and applied to different tasks with respect to the available information and robots' sensing abilities.

We hope that the book will guide the reader over different approaches to mobile robot systems and will inspire further developments in the field of navigation of mobile robots in unstructured environments.

Eugene Kagan, Nir Shvalb and Irad Ben-Gal
November 2018, Ariel – Tel-Aviv, Israel

---

4 Lumelsky, V. (2006). *Sensing, Intelligence, Motion. How Robots and Humans Move in an Unstructured World.* Wiley-Interscience: Hoboken, NJ.

Acknowledgments

The idea of preparing this book was inspired by the brief course given in 2013 by Prof. Vladimir Lumelsky at the Shenkar College, Ramat-Gan, Israel. The editors thank Prof. Lumelsky for this inspiration and for the ideas that form a basis for the book.

The authors and editors of the book are in great debt to the many people who influenced our understanding of mobile robot systems and contributed to the algorithms and projects. Especially, we thank Prof. Boaz Golany, Prof. Zvi Shiller, Prof. Hava Siegelmann, Dr. Nahum Kogan (ז"ל), Dr. Alexander Rybalov, and Mr. Sergey Khodorov for numerous discussions.

Certainly, nothing could be done without our students Rottem Botton, Rakia Cohen, Hadas Danan, Shirli Dari, Dikla El-Ani, Liad Eshkar, Chen Estrugo, Gal Goren, Idan Hammer, Moshe Israel, Mor Kaiser, Stav Klauzner, Sharon Makmal, Yohai Maron, Harel Mashiah, Noa Moran, Elad Mizrahi, Eynat Naiman, Alon Rapoport, Amir Ron, Eynat Rubin, Emmanuel Salmona, Alon Sela, Michal Shor, Jennie Steshenko, Tal Toibin, Dafna Zamir, and Hodaya Ziv. We thank them all for their practical work and for their questions, remarks, and ideas.

About the Companion Website

This book is accompanied by a companion website:

www.wiley.com/go/kagan/robotsystems

The website includes:

- C code used for navigation of Kilobots
- C and C++ code for navigation of Lego NXT robots
- Matlab code used in simulations and figures

Introduction

Eugene Kagan, Nir Shvalb, and Irad Ben-Gal

I.1 Early History of Robots

The first known robot, a mechanical bird known as the "the Pigeon" actuated by a water stream, was built in the fourth century BCE by the Greek mathematician Archytas of Tarentum.[1] Archytas constructed his robo-bird out of wood and used steam to power its movements. A century and a half after that, in 250 BCE, the Greek physicist Ctesibius of Alexandria proposed another machine "clepsydra," – literally, a *water clock* – that formed the main driving principles for mechanical devices of the ancient world. At about the same time (150–100 BCE), analog computation mechanisms were used to accurately compute astronomical events. The Antikythera clockwork mechanism was composed of at least 30 meshing bronze gears. At the beginning of the Current Era, the Greek mathematician and engineer Heron of Alexandria (approximately 10–70 CE) described most of mechanisms of this kind and his own inventions, including the rocket-reaction engine "Aeoliple"[2] and the wind-powered organ, in the books *Pneumatics* and *Automata*[3] (Hero of Alexandria 2009). These books predefined the development of the automated mechanical tools for more than a thousand years, even the Leonardo da Vinci (1452–1519) inventions.

Credit for the first sign of changing the ancient Greek paradigm of robotics is often given to the mathematician and inventor Al-Jazari (1136–1206), an engineer of the Artuqid Palace in Diyarbakir, Turkey.[4] In the *Book of Knowledge of Ingenious Mechanical Devices* (al-Jazari 1974), among many novel automata he described a boat with four automatic musicians such that the rhythms and drums could be reprogrammed by changing the configuration of pegs or cams. In spite of the elementary nature of such programming, this band can be considered as a prototype of the further programmable devices, including machines with numerical control and programmable manipulators.

1 The town in southern Italy, which in that time was the Greek colony of Taras.
2 That is, the ball of the god of the air and wind.
3 The word *automaton* (pronounced "automatos" in Greek) was suggested by Homer in *Iliad* and literally means "acting by itself."
4 At that time, the regional residence of the Artuqid (pronounced "Artuklu" in Turkish) dynasty who ruled eastern Anatolia and Jazira.

Autonomous Mobile Robots and Multi-Robot Systems: Motion-Planning, Communication, and Swarming, First Edition. Edited by Eugene Kagan, Nir Shvalb and Irad Ben-Gal.

Companion website: www.wiley.com/go/kagan/robotsystems

With the development of mechanics and, in particular, of the mechanical watch technology in the eighteenth century, automated devices became more precise and stable and earned the relative freedom of movement. However, the most significant innovation in robotics of this time was the invention of the automated loom, programmed using the punched cards, made in 1801 by the French weaver and merchant Joseph Marie Charles, known as Joseph Jacquard (1752–1834). His mechanical loom was, certainly, the first comprehensive version of the programmable machine. Twenty years later, in 1822, the English mathematical and engineer Charles Babbage (1791–1871) applied the same programming principle based on punched cards in the first programmable computer "Difference Engine," while Babbage's collaborator Augusta Ada King, Countess of Lovelace, known as Ada Lovelace (1815–1852), suggested the first algorithm[5] of mechanical computation, commonly recognized as the first computer programmer.

In parallel with the development of the automation, steam engines improved dramatically during the eighteenth to nineteenth centuries.[6] A crucial point in these investigations was a steam engine patented in 1781 by the Scottish engineer James Watt (1736–1819). In contrast to its predecessors, Watt's engine produced continuous rotary motion that formed a basis for wide application of the steam power. In addition, in 1788, Watt designed the first automatic regulator or moderator, also known as a centrifugal governor, for preserving the motion speed using mechanical feedback control. In 1868, the English physicist James Clerk Maxwell (1831–1879), in his report to the Royal Society "On governors" (Maxwell 1868), suggested mathematical principles of the theory of automatic regulation, or the theory of automatic control that provides the methods for preserving stable operating of the machines. These methods, together with the Jacquard's principles of programmable operations, form a basis of modern robotics.

An invention of electricity led to substitution of the mechanical and heat engines by electric motors and quick changes in the technology of automated tools, and even a possibility of remote control demonstrated in 1898 by the Serbian American inventor Nikola Tesla (1856–1943). However, the main principles of automatic machines inheriting the inventions of Heron, Jacquard, and Maxwell were still unchanged, up to the middle of the twentieth century. That changed with the publication in 1942 of the Isaac Asimov story "Runaround" and the later book *I, Robot* (Asimov 1950), which complicated things.

1.2 Autonomous Robots

In 1945, while developing of the digital electronic computer, John von Neumann (1903–1957) prepared the "First Draft of a Report on the EDVAC" (von Neumann, First Draft of a Report on the EDVAC, 1945),[7] in which he suggested the model of

---

5 The term *algorithm* is a transliteration of the name of the Persian Uzbek mathematician and astronomer Al Khwarizmi (approximately 780–850) who, in his *Compendious Book on Calculation by Completion and Balancing* suggested formal procedures of solving linear and quadratic equations and recognized the difference between nonexistence and zero quantity.

6 The first steam engine was patented in 1606 by the Spanish inventor Jeronimo de Ayanz y Beaumont (1553–1613), and a commercial version of the steam engine with a steam pump was patented in 1698 by the English engineer Thomas Savery (1650–1715).

7 The abbreviation EDVAC stands for the Electronic Discrete Variable Automatic Computer.

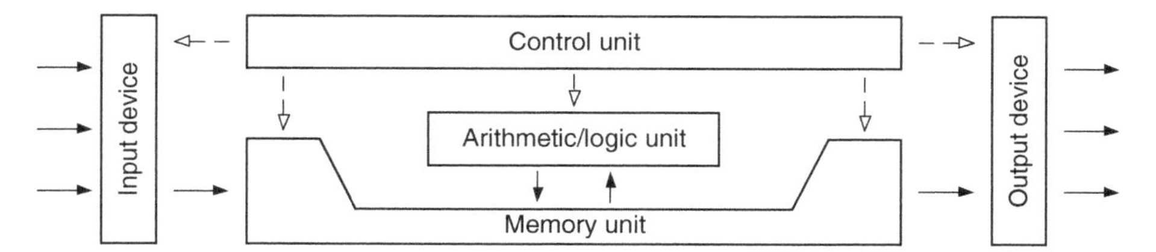

Figure I.1 The von Neumann architecture of digital computer.

the computer currently known as the von Neumann architecture or the Princeton architecture.[8] The scheme of this model is shown in Figure I.1.

According to this scheme, computations are executed by the arithmetic/logic unit following the instructions and using the data, which are received via the input device and stored in the memory unit. The results of the computations are passed to the user via the output device. The control unit synchronizes the activities of the parts of the computer. Later, another model known as the Harvard architecture was suggested, in which the instructions and the data are stored in separate memory units.

In the original model, it was assumed that the input device would be able to receive numerical or other information that "can be produced more or less directly by human action (typing, punching, photographing light impulses produced by keys of the same type, magnetizing metal tape or wire in some analogous manner, etc.)," and the output device would be able to transfer "presumably only numerical information" (von Neumann, First Draft of a Report on the EDVAC, 1945, p. 7).

With the development of the computers, the functions of the computer parts were strongly improved, and, in particular, the input and output devices were enriched by various sensing and actuating tools, which allowed receiving information both about internal states of the computer and external states of the environment and transmitting non-numerical information and actuating certain devices both inside and outside of the computer. Nowadays, the sensors that obtain information regarding internal states are often called proprioceptive sensors, and the sensors that obtain information regarding the states of the external environment are called exteroceptive sensors (Bekey 2005).

Usually, the proprioceptive sensors are implemented together with the internal effectors that, by respective feedback control, change the internal states of the system and return it as close as possible to the functionally steady state. The simplest example of the proprioceptive sensor is a thermometer that measures the temperature of the processor and leads to starting the cooler if the temperature becomes higher than a certain threshold value. A more complex example of the feedback control is the performance manager, which itself is a program that initiates swapping data when the memory usage overpasses the allowed percentage. After publication (1948) of the famous *Cybernetics* by Norbert Wiener (1894–1969), the property of the system to preserve its steady state by slow feedback control became known as homeostasis, and nowadays it is considered as an essential principle in the design of autonomous machines.

---

8 Following Norbert Wiener (1948), the similar principles of the computer architecture were formulated in the report sent by the Wiener group to Dr. Vannevar Bush in 1940.

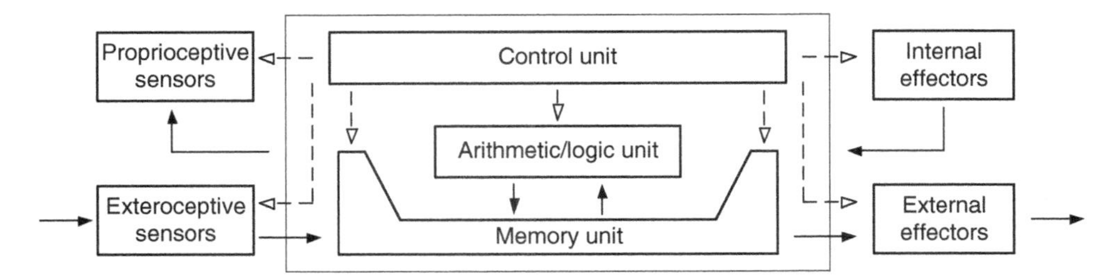

Figure I.2 The von Neumann-based architecture of robot.

Excluding user-oriented microphones and cameras, typical computers are not equipped with specific exteroceptive sensors, and, certainly, are not equipped with the external effectors. The use of such effectors is a facility of computer numerically controlled machines (CNC machines) and controlled manipulators, while the presence of both exteroceptive sensors and external effectors is a privilege of robots. The scheme of the robot based on the von Neumann model of the digital computer is shown in Figure I.2.

It is clear that the only difference between the presented architecture of the robot and the architecture of computer is the presence of the proprioceptive and exteroceptive sensors and of the internal and external effectors. Proprioceptive sensors and internal effectors provide the stable functionality or homeostasis of the robot, and exteroceptive sensors and external effectors, like wheels and arm manipulators, allow the robot to change its location in the environment and/or to change an environment according to the sensed information and programmed task. Moreover, as stressed by Lumelsky, "reacting to sensing data is essential for robot's being a robot" (Lumelsky 2006, p. 15).

The indicated roles of the sensors and effectors imply the following properties (Brooks 1991, p. 1227):

- *Situatedness*. The robots are situated in the world – they do not deal with abstract descriptions, but with the "here" and "now" of the environment that directly influences the behavior of the system.
- *Embodiment*. The robots have bodies and experience the world directly – their actions are part of a dynamic with the world, and the actions have immediate feedback on the robots' own sensations.

Later, these two properties were enriched by the principle of emergence (Arkin 1998, p. 27):[9]

- *Emergence*. Intelligence arises from the interactions of the robotic agent with its environment. It is not a property of either the agent or the environment in isolation but rather a result of the interplay between them.

The third property directly separates the robotic system from the pure computer system. As any material entity, a computer has bodily parameters and limits, but these characteristics and physical location of the computer do not influence the computation

9 This principle follows general ideas of cybernetics and bioinspired systems and appears in the works by Michael Tsetlin (1973).

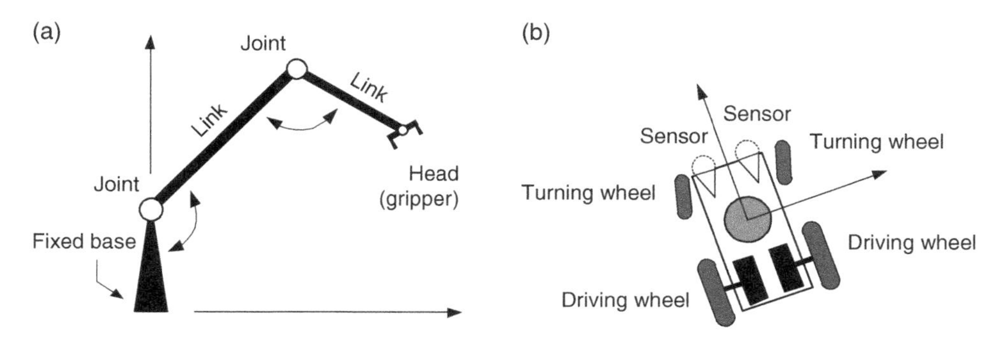

Figure I.3 (a) Fixed-base robot arm manipulator; and (b) mobile robot.

itself. In robotics, by contrast, physical characteristics of the robot, its location and environment, have the same or even higher importance than the computation abilities of the robot's controller or its on-board computer.

During the last decades, these three properties have obtained a status of general principles and have given rise to the behavior-based methodology (Arkin 1998) that introduces the main ideas of cybernetics into robotics and uses biological inspirations in the robots' design (for general and even philosophical considerations of this methodology see e.g. the book by Clark [1997]).

With respect to the purposes and conditions of activity, the robots are equipped with different sensors and effectors. However, while the proprioceptive sensors and internal effectors provide the stable functionality of the robot as a unit, the exteroceptive sensors and, furthermore, external effectors define the type of the robot and the methods of planning its motion.

The robots of the first type, often called the fixed-base robot arm manipulators (Choset et al. 2005), are used in highly structured spaces, like industrial production lines or storerooms, and their motion is considered in global or world Cartesian coordinate systems defined by the surrounding space with the origin fixed in the base of the robot (Lumelsky 2006). The required motion for such robots is the motion of the end-point of the arm or the head, and the main external effectors are motors, which are placed in the joints of the arm and change the angles between the links. The scheme of such robot is shown in Figure I.3a.

The robots of the second type are mobile robots that act in unstructured or unknown environment, like arbitrary terrains or randomly changing surroundings. In such robots, the local Cartesian coordinate system is related with the robot, and the main problem is a motion-planning of the origin of this system with respect to the environment conditions. The main external effectors of such robots are the motors, which provide the movement of the robot's body in the environment with respect to the information obtained by the exteroceptive sensors. The scheme of such robot is shown in Figure I.3b.

Certainly, such a division of robot types is rather provisional. In some applications, as in the case of surgery robots, the fixed-base arm manipulators can act in unstructured and changing environment, and the motion is planned with respect to environmental conditions and robot's limitations. At the same time, mobile robots can be applied in highly structures environments and their planned trajectories do not depend on the sensed data like for unmanned ground vehicles (UGVs) that typically use the predefined roadmaps. However, the ability of the automatic device to act in the unstructured environment is often considered as crucial to robotics and even leads to the following operational definition of the robot (Lumelsky 2006, p. 15):

> A robot is an automatic or semiautomatic machine capable of purposeful motion in response to its surroundings in an unstructured environment.

Starting in the 1960s, there were designed and produced different types of CNC machines and industrial fixed-base robot arm manipulators (for particular consideration of such robots, see, e.g., the books [Asfahl 1985 and Selig 1992]) and a wide variety of mobile robots (for overviews, see, e.g., the books [Arkin 1998; Bekey 2005; Choset et al. 2005; Dudek and Jenkin 2010; and Siegwart, Nourbakhsh, and Scaramuzza 2011]; the book by Dudek and Jenkin (2010) also includes a review of the appearance of the mobile robots in the fiction and movies). The detailed information regarding the most types of contemporary robots is presented in Siciliano and Khatib (2008).

I.3 Robot Arm Manipulators

The first remotely controlled arm manipulator, Master-Slave Manipulator Mk. 8 (MSM-8), was developed in 1945 by the US company Central Research Laboratories for the US Argon National Laboratory and was aimed to transfer radioactive materials. In this manipulator, the master human operator located in one room governed the servomotors of the slave manipulator located in the other room and controlled the movements by direct vision via a glass window. This manipulator gave rise to the development of different remotely controlled manipulators, from industrial and surgery robots via military unmanned ground and aerial vehicles and up to the Moon and Mars explorers. In most cases, both governing and control of these devices were conducted by radio and video streaming, but the idea of remote control is still the same as in the Mk. 8 manipulator.

In 1954, the American inventor George Charles Devol Jr. (1912–2011) applied for the patent titled "Programmed article transfer" (issued in 1961) that was defined as a universal automation and was characterized as follows (Devol 1961):

> Universal automation, or Unimation, is a term that may well characterize the general object of the invention. It makes article transfer machines available to the factory and warehouse for aiding the human operator in a way that can be compared with business machines as an aid to the office.

In the patent, the functionality of the programmed article transfer was specified as follows (Devol 1961):

> … the article transfer head (whether it takes the form of jaws, a suction gripper or other comparable article-handling tool) is moved by a mechanical power source through a sequence of strokes whose lengths and directions are determined by a program controller…

> ... According to this concept, the transfer mechanism operates the transfer head (...) and at the same time it displaces a position detector or position representing device; and the position detector is compared through a feedback loop with the program controller, until the position detector of the transfer head is displaced into coincidence or matching.

Two years later, the term *unimation* was used as a name of the first robot manufacturing company Unimation Inc., which was founded in 1956 by the American engineer Joseph F. Engelberger (born 1925). The first industrial robot Unimate was sold in 1960 to General Motors, where it was installed in 1961 and used for transferring hot metal parts.

Finally, in 1958 Claude Shannon and Marvin Minsky originated the idea of building an arm manipulator operated by a digital computer. This work was assigned to Heinrich Arnold Ernst, at that time the Shannon's D.Sc. student at MIT, who accomplished the research in 1961 by building the mechanical hand MH-1. In the abstract to his D.Sc. thesis, Ernst characterized the work as follows (Ernst 1961, p. 2):

> Digital computers are used in a human environment in almost all of their applications. A man translates the physical problem into numerical form, a programmer provides the machine with all the information needed, and a human user interprets and applies the results to the real world. In this study, instead of using the digital computer as a tool during the execution of human thought processes that are ultimately concerned with the real world we wish to let the computer deal directly with the real world by itself, beginning with the perception of the real world and the appreciation of it and ending with performance of a purposeful, active task in real world.

In the Ernst's MH-1, gripper of the manipulator was equipped with binary touch sensors placed in different parts of the gripper and with pressure sensors placed inside the gripper. The touch sensors were used to indicate the contact between the gripper and the obstacles in the environment and between the gripper and the object, which has to be transferred; the pressure sensors represented the size of the object. The actions of the manipulator were controlled by a digital controller programmed using specific programming language, in which the motions of the manipulator were defined in global Cartesian coordinates with respect to the information obtained by the touch and pressure sensors.

In light of the presented above definition, the MH-1 manipulator was certainly the first robot in the exact meaning of this word, and all subsequent robots, both fixed-base robot arms and mobile robots, inherited its properties and underlying ideas. However, the problems and the methods used in the studies and developments of the fixed-base robot arms and of the mobile robots are rather different. Limitations of possible motions and complete information about location of the arm in any given moment allow its motion planning in the global Euclidian space with definite origin, and, in the worst case, the lack of information results from incomplete knowledge of the obstacle locations. For mobile robots, in contrast, it is hard to determine a fixed origin and Cartesian coordinate frame and motion planning is conducted using relative coordinates with respect to the sensed information.

I.4 Mobile Robots

Automatic devices that, in certain sense, can be considered as the first remotely controlled mobile robots were demonstrated in the 1870s. These devices – the self-propelled wire-controlled torpedo boats – combined the properties of automatically regulated machines that allowed a stable motion in the sea and the properties of human-controlled mechanisms for navigation toward the enemy vessels. As already indicated, in 1898, Nicola Tesla invented the wireless radio-controlled torpedo, and in 1917 the English engineer Archibald Montgomery Low (1888–1956) demonstrated the first radio-controlled airplane and then the radio-controlled rocket that gave a rise to the further developments of remotely controlled UGVs and unmanned aerial vehicles (UAVs).

Following this approach, in 1912 the American engineers John Hays Hammond Jr. (1888–1965) and Benjamin Franklin Miessner (1890–1976) designed the first mobile robot in its exact modern sense. The construction of this robot called "electric dog" was introduced in 1919 in the paper by Miessner (1919). This was the first autonomous electronic machine that acted and changed its location in the environment according to the sensed information only. In the description of the robot, Miessner wrote (Miessner 1919, p. 376):

> In its present form the electric dog consists of a rectangular box about three feet long, one and one-half feet wide and one foot high. This box contains all the instruments and mechanism, and is mounted upon three wheels, two of which are geared to a driving motor; the third, on the rear end, is so mounted that its bearings can be turned in a horizontal plane for steering, much like the front wheel of a child's velocipede. Two large, glass lenses on the forward end, separated by a protruding, nose-like partition, appear very much like huge eyes.

The electronics and mechanical mechanism of this machine were constructed in such a way that the steering wheel turned the robot in the light direction. This Miessner describes this behavior as follows (Miessner 1919, p. 376):

> If a pocket flash light be turned on the machine it will immediately spring into action, but will stop as suddenly if the light be snapped off or turned away. If the light be held stationary and directed upon the dog, it will amble up until its own motion causes it to come directly under the light, and therefore into such a position that the light will not shine in its glass eyes; there it stops and the whining of its driving motor also ceases. If now one turns the flashlight into its eyes and walks around the room the dog will immediately respond and follow the moving light wherever it goes, with a loud metallic clank at each wag of its steering-wheel tail.

Twenty years later, the English neurophysiologist William Grey Walter (1910–1977) in the paper "An Electromechanical Animal" (Walter 1950a,b) published in 1950 and two next papers "An imitation of life" (Walter 1950a,b) and "A machine that learns" (Walter 1951) presented another version of mobile robot that followed the light. The main goal of this study was to check the possibility of mimicking "the certain aspects of brain function by means of electro-mechanical devices in order to discover how far the neuroelectronic analogy can be pushed" (Walter 1950a,b, p. 208).

Presenting his robots, Walter wrote (Walter 1950a,b, p. 43):

> We have given them the mock-biological name Machina speculatrix, because they illustrate particularly the exploratory, speculative behavior that is so characteristic of most animals. The machine on which we have chiefly concentrated is a small creature with a smooth shell and a protruding neck carrying a single eye which scans the surroundings for light stimuli; because of its general appearance we call the genus "Testudo," or tortoise. The Adam and Eve of this line are nicknamed Elmer and Elsie, after the initials of the terms describing them – ELectro MEchanical Robots, Light-Sensitive, with Internal and External stability. Instead of the 10,000 million cells of our brains, Elmer and Elsie contain but two functional elements: two miniature radio tubes, two sense organs, one for light and the other for touch, and two effectors or motors, one for crawling and the other for steering.

The control circuits of the robots were built in such a manner that without the light stimulus the robot continuously explores the environment and its driving mechanism moves it forward, and in presence of the light stimulus the robot moves toward the light. However, since while seeking for a light source the light sensor turns around, the trajectory of the robot is a cycloid. In the experiments with two robots, each of which was equipped with the lamp, the motion of the robots was highly unpredictable and looked like the behavior of living organisms (Walter 1950a,b). In the further versions (Walter 1951), the robots were endowed with the simplest learning ability mimicking the Pavlov stimulus-reaction mechanism that resulted in the close correspondence between the robots' behavior and the simplest behavior of animals.

Following the Walter's cybernetics approach to the mobile robots, in 1984 the Italian Austrian neuroscientist Valentino Braitenberg (1926–2011) returned to the idea of Hammond and Miessner's "electric dog" and considered the wheeled mobile robots with two light sensors with different connections with the motors. Introducing these robots, widely known as Braitenberg vehicles, he wrote (Braitenberg 1984, p. 2):

> We will talk only about machines with very simple internal structures, too simple in fact to be interesting from the point of view of mechanical or electrical engineering. Interest arises, rather, when we look at these machines or vehicles as if they were animals in natural environment.

In the simplest configuration without control, there are two possible variants of connectivity between the sensors and the motors. In direct connectivity, the left sensor is connected with the left motor and right sensor is connected with the right motor, and in cross connectivity, left sensor is connected with the right motor, and the right sensor is connected with the left motor. Two different behaviors of the Braitenberg vehicle interpreted as fear (direct connectivity) and aggression (cross connectivity) are shown in Figure I.4 (following Figure 3 from [Braitenberg 1984]).

In direct connectivity, a stronger signal received by the left sensor results in a right turn and a stronger signal received by the right sensor results in a left turn of the robot. By contrast, in cross connectivity a stronger signal received by the left sensor results in a left

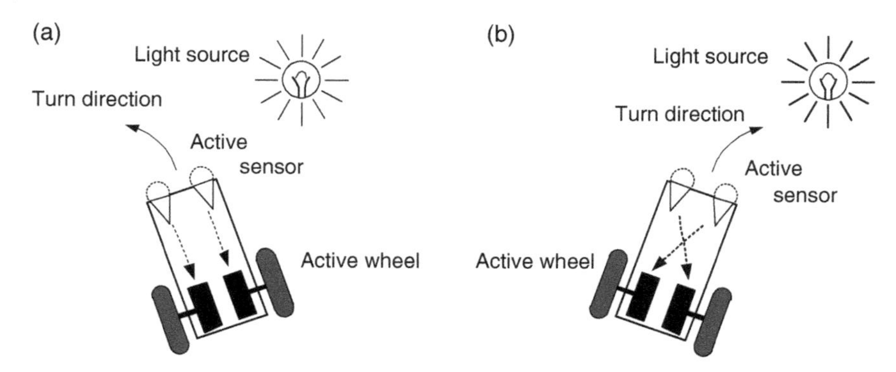

Figure I.4 The Braitenberg vehicle with "fear" (a) and "aggression" (b) behaviors.

turn and a stronger signal received by the right sensor results in a right turn of robot. Certainly, application of additional control and sensors as well as the activity of the pairs of robots result in more complicated behavior, which can be interpreted in different psychological terms (Braitenberg 1984).

Starting from the Walter's Elmer and Elsie robots, the development of mobile robots followed several directions, which deal with different abilities of the mobile robots. In particular, an improvement of locomotion resulted in creating legged robots and humanoid robots; an implementation of advanced sensing abilities and corresponding pattern recognition techniques allowed application of on-board cameras for exploring environment and image and video processing for reacting to the observed situations; and further achievements in automatic regulation and control methods allowed building the domestic service robots, like floor-cleaners and grass-cutters, and the semiautonomous mobile robots for Moon and Mars exploration. For brief consideration of the types of mobile robots, see, e.g., Arkin (1998); Bekey (2005); Siegwart, Nourbakhsh, and Scaramuzza (2011); the illustrated history of mobile robots is presented in a recent book by Foran (2015).

Certainly, in most modern mobile robots, the control is conducted by the on-board programmable controllers that are built following the same principles as the controllers of robotic arm manipulators, but instead of governing the motions of the arm, the controllers govern the movements of the robot around the environment. The motion planning in this case requires defining a path or a trajectory of the robot in the environment such that the robot avoids collisions with obstacles and the other agents and fulfills the definite task. In addition, global Euclidian space with Cartesian coordinates is usually considered as physical space, and relation between physical space and configuration space differs from such relation used in path planning for fixed-base robots. Notice that usually the path of the robot is considered as a sequence of positions or a curve in the configuration space of the robot, while the trajectory is defined as a path together with the robot's velocity (Choset et al. 2005; Medina, Ben-Moshe, and Shvalb 2013; Shiler 2015).

In general, motion planning for mobile robots addresses several different tasks. The classical task deals with defining the path of the robot from a given initial or starting position $\mathfrak{x}_{init}$ to a given final or target position $\mathfrak{x}_{fin}$ such that while the motion the robot avoids collision with obstacles. For the predefined locations of the obstacles and their size

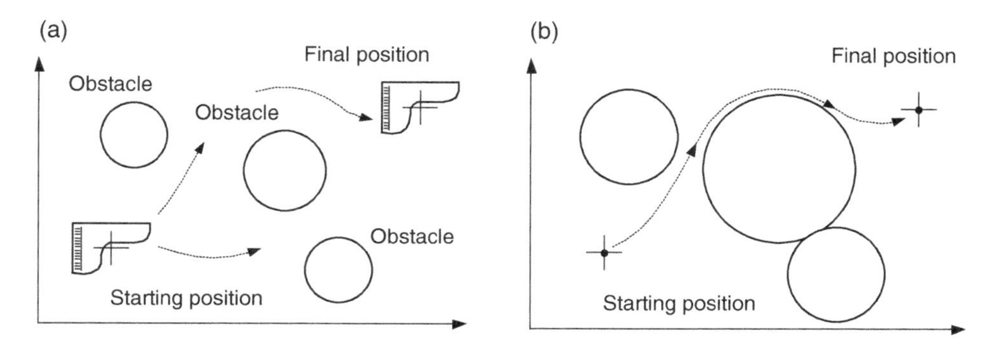

Figure I.5 Illustration of the piano mover problem: (a) physical space and (b) configuration space.

and form, such a problem is known as the piano-mover problem (Choset et al. 2005; Lumelsky 2006). The piano-mover problem is illustrated in Figure I.5.

In this problem, an object of some complex form (a piano) should be transferred from the position x_{init} to the position x_{fin} over the space with obstacles. Then, at the first stage, the object is shrunk to point and the sizes of obstacles are grown, respectively, and at the second stage the path between the positions x_{init} and x_{fin}, if it is exists, is defined.

The piano-mover problem illustrates the motion planning in the known environment following a global coordinates system. Such motion planning can be executed offline and, if the required path exists, the robot should follow the path without using any information sensed online during the movement. If, contrast, locations or form of the obstacles are unknown, then the motion planning is conducted online with respect to the information sensed by the robot. The basic algorithms that solve this problem are the Bug algorithms. In these algorithms, it is assumed that the robot is equipped with the sensors, which perceive the obstacle (locally or at definite distance), and the sensor, which perceive the target point at a distance. In addition, it is assumed that the controller of the robot has a memory for storing the intermediate points of the path and a processor for calculating the distances.

More complicated problems appear in motion planning with moving obstacles. With respect to the available information and sensors, as well as depending on the exact task, the methods of potential function (Shahidi, Shayman, and Krishnaprasad 1991; Koditschek and Rimon 1990; Hacohen, Shraga, and Shvalb 2014) and the methods of dynamic trajectory planning (Fiorini and Shiller 1998; Shiller and Gwo 1991; Shiller et al. 2013; Shiler 2015) are applied.

In the indicated above problems, the task was to define a path of the robot from the starting point to the final point. The other group of problems appear in the tasks, in which mobile robot is aimed to screen the domain or to cover as greater as possible part of the domain. In general, such tasks follow the framework of search and foraging problems (Viswanathan et al. 2011; Kagan and Ben-Gal 2015); however, certain implementation of the methods, like in the floor cleaning and grass cutting robots, can be rather far from the original formulation.

Finally, more complicated problems appear in the tasks of motion planning for aerial and underwater mobile robots. For such robots, the problem of obstacle avoidance becomes less important, and the main difficulties appear in positioning of the robots in physical space. In the simplest case, when the robots are able to communicate with a central unit, such positioning is conducted using global coordinates system (Tsourdos, White, and Shanmugavel 2011). However, if the robots are acting autonomously, positioning is based on local environment and relative coordinates and is conducted by the robot using certain mapping algorithms and sensed information about the surroundings.

Information obtained by the sensors about neighboring environment can include the knowledge both about the environment itself and about the other agents, robots or humans, acting together with the mobile robot. In many cases, such agents are considered as elements of environment and the robot is required to avoid collisions with these agents. In the last decades, the groups of mobile robots, interactions between them, and their collective behavior has attracted strong interest of researchers in different fields (Weiss 1999).

I.5 Multi-Robot Systems and Swarms

Formally, the multi-robot system is defined as a group of robots cooperating to execute a certain mission (Dudek and Jenkin 2010), and in such a sense a manufacturing line, which includes several robot arm manipulators and CNC machines, can be considered as a multi-robot system. In such a system, each robot and machine fulfills a definite task and the results of their functionality are combined in a final product. In the case of mobile robots, in contrast, cooperative behavior is considered in different way. In such systems, the robots are rather simpler than the ones used in manufacturing lines; however, since the configuration space of the swarm is embedded in a much higher-dimension ambient space, motion planning is computationally much harder (see, e.g., Ghrist 2010).

In most cases, all robots in the swarm are assumed to have equal sensing and functional abilities, and are designed and programmed in such a manner that they are able to create temporary groups with respect to current assignment and to act together in the same way as animals' societies (Sumpter 2010). Following the biological inspirations, such systems of mobile robots are called the swarm robotic systems (Hamann 2010; Trianni 2008).

Probably, the first system of mobile robots that demonstrated the basic properties of swarm was the pair of the Walter's Elmer and Elsie tortoises equipped with the lamps and light sensors (Walter 1950a,b, 1951). However, swarm robotic systems attracted much greater interest in later 1980s, when the term *swarm robotics* appeared. Recalling this moment, Gerardo Beni writes (Beni 2005, pp. 1–2):

> With Jing Wang, we presented a short paper on cellular robots at one of the Il Ciocco conferences.[10] The discussion was quite lively and I remember Alex Meystel saying that "cellular robot" was an interesting concept, but the name was not appealing; a buzz word was needed to describe "that sort of 'swarm,'" as he put it. I agreed that the term "cellular robot" was not very exciting and, besides, it had already been used by Fukuda... Anyway, in thinking of how to call the "cellular robots" with better term, I did not make any leap of imagination but simply used the word "swarm" that Alex had mentioned causally.

10 The NATO Advanced Workshop on Highly Redundant Sensing in Robotic Systems, Il Ciocco, Italy, June 1988; the indicated paper is Beni and Wang (1990).

Stressing the reasons for the use of such term, Beni indicates (Beni 2005, p. 2):

> The fact is that the group of robots we were dealing with was not just a "group." It had some special characteristics, which in fact are found in swarms of insects, i.e., decentralized control, lack of synchronicity, simple and (quasi) identical members.

By these reasons, the swarm robotic systems are currently considered in the framework of the studies multi-agent systems and of the biological researches of the societies of living organisms, especially of insects, fishes, and birds. In this framework, the following definition of the agent is widely accepted (Wooldridge 1999, p. 29):

> An agent is a computer system that is situated in some environment, and that is capable of autonomous action in this environment in order to meet its design objectives.

It is clear that if an "environment" means "physical environment," and the action assumes changes of the environmental states or the robot's locations, then this definition has a direct link with the indicated above Brooks properties of the robot. Consequently (Sahin 2005, p. 12),

> Swarm robotics is the study of how large number of relatively simple physically embodied agents can be designed such that a desired collective behavior emerges from the local interactions among agents and between the agents and the environment.

The multi-robot systems and robot swarms have several advantages that make them rather attractive in different tasks (Yogeswaran and Ponnambalam 2010). In particular, because of the equivalence and concurrent activity of the swarm members, the multi-robot systems, in comparison with single robots, demonstrate the following features:

- *Parallelism.* Quicker fulfilling of decomposable missions by the parallel execution of the subtasks.
- *Robustness*: Higher tolerance to the faults of the individual robots or their small subgroups.
- *Flexibility.* Easier adaptation to different applications and missions.

However, to achieve these and the other benefits of robotic swarms, we must solve several specific problems that do not appear in the considerations of fixed-base robots and of the mobile robots acting individually – in particular, the problem of preserving activity of the swarm as a unit, communication and interactions between the agents, collision avoidance, and division of labor. In addition, the problems of optimization and control, which are common for all robotic systems, require strongly different approaches in the case of swarms (Bayindir and Sahin 2007).

Fortunately, a lot of insights regarding the robots' swarms can be obtained from the studies of natural swarms, like colonies of ants, schools of fishes, flocks of birds, and herds of land animals (McElreath and Boyd 2007; Holldobler and Wilson 2009; Gordon 2010; Sumpter 2010). Such considerations resulted in the researches of swarm intelligence (Weiss 1999; Kennedy, Eberhart, and Shi 2001; Panigrahi, Shi, and Lim 2011), swarm optimization (Dorigo and Stutzle 2004; Passino 2005), and bio-inspired robotics (McFarland and Bösser 1993; Bekey 2005). The progress and results in these directions are represented by the series of conferences on simulation and adaptive behavior (From Animals to Animats 1991–2012) and on swarm intelligence (ANTS 1998–2014) and in the series of symposia on distributed autonomous robotic systems (Distributed Autonomous Robotic Systems 1992–2013) and are compiled in the detailed handbook (Kernbach 2013). The overviews of key concepts in the field appear in the inspiring books by Trianni (2008) and by Hamann (2010), and popular presentation of the main ideas is presented by Clark (1997, 2003).

The basic task of the mobile robots acting in swarm is to move together and to preserve geometrically distinguishable cluster of the swarm. An insight for solving this problem came from the behavioral model, which was suggested by the American computer graphics expert Craig Reynolds, who in 1987 formulated three famous rules to simulate flocking[11] (Reynolds 1987, p. 31):

1) *Collision avoidance.* Avoid collisions with nearby flockmates.
2) *Velocity matching.* Attempt to match velocity with nearby flockmates.
3) *Flock centering.* Attempt to stay close to nearby flockmates.

Consequently (Reynolds 1987, p. 26),

> flock refers generically to a group of objects that exhibit this general class of polarized, non-colliding aggregate motion, [and flockmates are considered as] simulated bird-like 'bird-oid' objects [and generically referred] as 'boids' even when they represent other sorts of creatures such as schooling fish.

Clarifying these rules, Reynolds writes (Reynolds 1987, p. 31):

> Static collision avoidance and dynamic velocity matching are complementary. Together they ensure that the members of a simulated flock are free to fly within the crowded skies of the flock interior without running into one another...
>
> Flock centering makes a boid want to be near the center of the flock. Because each boid has a localized perception of the world, "center of the flock" actually means the center of the nearby flockmates. Flock centering causes the boid to fly in a direction that moves it closer to the centroid of the nearby boids.

The indicated Reynolds rules provide the simplest techniques for preserving the swarm and achieving the key benefit of moving together that is an access to information (Sumpter 2010). The information exchange between the robots in swarms is conducted in two main ways (Iocchi, Nardi, and Salerno 2001):

1) Information is directly communicated between the agents according to definite communication protocols.
2) Information is indirectly communicated by changing the environmental states and perceiving these changes.

---

11 Currently, these rules are also known as *separation, alignment,* and *cohesion,* correspondingly (see, e.g., [Gazi and Passino 2011]).

In the second case, the resulting collective behavior is known as *stigmergy* (Grasse 1959), and such behavior is considered as a characteristic of the system. More precisely (Hamann 2010, p. 9),

> a stigmergic system has a process that changes the environment, and the system itself reacts again to the changes of the environment.

It is clear that this definition has direct link with both the Brooks properties of robot and the Ernst remark on robots dealing with the real world and reacting to its changes.

Following general taxonomy (Iocchi, Nardi, and Salerno 2001), dynamics of the mobile robot swarms is usually considered as a cooperative activity such that the robots are aware of the actions of the other robots, are influenced by these actions, and coordinate their activities with respect to their influence to the behavior of the other robots. Following the decentralized control scheme, decision-making regarding the robots' actions is conducted autonomously by each robot according to obtained information the activities of the other robots. A detailed review of the methods of coordination and control and corresponding task and motion planning is presented in the paper by Yan, Jouandeau, and Cherif (2013).

The decentralized control and autonomous activity of the mobile robots in swarms often result in the self-organization phenomenon that is a formation of ordered spatial or temporal structures. The basic ideas of such phenomenon in the collective behavior first appeared in the second edition of the mentioned above Wiener's "Cybernetics"[12] and in the "Theory of Self-Reproducing Automata" by von Neumann (von Neumann, Theory of Self-Reproducing Automata, 1966). General mathematical framework for consideration of pattern formation was suggested by the British mathematician and computer scientist Alan Turing (1912–1954) in his seminal paper "The chemical basis of Morphogenesis" (Turing 1952), in which he suggested a dynamical model describing formation of macroscopic spatial structures in the media of interacting agents. A new impulse to the studies of self-organization was given in 1977 by the book *Synergetics* (Haken 1977) by the German physicist Hermann Haken (born 1927) and the book *Self-Organization in Non-Equilibrium Systems* (Prigogine and Nicolis 1977) by the group led by the Russian Belgian chemist Ilya Prigogine (1917–2003).

Implementation of this approach to the swarm robotics resulted in evolutionary models of swarm dynamics (Trianni 2008; Hamann 2010) that allow an application of general models of collective behavior of active particles (Schweitzer 2003) and swarm optimization [Gazi and Passino 2011] to the swarms of mobile robots. The inspiring projects in this direction are the group of ALICE robots following the gradient field (Garnier et al. 2013), the self-assembling swarm of 1024 KBot robots (Rubenstein, Cornejo, and Nagpal 2014) and collective decision-making (Valentini, Hamann, and Dorigo 2014), which was implemented by the swarm of 100 KBot robots.

Additional historical reviews and methodological considerations of swarm robotics are presented in the books by Trianni (2008) and Hamann (2010). Reviews of the projects in swarm robotics can be found in the papers by Navarro and Matia (2012) and by Tan and

12 Published in 1961 by the MIT Press, Cambridge, MA.

Zheng (2013), and useful information on the software simulation platforms is presented by Shi et al. (2012). A complete review of the studies and results in multi-robot systems is presented in Kernbach (2013).

I.6 Goal and Structure of the Book

The main goal of the book is to provide theoretical and practical guide and ready-to-use algorithms of navigation of mobile robots, which can be directly implemented in the laboratory and can be used as a starting point for further research and as a basis for solutions of engineering tasks. The book is aimed to serve the following three purposes.

As a textbook, the book is aimed to support the course in mobile robotics for undergraduate and graduate students specializing in applied mathematics and engineering. It is structured with respect to the program of a single semester course and provides complete theoretical material and algorithmic solutions, which can be used for further laboratory training and students' projects.

As a research text, the book is aimed to provide a starting point for the research in different directions of mobile robots navigation and can be used as a guide in the field. In this concern, the book follows a direction "Sensing-Intelligence-Motion" and implements it for navigation of mobile robots and mobile robot systems.

As a practical guide, the book is self-contained and includes ready-to-use algorithms, numerical examples, and simulations, which can be directly implemented in both simple and advanced mobile robots and applied to different tasks with respect to the available information and robots' sensing abilities.

Methodologically, the book consists of four main parts. The first part considers the models and algorithms of navigation and motion planning in global coordinates system with complete information about the robot's location and velocity. In this part, robotic motion is defined by classical methods of kinematics and dynamics in two dimensions for ground vehicles and sea vessels and in three dimensions for aerial and underwater vehicles, and motion planning follows offline path-planning techniques, including the information obtained by the basic sensors.

The second part considers the motion of the robots in the potential field, which is defined by the environmental states of the robot's expectations and knowledge. In this case, the robot's motion is planned with respect to the sensed information about the neighboring environment, and the movements of the robots are specified in local coordinates (in two or three dimensions), which are defined relatively to the coordinates of the other robots or certain reference marks in the environment.

The third part considers the robot's motion in the unknown environments and the corresponding tasks of environment mapping using sensed information. If global positioning system is accessible, such tasks use obstacle avoidance algorithms based on local sensing, while the map of the environment is developed with respect to global coordinates system. In contrast, if the robots have no access to global coordinates, then the map is created using relative coordinates and global mapping is conducted by combination of local maps.

Finally, the fourth part of the book deals with the multi-robot systems and swarm dynamics in two and three dimensions. This includes models of motion in global and local coordinates systems with and without direct communication between the robots. The main attention in this part is concentrated on the specific problems and phenomena that appear in the collective behavior of the robots, namely on communication between the robots, aggregation, collision avoidance, cooperation and division of labor, activity in homogeneous and heterogeneous groups, and self-organization in large systems.

For convenience, the discourse is divided into 12 chapters (excluding Introduction and Conclusion) such that each chapter considers an issue in the field of motion planning of mobile robots. The chapters include theoretical considerations and models that are supported by ready-to-use algorithms and numerical examples and simulations, which can be used both in classes and laboratory trainings and in solutions of practical engineering tasks.

References

ANTS (1998–2014). *Proceedings of International Conference on (Ant Colony Optimization and) Swarm Intelligence.* Berlin: Springer-Verlag.

Arkin, R.C. (1998). *Behavior-Based Robotics.* Cambridge, MA: The MIT Press.

Asfahl, C.R. (1985). *Robots and Manufactury Automation.* New York: Wiley.

Asimov, I. (1950). *I, Robot.* New York: Gnome Press.

Bayindir, L. and Sahin, E. (2007). A review of studies in swarm robotics. *Turkish Journal of Electrical Engineering* 15 (2): 115–147.

Bekey, G.A. (2005). *Autonomous Robots. From Biological Inspiration to Implementation and Control.* Cambridge, MA: The MIT Press.

Beni, G. (2005). From swarm intelligence to swarm robotics. In: *Swarm Robotics. Lecture Notes in Computer Science,* vol. 3342 (ed. E. Sahin and W.M. Spears), 1–9. Berlin: Springer.

Beni, G. and Wang, J. (1990). Self-organizing sensory systems. In: *Proceedings of NATO Advanced Workshop on Highly Redundant Sensing in Robotic Systems, Il Cioco, Italy (June 1988),* 251–262. Berlin: Springer.

Braitenberg, V. (1984). *Vehicles: Experiments in Synthetic Psychology.* Cambridge, MA: The MIT Press.

Brooks, R.A. (1991). New approaches to robotics. *Science* 253 (5025): 1227–1232.

Choset, H., Lynch, K., Hutchinson, S. et al. (2005). *Principles of Robot Motion: Theory, Algorithms, and Implementation.* Cambridge, MA: Bradford Books/The MIT Press.

Clark, A. (1997). *Being There: Putting Brain, Body and World Together Again.* Cambridge, MA: The MIT Press.

Clark, A. (2003). *Natural-Born Cyborgs – Minds, Technologies, and the Future of Human Intelligence.* Oxford: Oxford University Press.

Devol, G. C. (1961). *Patent No. US 2988237 A. USA.*

Distributed Autonomous Robotic Systems (1992–2013). *Proceedings of International Symposia.* Berlin: Springer.

Dorigo, M. and Stutzle, T. (2004). *Ant Colony Optimization.* Cambridge, MA: The MIT Press/A Bradford Book.

Dudek, G. and Jenkin, M. (2010). *Computational Principles of Mobile Robotics*, 2e. New York: Cambridge University Press.
Ernst, H. A. (1961). *MH-1, A Computer-Operated Mechanical Hand*. D.Sc. Thesis, Massachusetts Institute of Technology, Dept. Electrical Engineering, Cambridge, MA.
Fiorini, P. and Shiller, Z. (1998). Motion planning in dynamic environments using velocity obstacles. *International Journal of Robotics Research* 17 (7): 760–772.
Foran, R. (2015). *Robotics: From Automatons to the Roomba*. Edina, MN: Abdo Publishing.
From Animals to Animats (1991–2012). *Proceedings of International Conference on Simulation and Adaptive Behavior*. Bradford Books/MIT Press/Springer.
Garnier, S., Combe, M., Jost, C., and Theraulaz, G. (2013). Do ants need to estimate the geometrical properties of trail bifurcations to find an efficient route? A swarm robotics test bed. *PLoS Computational Biology* 9 (3): e1002903.
Gazi, V. and Passino, K.M. (2011). *Swarm Stability and Optimization*. Berlin: Springer.
Ghrist, R. (2010). *Configuration Spaces, Braids, and Robotics, Lecture Notes Series*, vol. 19, 263–304. Institute of Mathematical Sciences. National University of Singapore.
Gordon, D.M. (2010). *Ant Encounters: Interaction Networks and Colony Behavior*. Princeton: Princeton University Press.
Grasse, P.-P. (1959). La reconstruction du nid et less coordinations interindividuelles chez bellicositermes natalensis et cubitermes sp. la theorie de la stigmergie: essai d'interpretation du comportement des termites constructeurs. *Insectes Sociaux* 41–83.
Hacohen, S., Shraga, S., and Shvalb, N. (2014). Motion-planning in dynamic uncertain environment using probability navigation function. In: *IEEE 28th Convention of Electrical and Electronics Engineers in Israel*. Eilat: IEEE.
Haken, H. (1977). *Synergetics: An Introduction. Nonequilibrium Phase transitions and Self-Organization in Physics, Chemistry and Biology*. Berlin: Springer-Verlag.
Hamann, H. (2010). *Space-Time Continuous Models of Swarm Robotic Systems: Supporting Global-to-Local Programming*. Berlin: Springer.
Hero of Alexandria (2009). *The Pheumatics* (trans. B. Woodcroft). CreateSpace.
Holldobler, B. and Wilson, E.O. (2009). *The Superorganism: The Beauty, Elegance, and Strangeness of Insect Societies*. New York: W. W. Norton & Company.
Iocchi, L., Nardi, D., and Salerno, M. (2001). Reactivity and diliberation: a survey on multi-robot systems. In: *Balancing Reactivity and Social Diliberation in Multi-Agent Systems. From RoboCup to Real-World Applications* (ed. M. Hannebauer, J. Wendler and E. Pagello), 9–32. Berlin: Springer.
al-Jazari, I.a.-R. (1974). *The Book of Knowledge of Ingenious Devices* (trans. D. R. Hill). Dordrecht/Holland: D. Reidel.
Kagan, E. and Ben-Gal, I. (2015). *Search and Foraging. Individual Motion and Swarm Dynamics*. Boca Raton, FL: Chapman Hall/CRC/Taylor & Francis.
Kennedy, J., Eberhart, R.C., and Shi, Y. (2001). *Swarm Intelligence*. San Francisco: Morgan Kaufmann.
Kernbach, S. (ed.) (2013). *Handbook of Collective Robotics: Fundamentals and Challenges*. Boca Raton: CRC Press/Taylor & Francis.
Koditschek, D.E. and Rimon, E. (1990). Robot navigation functions on manifolds with boundary. *Advances in Applied Mathematics* 11 (4): 412–442.
Lumelsky, V.J. (2006). *Sensing, Intelligence, Motion. How the Robots and Humans Move in an Unstructured World*. Hoboken, NJ: Wiley.
Maxwell, J.C. (1868). On governors. *Proceedings of the Royal Society of London* 16: 270–283.

McElreath, R. and Boyd, R. (2007). *Mathematical Models of Social Evolution: A Guide for the Perplexed.* Chicago: The University of Chicago Press.
McFarland, D. and Bösser, T. (1993). *Intelligent Behavior in Animals and Robots.* Cambridge, Massachusetts: The MIT Press.
Medina, O., Taitz, A., Ben Moshe, B., and Shvalb, N. (2013). C-space compression for robots motion planning. *International Journal of Advanced Robotic Systems* 10 (1): 6.
Miessner, B.F. (1919). The electric dog. *Scientific American Supplement (2267)* 376: 6.
Navarro, I. and Matia, F. (2012). An introduction to swarm robotics. *ISRN Robotics 2013* 1–10.
von Neumann, J. (1945). *First Draft of a Report on the EDVAC.* University of Pennsylvania, Moore School of Electrical Engineering. Contract No. W-670-ORD-4926 between the US Army Ordnance Department and the Univeristy of Pennsylvania.
von Neumann, J. (1966). *Theory of Self-Reproducing Automata* (ed. A.W. Burks). Champaign: University of Illinois Press.
Panigrahi, B.K., Shi, Y., and Lim, M.-H. (eds.) (2011). *Handbook of Swarm Intelligence: Concepts, Principles and Applications.* Berlin: Springer-Verlag.
Passino, K.M. (2005). *Biomimicry for Optimization, Control, and Automation.* London: Springer.
Prigogine, I. and Nicolis, G. (1977). *Self-Organization in Non-Equilibrium Systems: From Dissipative Structures to Order through Fluctuations.* New York: John Wiley & Sons.
Reynolds, C.W. (1987). Flocks, herds, and schools: a distributed behavioral model. *Computer Graphics Graphics 2.1 (ACM SIGGRAPH'87 Conference Proceedings)* 4: 25–35.
Rubenstein, M., Cornejo, A., and Nagpal, R. (2014). Programmable self-assembly in a thousand-robot swarm. *Science* 345 (6198): 795–799.
Sahin, E. (2005). Swarm robotics: from sources of inspiration to domains of application. In: *Swarm Robotics, Lecture Notes in Computer Science,* vol. 3342 (ed. E. Sahin and W.M. Spears), 10–20. Berlin: Springer.
Schweitzer, F. (2003). *Brownian Agents and Active Particles. Collective Dynamics in the Natural and Social Sciences.* Berlin: Springer.
Selig, J.M. (1992). *Introductory Robotics. Hertfordshire.* UK: Prentice Hall.
Shahidi, R., Shayman, M., and Krishnaprasad, P.S. (1991). Mobile robot navigation using potential functions. In: *IEEE International Conference on Robotics and Automation,* 2047–2053. Sacramento, CA.
Shi, Z., Tu, J., Qiao, Z. et al. (2012). A survey of swarm robotics system. In: *Advances in Swarm Intelligence. Lecture Notes in Computer Science,* vol. 7331 (ed. Y. Tan, Y. Shi and Z. Ji), 564–572. Berlin: Springer.
Shiler, Z. (2015). Off-line vs. on-line trajectory planning. In: *Motion and Operation Planning of Robotic System. Mechanisms and Machine Science,* vol. 29 (ed. G. Carbone and F. Gomez-Barvo), 29–62. Switzerland: Springer International.
Shiller, Z. and Gwo, Y.-R. (1991). Dynamic motion planning of autonomous vehicles. *IEEE Transactions on Robotics and Automation* 7 (2): 241–249.
Shiller, Z., Sharma, S., Stern, I., and Stern, A. (2013). Online obstacle avoidance at high speeds. *International Journal of Robotics Research* 32 (9–10): 1030–1047.
Siciliano, B. and Khatib, O. (eds.) (2008). *Springer Handbook of Robotics.* Berlin: Springer.
Siegwart, R., Nourbakhsh, I.R., and Scaramuzza, D. (2011). *Introduction to Autonomous Mobile Robots,* 2e. Cambridge, MA: The MIT Press.
Sumpter, D.J. (2010). *Collective Animal Behaviour.* Princeton: Princeton University Press.

Tan, Y. and Zheng, Z.-y. (2013). Research advance in swarm robotics. *Defence Technology* 9: 18–39.

Trianni, V. (2008). *Evolutionary Swarm Robotics: Evolving Self-Organising Behaviors in Groups of Autonomous Robots*. Berlin: Springer-Verlag.

Tsetlin, M.L. (1973). *Automaton Theory and Modeling of Biological Systems*. New York: Academic Press.

Tsourdos, A., White, B.A., and Shanmugavel, M. (2011). *Cooperative Path Planning of Unmanned Aerial Vehicles. Chichester*. UK: Wiley.

Turing, A.M. (1952). The chemical basis of morphogenesis. *Philosophical Transactions of the Royal Society of London, Series B* 237: 37–72.

Valentini, G., Hamann, H., and Dorigo, M. (2014). *Self-Organized Collective Decision-Making: The Weighted Voter Model*. IRIDIA, Institut de Recherches Interdisciplinaires et de Developpements en Intelligence Artificielle. Bruxelles: IRIDIA - Technical Report TR/IRIDIA/2014–005.

Viswanathan, G.M., da Luz, M.G., Raposo, E.P., and Stanley, H.E. (2011). *The Physics of Foraging*. Cambridge: Cambridge University Press.

Walter, G.W. (1950a). An electromechanical animal. *Dialectica* 4: 42–49.

Walter, G.W. (1950b). An imitation of life. *Scientific American* 182 (5): 42–45.

Walter, G.W. (1951). A machine that learns. *Scientific American* 185 (2): 60–63.

Weiss, G. (ed.) (1999). *Multiagent Systems. A Modern Approach to Distributed Artificial Intelligence*. Cambridge, MA: The MIT Press.

Wiener, N. (1948). *Cybernetics: Or Control and Communication in the Animal and the Machine*. Paris: Herman & Cie.

Wooldridge, M. (1999). Intelligent agents. In: *Multiagent Systems. A Modern Approach to Distributed Artificial Intelligence* (ed. G. Weiss), 27–78. Cambridge, MA: The MIT Press.

Yan, Z., Jouandeau, N., and Cherif, A.A. (2013). A survey and analysis of multi-robot coordination. *International Journal of Advanced Robotic Systems* 10 (399): 2013.

Yogeswaran, M. and Ponnambalam, S.G. (2010). Swarm robotics: an extensive research review. In: *Advanced Knowledge Application in Practice* (ed. I. Fuerstner), 259–278. Rijeka: InTech.

1

Motion-Planning Schemes in Global Coordinates

Oded Medina and Nir Shvalb

1.1 Motivation

During the robot's tasks performance, it may change its configuration frequently and gather information from its environment using its sensors. Changing the robot's configuration from one to another can turn out to be not an easy task. The robot might perform an intricate maneuver from its initial configuration to its goal configuration rather than proceeding in a straight line in order to avoid obstacles. Such maneuvers may require detailed information regarding the robot's environment. For example, in a robot competition that involves several mobile robots playing soccer autonomously (RoboCup competition, small size league), the information is collected by a camera that captures the entire game field from above the playing ground. Another example where motion is a complicated task is the case of hyper-redundant robots, especially when motion takes place in the vicinity of obstacles.

> ***The term "motion planning" indicates the algorithm that calculates the maneuver from a start configuration c_s to a goal configuration c_g.***

Many examples of motion-planning algorithms are known (see, e.g., [Canny 1988; Laumond, Sekhavat, and Lamiraux 1998] for robotic arm and mobile robots motion planning designed to avoid obstacles collisions). Some of these algorithms are deterministic (i.e., they assume all information is accurately known or calculable) and some are stochastic (i.e., known to some extent). Still, informally, these may be categorized into three main groups: potential-field algorithms, grid-based algorithms, and sampling-based algorithms. We shall exemplify one algorithm of each of these groups, yet this is a drop in the ocean since the motion-planning problem continues to be highly investigated in the literature.

1.2 Notations

This section defines some key terms related to the configuration and workspaces and the weight function.

Autonomous Mobile Robots and Multi-Robot Systems: Motion-Planning, Communication, and Swarming,
First Edition. Edited by Eugene Kagan, Nir Shvalb and Irad Ben-Gal.

Companion website: www.wiley.com/go/kagan/robotsystems

1.2.1 The Configuration Space

A mobile robot may independently be situated in different locations and orientations. A mechanism may acquire different configurations, which are defined as the set of postures of all the mechanism's parts. Mathematically, we define a configuration as the minimal ordered set of parameters that defines the robot's posture.

So, for the case of a planar moving object, three parameters are of importance – the coordinates of the object's center gravity and its orientation angle. In the case of a spatial moving object, we consider six parameters – three coordinates of the object's center gravity and three angles corresponding to the object's yaw, pitch, and roll axes (see Chapter 3).

Let us now think of a planar, serial robot – a concatenation of two links connected by rotational joints. The configuration space of such a robot is the set of vectors $(\theta_1, \theta_2) \in [0, 2\pi] \times (0, 2\pi)$. In some cases, the configuration space should also be endowed with a metric that defines the distance between two configurations. Here we assume that a joint can be activated continuously without collisions. This is notated in the literature as the product of two circles $T^2 = S^1 \times S^1$, which is a two-dimensional torus. For now, we shall use an extended example – a seven-link chain robot with its end link attached to the ground, as depicted in Figure 1.1a.

Any configuration of this mechanism can be represented by the set of its six degrees of freedom (DoF) so the configuration space is a six-dimensional torus T^6. Nevertheless, for practical purposes it is natural to consider the actuators' limitations – for example, to $-\frac{\pi}{2} < \theta_i < \frac{\pi}{2}$ for all $i = 1 \dots 6$. Thus, the configuration space in this case is reduced to $C = I^6$ – the six- dimensional cube. In the case of a spatial object such as a quadrotor (Figure 1.1b), the configuration space is $C = \mathbb{R}^3 \times SO(3)$ where $\mathbb{R}^3$ is the center gravity coordinates and $SO(3)$ corresponds to the algebraic structure created by the yaw, pitch, and roll axes.

Note that a C-space may have more than one connected component. For example, the configuration space of a four-bar mechanism (robot) with links' lengths (1, 0.5, 1.75, 2) is

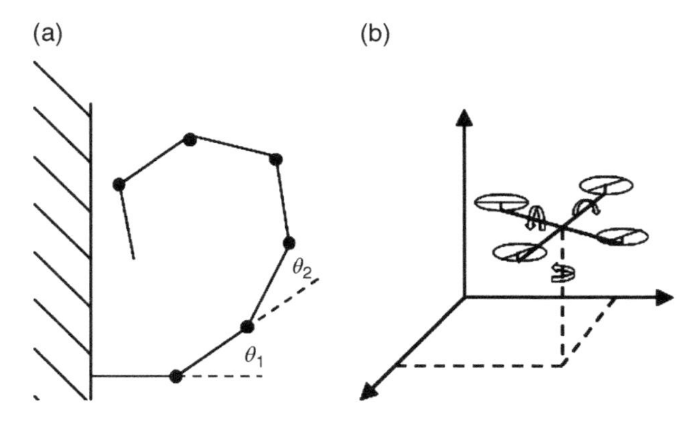

Figure 1.1 (a) A six-DOF planar robot. Each configuration $\vec{\theta} \in \mathbf{C}$ is presented by $\vec{\theta} = (\theta_1, \dots, \theta_6)$. (b) A quadrotor. Each configuration $\vec{c} \in \mathbf{C}$ is presented by $\vec{c} = (x, y, z, \theta_1, \theta_2, \theta_3)$.

congruent to two disjoint circles. For a mobile robot, the C-space may also have more than one connected component (e.g., when two sets of positions cannot be reached from one to another). Nevertheless, in some cases when discussing motion planning, one may pick a single connected component and apply a motion-planning scheme solely on it (Shvalb, Shoham, and Blanc 2007).

1.2.2 The Workspace

The workspace of a robot is defined as the geometrical set of points that it can reach. For example, for a planar mobile robot, this is the two-dimensional space where it operates, while for a robotic arm the workspace is usually taken as the set of points that can be reached by its end-effector. The workspace for the robot depicted in Figure 1.1a is the planar disc with a radius equal to the sum of its six links' lengths centered at the lower joint. For practical reasons, a robot should avoid collisions with obstacles located in its working space. We define the subspace $C_{free} \subset C$ as the set of obstacle-free configuration.

1.2.3 The Weight Function

Occasionally, C_{free} is coupled with a weight function (referred to in literature as a cost function). This may be the energy function, a height function, or the distance from a target configuration, for example. Integrating the cost function through a path connecting two configurations yields the maneuver cost. It is mostly common that the robot is physically limited to a given range of values of the weight function. In such a case, the configurations that exceed this range may be treated as C-space obstacles. The motion planning is the action of calculating the path (i.e., set of configurations) from the start configuration c_s to the goal configuration c_g avoiding obstacles.

In order to visually exemplify the motion-planning problem, the reader is referred to Figure 1.2a, which depicts a planar robot arm having three DoF, fixed to the ground in its one end, and Figure 1.2b, which shows a circular-shaped planar mobile platform. In both examples, obstacles may be placed between the start and the goal configurations such that motion from c_s to c_g cannot be performed in a straight line in C.

Note that for the planar robot arm case, in addition to avoiding the obstacles in the workspace the robot should avoid self-collisions, i.e., avoid mutual collisions between its rigid links. For simplicity of the discussion, we shall make sure that only the end effector will avoid collisions. In addition, we assume that the motors' torques are limited. The base motor is carrying most of the robot's mass, so in order to be on the safe side, we limit the torque at the base joint. In order for the robot to have a safe distance from the maximal torque configurations, we define a new obstacle in C as the level set where the torque equals a threshold (maximal) torque.

Figures 1.3 depict the C-space of the exemplified robots. For the planar robot arm, there are two manifolds. One represents the boundary of the configurations where the robot collides with the workspace physical obstacle and the other represents the torque threshold as already defined.

Figure 1.2 Two mobile robots. (a) The serial (climbing) planar robot. Its goal is to traverse a vertical wall while avoiding obstacle collisions, collision with wall, and self-collisions. (b) A mobile planar robot aims to traverse the plane toward the goal configuration while avoiding obstacles in the plane.

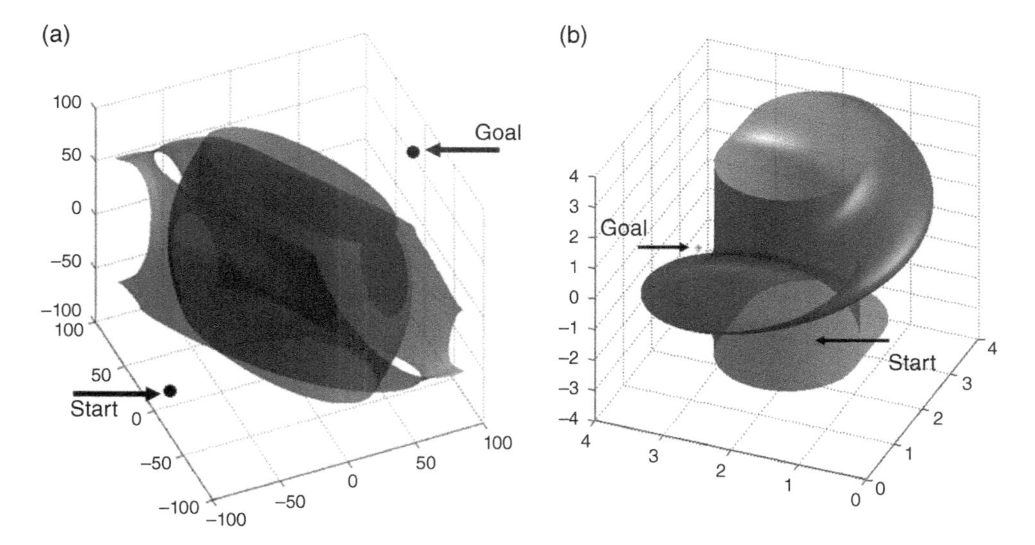

Figure 1.3 The C-spaces and its associated obstacles for a three-DoF planar robot arm (a) and the mobile platform (b). The manifolds are the workspace physical obstacle embedded in *C* and the torque threshold (a).

1.3 Motion-Planning Schemes: Known Configuration Spaces

There are many examples of motion-planning algorithms. Informally, they can be categorized into three main groups: potential-field algorithms, grid-based algorithms, and sampling-based algorithms, which we shall briefly review.

1.3.1 Potential-Field Algorithms

Consider a simple case with a two-dimensional C-space, start and goal configurations with a round obstacle situated within. In order to solve for motion planning, one may intuitively think of a negative electrically charged particle fixed in the goal configuration c_g, while in c_s one may imagine a positive electrically charged particle. A positive-charged particle that represents the robot is initially situated at c_s. The robot will be repelled from c_s and attracted to c_g. Following the same idea, we may place another positive-charged particle in the center of the obstacle, so c_s moves in the gradient direction of both potential fields, consequently avoiding the obstacle while being attracted to the goal configuration.

The electrostatic potential energy of a pair of charged particle q_1, q_2 situated at a distance r from each other is given by $\Phi = k\frac{q_1 q_2}{r}$ (with constant k). The resulting force on the charged particle is $\nabla\Phi = k\frac{q_1 q_2}{r^2}$. Nevertheless, for the sake of motion planning, different potential functions may also do. Note that since an obstacle may have various geometries, a simple solution would be to model it by a set of charged particles. It is not uncommon for the resultant field around the obstacle to deviate somewhat from the real obstacle boundary, but for reasonable cases, this will do.

Algorithm 1.1 Potential Field Pseudocode

Definitions:
Obs_i: The coordinate of the i-th obstacle center
k_i: The force coefficient of the i-th obstacle
ε: The desired proximity to c_g
c_k: the current configuration
Define $f(a,b) := \frac{a-b}{|a-b|^2}$
While $|c_k = c_g| < \epsilon$ **do**
set $V_k = f(c_k, c_g) + \sum k_i f(c_k, Obs_i)$
Proceed motion by $c_{k+1} = c_k + v_k$
End while

Although potential field algorithms are simple, since no prior calculation is required (apart from fixing the charged particle), their main vulnerability is the appearance of undesirable configurations where the resulting force vanishes. In such configurations,

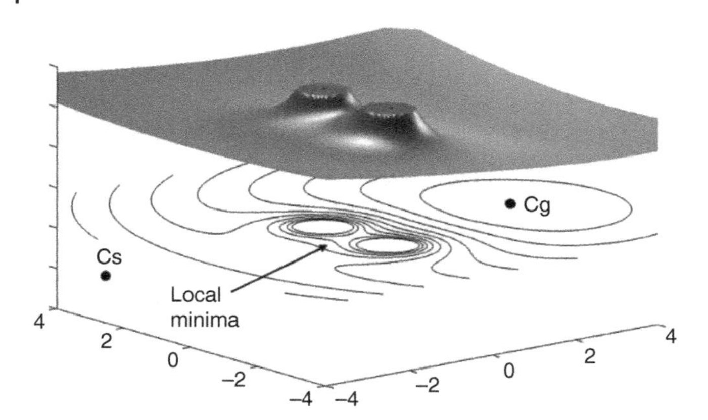

Figure 1.4 A local minima configuration with relation to $\boldsymbol{c_g}$.

the motion planner may be not definitive (a saddle point), while in other cases, the robot may altogether stop (local minima) (Figure 1.4). This may occur, for example, when multiple obstacles surround the robot (see, e.g., Figure 1.5). In order to avoid this problem, one should choose the potential function properly (Kim and Khosla 1992).

Recall that harmonic functions f are those that satisfy Laplace equation $\Delta f = \nabla^2 f = 0$ (i.e., the trace of the Hessian matrix vanishes), and therefore a local minima within the region is impossible. In other words, a function f that satisfies the Laplace equation on some region $\Omega \subset \mathbb{R}^n$ will attains its minimum and maximum values only on the boundary $\partial\Omega$ of Ω. Any other critical points of f in the interior of this region must be saddle points. We can now apply the principle of superposition, which follows from the linearity of the Laplace equation. If f_1 and f_2 are harmonic, then any linear combination of these functions is also harmonic (i.e., it is a solution of the Laplace equation).

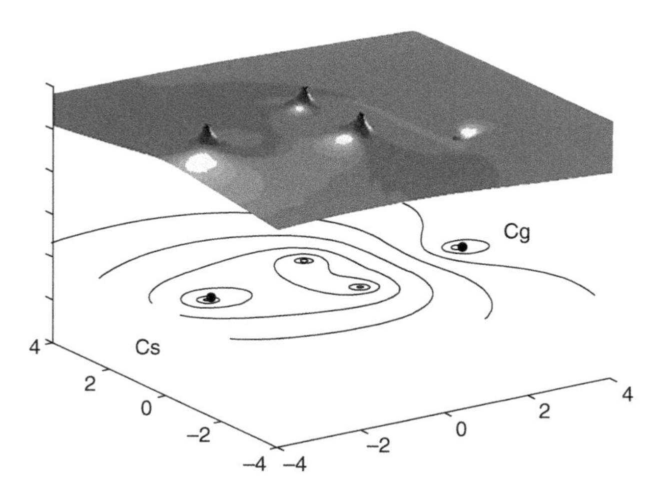

Figure 1.5 Harmonic potential field with a maximal value at $\boldsymbol{c_s}$ and a minimal value at $\boldsymbol{c_g}$.

1.3.2 Grid-Based Algorithms

Most mechanisms are continuously actuated (but discrete actuated robots are also being researched). A grid-based algorithm would first sample the space to create a discrete C-space. Each configuration is surrounded by neighboring configurations, the exact definition of which may vary – e.g., one may consider the Cartesian vicinity, including the diagonal vicinities, etc. The simplest definition would include only the Cartesian vicinities, i.e., each configuration is surrounded by $2d$ neighboring configurations where d is the space dimension.

To solve the motion-planning problem, let us consider the two-dimensional discrete C-space depicted in Figure 1.6. We would like to find the shortest path from c_s to c_g, avoiding all obstacles on the way (dark cells). The simplest algorithm is breadth-first-search (BFS), which starts by tagging c_s surrounding configurations with a "1" tag. The neighboring configurations of these eight cells would be given "2" tags, and so on (Figure 1.7).

This procedure is continued until reaching c_g. The shortest path from c_s to c_g is found by moving from c_g back to c_s in such a way that in each step, one advances to a configuration cell that is tagged smaller than the current configuration. Commonly, this will result in several paths from which one needs to choose. Note that this algorithm requires storing all the tagged configurations. The amount of data saved (strongly) depends on the configuration space dimension and (linearly) depends on the resolution, so for high-dimensional C-space, this algorithm is not practical.

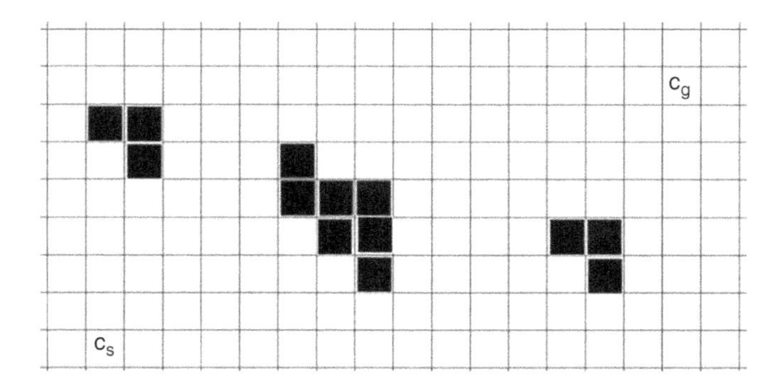

Figure 1.6 A two-dimensional discretized C-space. Obstacles are indicated by dark cells.

2	2	2	2	2	
2	1	1	1	2	
2	1	c_s	1	2	
2	1	1	1	2	
2	2	2	2	2	

Figure 1.7 Tagging the configurations from c_s toward c_g.

The A* algorithm solves this problem. The A* (reads A star) algorithm applies a different tagging method. The tagging is performed only to configurations c_k that are "more-likely" to be a portion of the path and is given by a function $f = h + g$ where g is the distance of the current configuration from c_s, and h is the estimation of the distance to c_g (commonly taken as the Euclidean distance to c_g, ignoring obstacles).

The A* involves two cell "bookkeepings": the open list OL is the *set* of cells that are currently under inspection and the close list CL is the set of cells that are already been inspected. Additionally, in order to eventually construct the path, each cell is equipped with information of its parent cell – the cell adjacent to it on the path and is closer to the starting point (i.e., if eventually this is where the path will pass). The last stage of A* is constructing the path from c_g back to c_s by tracking the parent cells.

Algorithm 1.2 A* Pseudocode

Initialize an open set OL.
Initialize a closed set CL.
Mark $c.g$ and $c.f$, $c.parent$ the g, f values of a cell c and its parent cell.
$OL \leftarrow c_s$
Set $c_s.f$ to zero.
While OL is not empty
  Pick q to be the node $q \in OL$ with q, f minimal.
  Drop q from OL
  Add q to CL
  If q is the goal cell **return**.
  For all neighbors n_i of q check
    if n_i is an obstacle cell or is in CL skip to next neighbor.
$g_{temp}, f_{temp} \leftarrow$ calculate g, f of n_i
    If $n_i.g > g_{temp}$
        Update $n_i.g \leftarrow g_{temp}$ and $n_i.f \leftarrow f_{temp}$
        Set $n_i.parent \leftarrow q$
    Add n_i to the set OL
  end for
end while
Construct a path from c_g back to c_s by tracking the parent cells.

Note that in order to perform smooth motions in the configuration space, one should increase the grid's density. Obviously, doing so will cause the data volume to increase. Note that the C-spaces volume increases exponentially with the configuration space dimension. For example, for the serial robot depicted in Figure 1.1a, the configuration space volume will exceed 2 Gb of computer data if we consider a resolution of 5° in each joint and only 8-bit storage for each cell. To overcome this, one may consider a compression of the data. The weight functions are usually smooth; therefore their corresponding entropy can be assumed to be low. So, one may compress the weight function supported by the configuration space in the same manner JPEG format compresses images. During the motion-planning process, different sectors of the C-space may be uncompressed as required for the motion-planning calculation (Medina , Taitz, Moshe, and Shvalb 2013). A different solution would be to store just a small portion of nodes (rather the entire grid); such a method is called a *sampling-based algorithm*.

1.3.3 Sampling-Based Algorithms

The basic idea of these algorithms is to build a roadmap between c_s and c_g passing through several intermediate configurations, usually randomly sampled (Kavraki and Latombe 1998). We begin by randomly sampling N configurations in C_{free}, (see Figure 1.9a). Each two configurations are checked for connectivity. This can be done, for example, by calculating whether the straight line connecting them is entirely in C_{free}. Alternatively, one can use potential-field algorithms (Koren and Borenstein 1991) for the connectivity check, which may result in a curved path. A third approach would be to crawl along the obstacle boundary rather than traversing a straight path (see [Shvalb, Moshe, and Medina 2013] see Figure 1.8). Anyway, for each connected pair, one calculates the cost for moving from one configuration to the other. Usually, this is done by integrating a predefined weight function (e.g., the required work (Borgstrom et al. 2008), path length (Ismail, Sheta, and Al-Weshah 2008), and maximal torque (Shvalb, Moshe, and Medina. 2013), etc.). This results in a weighted abstract graph Figure 1.9b. The algorithm proceeds with a path search on the graph from node c_s to node c_g while minimizing the total path cost.

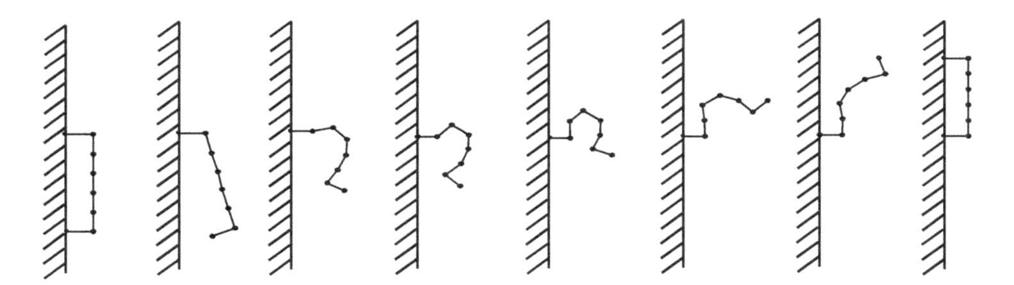

Figure 1.8 The maneuver of a six-DoF climbing robot using CPRM algorithm. The robot avoids configurations that exceed the predefined maximal power consumption. The robot also avoids collisions with the wall, as well as self-collisions (i.e., collisions of two of its links).

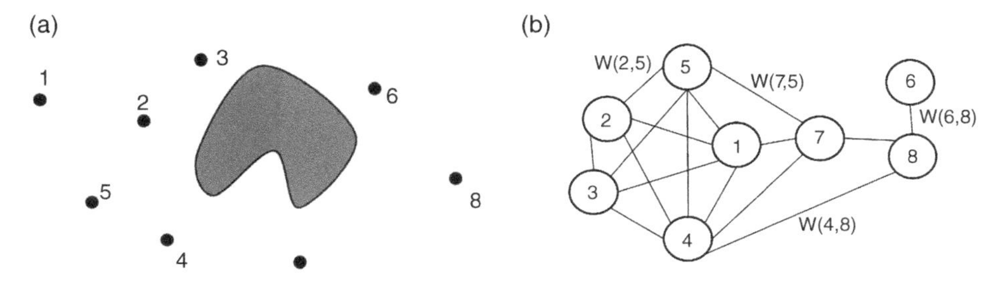

Figure 1.9 (a) The configuration space and (b) the weighted connectivity graph.

Note that configurations c_s and c_g may be added at the roadmap construction stage as waypoints or, preferably, may be added as new waypoints after the roadmap is calculated. This way the roadmap is calculated once and used for all pairs c_s and c_g.

Finding the path is a well-known problem in graph theory. The most common algorithm is Dijkstra, (1959), which is described in the following pseudocode.

Algorithm 1.3 Dijkstra Pseudo Code

```
Create a vertex set v_i ∈ V.
Mark dist(v_i) the distance from c_s to v_i .
Mark v_i. parent  the parent node of v_i.
Set dist(c_s) = 0
Set  dist(v_i) = ∞  for all v_i ≠ c_s

While V is not empty do
  Pick u from V such that dist(u) = min {dist(V)}.
  remove u from V
    for each neighbor v_i of u calculate
       temp = dist(u) + cost(u, v_i)
       if  temp > dist(u)
       dist(v_i) ← temp
       v_i.parent = u
       End if
    End for
End while
return dist, parent
```

Figures 1.8 and 1.10 exemplify the use of the sampling-based algorithm coupled with the Dijkstra graph search algorithm. In Figure 1.10, the mobile platform is assumed to have omni-wheels, i.e., it is capable of moving independently in the plane as well as changing its heading. Since the workspace is dense with obstacles, a sampling-based algorithm is the reasonable choice.

1.4 Motion-Planning Schemes: Partially Known Configuration Spaces

Commonly, the robot's available information is the data it gathers from its sensors during motion. These may include inertial measurement unit (IMU) orientation sensors, vision or distance sensors such as cameras, sonars, and infrared (IR) sensors, and location sensors such as GPS or optic-flow. These are usually fused for higher accuracy (Moravec 1988; Morer et al. 2016; Xiao, Boyd, and Lall, 2005). For most real-life motion-planning problems, there is not sufficient time or information for precalculations. The problem arises when the location and shape of the obstacles are unknown (e.g., autonomous vehicles that should identify pedestrians, vehicles, and street signs). Moreover, the obstacles

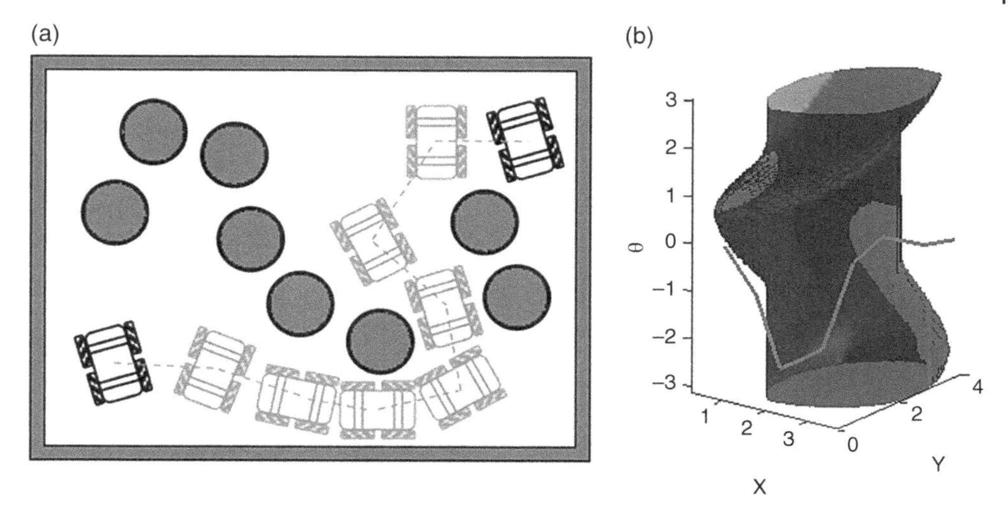

Figure 1.10 Probabilistic roadmap motion planning for a planar mobile robot. (a) Scattered obstacles and a rectangle bounding obstacle in the work space. (b) The resulting path in the configuration space.

can also be dynamic. In these situations, the problem is usually tackled by using local motion planners rather than global ones.

Assume a scenario where a robot searches for survivors in a disaster area. In such a case, in addition to the environment being unknown, the motion direction is indeterminate, i.e., c_g is not known (see, e.g. Kagan et al. 2014; Chernikhovsky et al. 2012). These kinds of problems require sensory information gathering of the surrounding environment as well as decision-making, which usually requires heuristics (Masehian and Sedighizadeh 2007; Kondo 1991; Kuffner and LaValle 2000). We shall now present the most basic motions applicable for planar C-spaces (i.e., planar mobile robots) (Lumelsky and Stepanov 1986):

1.4.1 BUG0 (Reads Bug-Zero)

Motion in c_{free} is set as a straight line toward c_g. If the robot meets an obstacle, motion is continued by following the obstacle boundaries, while the direction in which to turn (left/right) is random or predefined. The robot proceeds until the line segment between the current configuration and the goal configuration ceases to intersect the obstacle. Motion is then continued toward c_g in a straight line. Note that convergence is not certain, since the choice of motion direction as the robot meets the obstacle is not wisely chosen. Figure 1.11 depicts the potential outcomes from such an algorithm, where Figure 1.11a shows an efficient detour around the obstacle and Figure 1.11b shows an unfortunate possible outcome.

Note that applying such an algorithm to the plane is straightforward, since the obstacle's boundary tangent space is one-dimensional (i.e., the robot is left with the binary choice – turning left or right). When the C-space is three- or more-dimensional, the motion direction possibilities are infinite. In such a case, one may use a generalized approach – a gradient descent method – to move on the obstacle's boundary.

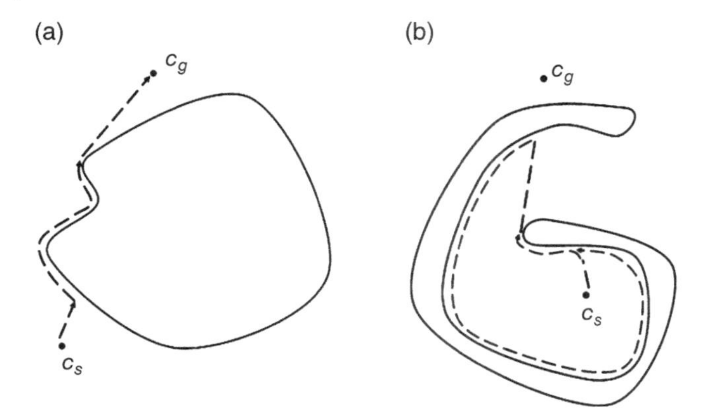

Figure 1.11 The BUG0 algorithm. (a) A scenario where the algorithm converges to $\boldsymbol{c_g}$. (b) A scenario where applying BUG0 may result in a loop (if the direction is chosen as depicted).

We shall first define a map from the C-space $Obs : C \rightarrow \mathbb{R}$ that for each configuration $c \in C$ calculates the distance to the i-th obstacle. During motion in C_{free}, the algorithm proceeds in the direction $\overrightarrow{c_g - c_k}$ where c_k is the current configuration. As the algorithm reaches an obstacle i,it calculates the obstacle gradient vector $\nabla Obs_i(c_k)$, which is perpendicular to the obstacle at c_k. Next, the algorithm calculates $null(\nabla Obs_i(c_k))$, which is the tangent space of Obs_i in c_k. Motion proceeds in the direction of the projection of $\overrightarrow{c_g - c_k}$ on $null(\nabla Obs_i(c_k))$ by small steps. This procedure continues until $\overrightarrow{c_g - c_k}^T \nabla Obs_i(c_k) > 0$, i.e., the line segment connecting the current configuration and the goal configuration ceases to intersect the obstacle. Motion is then continued toward c_g in a straight line. Obviously, the success of this algorithm depends on the convex geometry of the obstacle, yet, in some cases it may perform well and enable a fast solution, even in multidimensional spaces (Medina, Shapiro, and Shvalb 2016; Chernikhovsky et al. 2012).

1.4.2 BUG1

As long as the robot is in the interior of C_{free}, the BUG1 algorithm is similar to the aforementioned algorithm. Denote the configuration where the robot meets an obstacle by C_{Obs}. The algorithm encircles the obstacle looking for the minimal distance position to c_g. Next, the robot returns to the configuration that is the closest to c_g and proceeds toward c_g in a straight line.

1.4.3 BUG2

This algorithm differs from BUG1 in the decisions it makes as it meets an obstacle. As the robot reaches c_{free}, it marks a straight line l toward c_g. The algorithm encircles the obstacle until it encounters l again. The algorithm then proceeds to c_g in a straight line, as illustrated in Figure 1.12b.

By this algorithm, we finalize this introduction to motion planning in global coordinates. Certainly, there exists a wide variety of algorithms that either extend the presented one or address the motion-planning problem from the other points of view. Some such

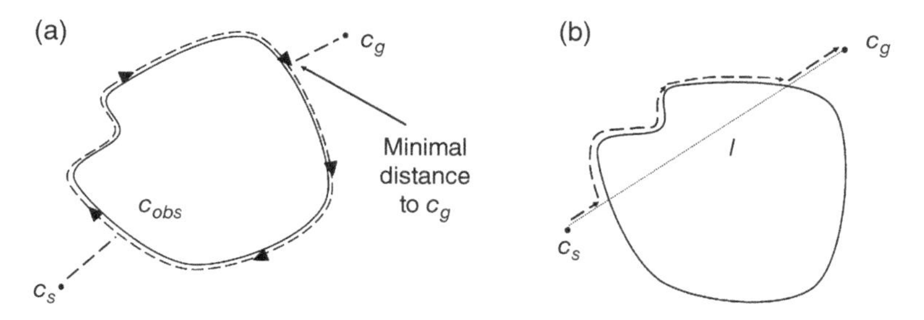

Figure 1.12 (a) The BUG1 algorithm. The algorithm encircles the obstacle looking for the minimal distance position. After returning to the minimal distance position, motion proceeds toward c_g in a straight line. (b) illustrates BUG2 algorithm. The robot circles the obstacle until it encounters the line l. The robot then proceeds toward c_g in a straight line.

algorithms are considered in the following chapters, and others can be found in the references listed in the chapters.

1.5 Summary

In this chapter, we considered the basic concepts of motion planning of mobile robots acting in global coordinates. In particular, we addressed the widely accepted concept of configuration space and its corresponding notation. Motion-planning schemes were presented for two settings: completely known configuration space and partially known configuration space.

1) For motion planning in the completely known configuration space, we considered three algorithms: (i) potential-field algorithms, which define and utilize certain artificial functions over the domain; (ii) grid-based algorithms, which assume that the domain is divided into a finite number of cells, and then apply the path-planning algorithms of the A* family; and (iii) sampling-based algorithms, which act using the intermediate configurations.
2) For motion planning with partially known configurations space, it is assumed that when it starts moving, the robot does not know about the obstacles and partially observes them only when it is close enough. For such problems, the chapter considered three basic algorithms of the same family – BUG0, BUG1, and BUG2.

The considered algorithms form the basis of motion-planning methods widely used for both mobile robots and also other tasks that deal with the activity of autonomous agents. Most of the methods presented in the next chapters will extend these algorithms for more complicated motion-planning problems and for navigation of the robot's swarms.

References

Borgstrom, P., Singh, A., Jordan, B. et al. (2008). Energy-based path planning for a novel cabled robotic system. In: *IEEE/RSJ International Conference on Intelligent Robots and Systems, 2008. IROS 2008*, 1745–1751. Institute of Electrical and Electronics Engineers (IEEE).

Canny, J. (1988). *The Complexity of Robot Motion Planning*. Cambridge, MA: MIT press.
Chernikhovsky, G., Kagan, E., Goren, G., and Ben-Gal, I. (2012). Path planning for sea vessel search using wideband sonar. In: *2012 IEEE 27th Convention of Electrical & Electronics Engineers in Israel (IEEEI)*, 1–4. Institute of Electrical and Electronics Engineers (IEEE).
Dijkstra, E. (1959). A note on two problems in connexion with graphs. *Numerische Mathematik* 1: 269–271.
Ismail, A., Sheta, A., and Al-Weshah, M. (2008). A mobile robot path planning using genetic algorithm in static environment. *Journal of Computer Science* 4 (4): 341–344.
Kagan, E., Rybalov, A., Sela, A. et al. (2014). Probabilistic control and swarm dynamics in mobile robots and ants. In: *Biologically-Inspired Techniques for Knowledge Discovery and Data Mining* (ed. E. Kagan, A. Rybalov, A. Sela, et al.), 11–47. IGI Global.
Kavraki, L., and Latombe, J.-C. (1998). Probabilistic roadmaps for robot path planning.
Kim, J.-O. and Khosla, P. (1992). Real-time obstacle avoidance using harmonic potential functions. *IEEE Transactions on Robotics and Automation* 8 (3): 338–349.
Kondo, K. (1991). Motion planning with six degrees of freedom by multistrategic bidirectional heuristic free-space enumeration. *IEEE Transactions on Robotics and Automation* 7 (3): 267–277.
Koren, Y. and Borenstein, J. (1991). Potential field methods and their inherent limitations for mobile robot navigation. In: *Proceedings of the 1991 IEEE International Conference on Robotics and Automation, 1991*, 1398–1404. IEEE.
Kuffner, J. and LaValle, S. (2000). RRT-connect: an efficient approach to single-query path planning. In: *Proceedings of the ICRA'00. IEEE International Conference on Robotics and Automation, 2000*, vol. 2, 995–1001. IEEE.
Laumond, J.-P., Sekhavat, S., and Lamiraux, F. (1998). Guidelines in nonholonomic motion planning for mobile robots. In: *Robot Motion Planning and Control* (ed. J.-P. Laumond, S. Sekhavat and F. Lamiraux), 1–53. Berlin, Heidelberg: Springer.
Lumelsky, V. and Stepanov, A. (1986). Dynamic path planning for a mobile automaton with limited information on the environment. *IEEE Transactions on Automatic Control* 31 (11): 1058–1063.
Masehian, E. and Sedighizadeh, D. (2007). Classic and heuristic approaches in robot motion planning-a chronological review. *World Academy of Science, Engineering and Technology* 29 (1): 101–106.
Medina, O., Taitz, A., Moshe, B., and Shvalb, N. (2013). C-space compression for robots motion planning. *International Journal of Advanced Robotic Systems* 10 (1): 6.
Medina, O., Shapiro, A., and Shvalb, N. (2016). Kinematics for an actuated flexible n-manifold. *Journal of Mechanisms and Robotics* 8 (2): 021009.
Moravec, H. (1988). Sensor fusion in certainty grids for mobile robots. *AI Magazine* 9 (2): 61.
Morer, R., Hacohen, S., Ben-Moshe, B. et al. (2016). *IEEE International Conference on the Improved GNSS Velocity Estimation Using Sensor Fusion. Science of Electrical Engineering (ICSEE)*, 1–5. Institute of Electrical and Electronics Engineers (IEEE).
Shvalb, N., Shoham, M., and Blanc, D. (2007). The configuration space of a parallel polygonal mechanism. *JP Journal of Geometry and Topology* December 17: 1–21.
Shvalb, N., Moshe, B., and Medina, O. (2013). A real-time motion planning algorithm for a hyper-redundant set of mechanisms. *Robotica* 31 (8): 1327–1335.
Xiao, L., Boyd, S., and Lall, S. (2005). A scheme for robust distributed sensor fusion based on average consensus. In: *Fourth International Symposium on Information Processing in Sensor Networks, 2005. IPSN 2005*, 63–70. Sringer-Verlag.

2

Basic Perception

Simon Lineykin

Sensors are required robot accessories that enable the robot to collect and process external as well as internal information. External information may include temperature, illumination conditions, distance to obstacle, magnetic field intensity, and more, while internal information has to do with the motor's electrical current, wheel revolution speed, battery voltage, arms position, etc. There are dozens of different types of sensors. A good classification is given by Tönshoff et al. (2008), Siegwart et al. (2011), and Dudek and Jenkin (2010). This chapter will only describe those types of sensors that are commonly used in robotics.

2.1 Basic Scheme of Sensors

Robot developers refer to both a sensing element and a sensing unit as a sensor. Sensing units are used in robotics for measuring, alarming, and control of processes. The sensing element, which is the principal component of a sensing unit, converts a controlled parameter (pressure, temperature, flow rate, concentration, frequency, velocity, displacement, voltage, current, etc.) into a signal (e.g., electrical, optical, pneumatic, etc.); see Zumbahlen (2007). Since the electrical signal is the most convenient for processing and digitizing, the vast majority of sensitive elements have electrical signal at the output.

Sometimes a sequence of sensing elements is used to provide an electrical output signal. The use of such sequences of sensing elements is common in exotic sensors as well as in popular ones. Such a sequence of sensing elements is best illustrated by magnetoelectric sensor (Jahns et al. 2012). This sensor consists of a pair of magnetostrictive and piezoelectric sensing elements coupled mechanically. The magnetic field that needs to be measured leads to mechanical deformations of the magnetostrictive element. This leads to strains in a piezoelectric element connected to it, which, in turn, generates an electrical field that is measured as a voltage.

Any sensing unit includes one or more sensing elements, signal conditioning circuitry, and analog-to-digital converter. Such sensors are called electronic sensors. Figure 2.1 shows the structure of an electronic sensor schematically. It is important to note that the market offers both entire measuring units and individual electronic components that

Autonomous Mobile Robots and Multi-Robot Systems: Motion-Planning, Communication, and Swarming, First Edition. Edited by Eugene Kagan, Nir Shvalb and Irad Ben-Gal.

Companion website: www.wiley.com/go/kagan/robotsystems

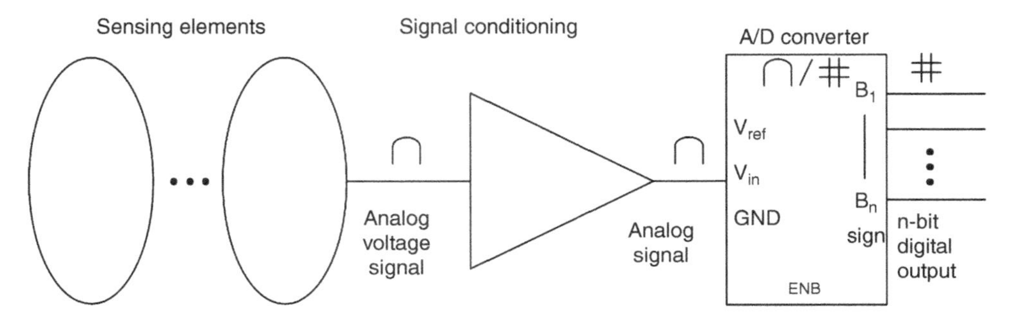

Figure 2.1 Schematic representation of the sensing unit.

allow the developer to assemble cost-effective sensor units tailored for the specific application. In the next section, we will consider each element of the sensor unit in detail.

A signal conditioning circuitry includes different types of amplifiers, filters, integrators and differentiators, frequency-to-voltage convertors, phase-to-voltage converters, avg-to-DC (direct current) and rms-to-DC voltage converters, comparators, Schmidt triggers, latch or flip-flop (FF) memory units, and more (Zumbahlen 2007). Some examples of signal conditional circuits will be discussed in further sections. The goal of the signal conditioning is avoiding unwanted noises and interferences, amplifying an analog signal to match its amplitude with the input voltage range of the analog-to-digital converter to provide a clear signal corresponding to the measured value.

An analog-to-digital converter (ADC) is selected in accordance with the required resolution, speed, and price. It is obvious that the bit depth, as well as the processing speed increase the cost of the ADC and the product as a whole. In some cases, a bit depth of one bit is sufficient. Such a bit depth is common in emergency sensors that require switching operating mode from one to another. An overheating sensor or sensors of clash with obstacle are typical examples. In this case, no special ADC is unnecessary. The single-bit digital signal can be formed using logic gates or comparator.

Processing the measured signal can be both analog and digital. Digital signal processing is more convenient and simple. However, sometimes it's easier to use the analog signal processing. The frequency-to-voltage conversion with subsequent digitization could be an example. It will be shown later by examples, such as the measurement of delay time of the light propagation. This delay of sub-nanoseconds is very difficult to process digitally by means of a super-fast processor only. At the same time this processing can be done easily in an analog way and the resulting low-frequency signal may be digitized by means of the basic ADC.

2.2 Obstacle Sensor (Bumper)

Bumper is one of the most basic and commonly used sensors (Dudek and Jenkin 2010). The main purpose of the bumper is to signalize that the movement zone's boundary is reached and no further movement in this direction is possible. The sensor of this type has

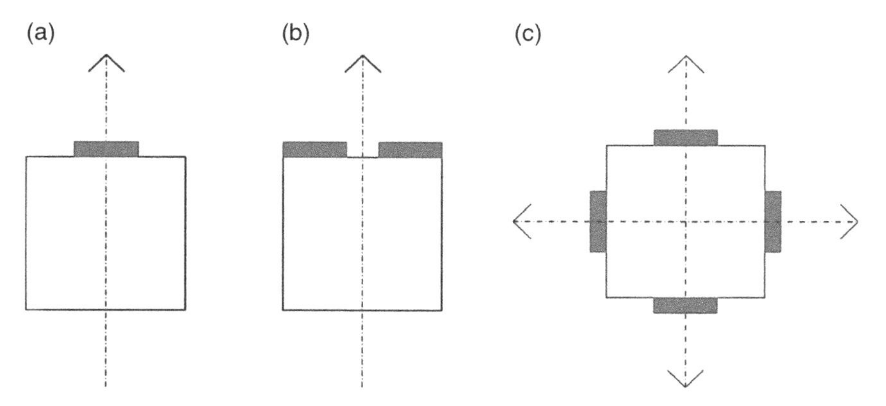

Figure 2.2 Typical installation methods of bumpers (shown as gray rectangular): (a) A sensor mounted on the axis of movement on front edge of the robot; (b) sensors are distributed along the front edge; and (c) sensors are mounted along each of the axes of movement in the case when the robot is able to move along multiple axes.

a mechanical input and electronic output. The resolution of the sensor is one-bit. One robot can be equipped with one or more bumper-sensors. The resolution of the sensor does not allow estimating the force of clash, preventing a clash, or determining the nature of the obstacle.

Figure 2.2 depicts the common methods of bumper-sensors installation. The single bumper-sensor is typically installed on the leading edge of the robot at the axis of its movement, as illustrated in Figure 2.2a. A group of such sensors can be distributed along its leading edge in the direction of its movement, as shown in Figure 2.2b. In the last case (Figure 2.2c), the robot is designed to move in multiple directions, and the bumper sensors in this case should be placed in each one of directions on the possible axes of movement, or distributed over the relevant edges. Installation of two or more sensors along the front edge of the robot allows collecting additional information about the location of the obstacle relative to the traveling direction.

The bumper-sensor consists of mechanical and electrical or electro-optical parts. The mechanical part (known as the bumper itself) is the one that involved in the impact, and gets an offset or elastic deformation as a result. There are several common topologies of such a sensor. Some of the topologies are shown in Figures 2.3–2.5.

The mechanical part of the sensor has to be massive enough to withstand any mechanical shocks. At the same time, the mass of the mechanical parts has to be small compared to the mass of the robot. Otherwise, it affects the robot's dynamics and introduces an undesirable delay between the time of impact and the time of its registration. When the mechanical part is displaced or elastically deformed, it turns on a button of micro switch or interrupts an optical beam.

A spring or built-in elastic component seeks to bring the bumper back into the original position after the cessation of contact between the robot and the obstacle.

The simple obstacle sensor is commonly used for small-mass micro- and mini-robots shown in Figure 2.3. The sensor includes a lightweight, sometimes elastic arm pivoting about the axis. The spring holds the arm in its end position. As soon as the obstacle is reached by the free end of the arm, the arm rotates about the axis and presses the button

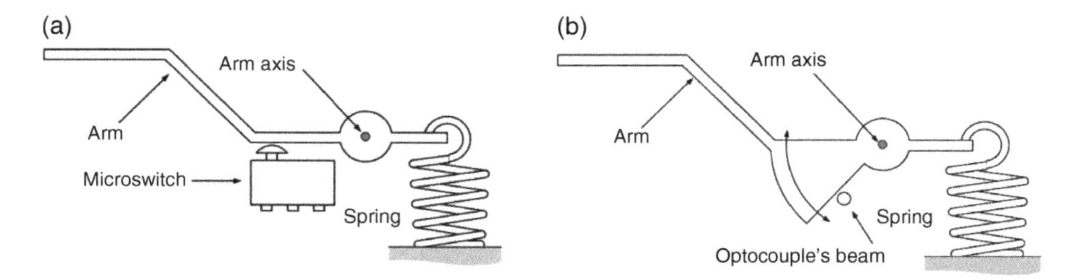

Figure 2.3 The basic mechanical construction of bumper: (a) with micro switch; and (b) with reflecting optocouple.

of the microswitch as in the case (a) of Figure 2.3. In the case of (b) shown in the same figure, the bumper is equipped with an optical switch instead of the microswitch. Advantages and disadvantages of each type of switches will be discussed below.

Figure 2.4 shows an example of arrangement of several sensors along the front of the robot in the direction of movement. With this arrangement, the obstacle located out of the axis of movement activates only one corresponding sensor. The figure shows a case when an obstacle arose ahead and to the left of travel direction. In this case, only MS 1 microswitch is actuated by means of the arm 1. In the opposite case, the MS 2 will be activated when the obstacle appears ahead and to the right of travel trajectory. In the case when the obstacle is straight ahead, both switches will be turned on.

The topology shown in Figure 2.5 is more commonly used in medium and large robots. The robotic vacuum cleaner is a good example; see Hasan and Reza (2014). A massive mechanical bumper is better suited to withstand shocks. Lack of mechanical pivots also facilitates reliability. The mechanical bumper translationally shifts into the slot due to clash and breaks (or reflects) an optical beam. A modification shown in Figure 2.5 has the ability to provide both translational and limited rotational movement. Thus, the sensor having this topology serves as two sensors mounted on the front end. As in the case described above, this sensor signalizes both the presence of the obstacle contact distance and roughly points out a direction to the obstacle relative to the direction of travel. Both the fact of impact and the direction to the obstacle with respect to the axis of motion are

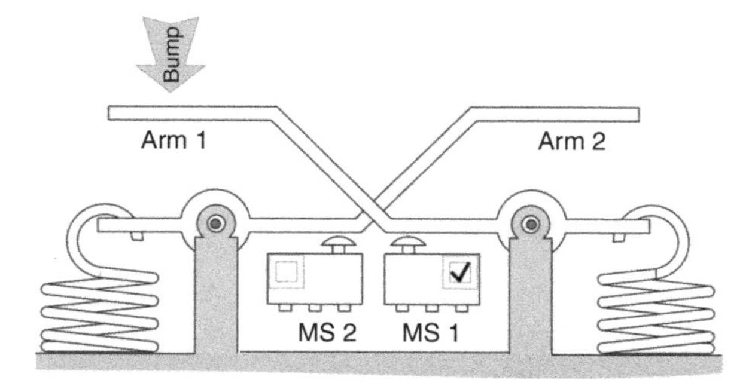

Figure 2.4 Example of two bumper sensors, which are installed in parallel on a leading edge of the robot. The obstacle on the left of the course activates the microswitch MS 1 by turning arm 1 about its axis. Vice versa, the obstacle ahead on the right-hand-side turns on MS 2 by arm 2.

Figure 2.5 Example of opto-electro-mechanical bumper. The mechanical bumper is suitable for translational and rotational shift. Thus, it first interrupts the ray on the side where the robot meets the obstacle.

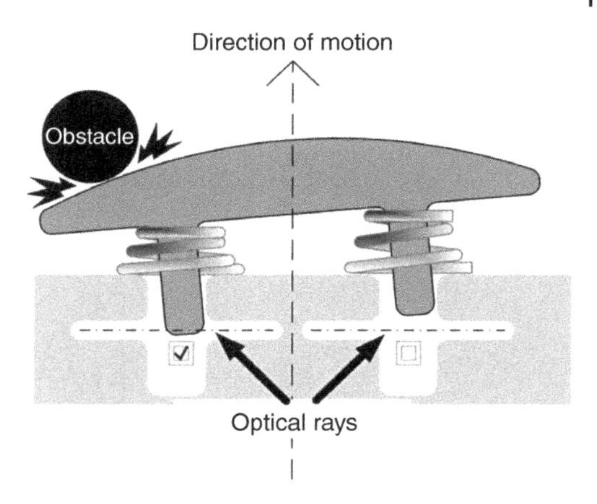

logged. The example depicted in Figure 2.5 shows that in case of an asymmetric kick, the bumper gets an additional rotational movement, and one of the optical rays is interrupted earlier than the second one. Such a scheme can be realized also by means of the microswitches.

Even in the case where the bumper's mass is negligible compared to the mass of the entire system and its dynamics due to inertia can be neglected, this type of sensor does not react immediately at the time when the bumper touches the obstacle. The mechanical part of the bumper has some backlash. It means that the mechanical part has to be displaced by a finite amount ΔX prior to activating the impact alarm. The value of ΔX helps to avoid the small mechanical vibrations effect and impact of lightweight obstacles that the robot can move out of its way without interruption, but it leads to some delay time Δt between the moment where the obstacle is already reached and the time when interruption triggers signal. The Δt delay can be calculated as

$$\Delta t = \frac{\Delta X}{v} \tag{2.1}$$

Where v is the robot's velocity at the moment when an obstacle is reached.

A mechanoelectrical/optoelectrical part of the systems converts the mechanical displacement of the bumper into an electrical signal, as an analog signal with subsequent conversion to digital form. The resolution of the output is 1 bit only.

The microswitch has two positions: "ON" and "OFF." In the "ON" position, the resistance R_{on} between its terminals is practically zero. In the position "OFF" the resistance (R_{off}) can be considered as infinite. There are two main types of microswitches: normally ON and normally OFF. The normally ON switch is short-circuited continuously and breaks the circuit when the button is pressed. The normally OFF microswitch operates in an opposite mode. It is continuously open-circuited until the button is pressed.

An auxiliary resistor together with a switch forms a widespread scheme that is called a *half-bridge* (Horowitz and Hill 1989). The four possible connection topologies of switching circuits are shown in Figure 2.6. If the auxiliary resistor is located on the high potential side of the half-bridge circuit, it is called a pull-up resistor (see Figures 2.6b,d). Otherwise, it is called a pull-down resistor, as in Figure 2.6a,c.

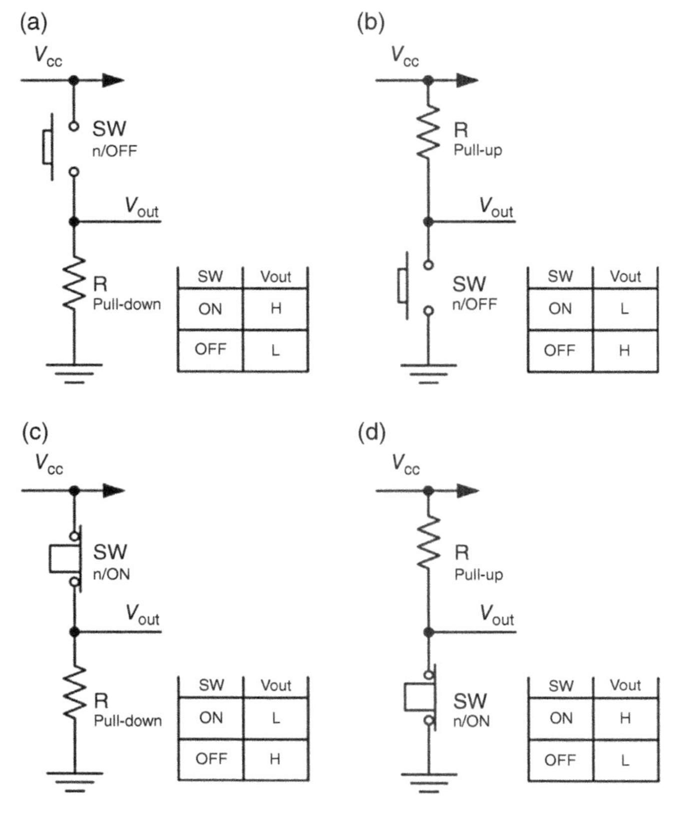

Figure 2.6 Possible variants of microswitch connection: (a) normally ON switch with pull-up resistor; (b) normally ON switch with pull-down resistor; (c) normally OFF switch with pull-up resistor, (d) normally OFF switch with pull-down resistor. The V_{cc} is a power supply voltage, H and L are logical values for high (V_{cc}) and low (ground) voltages.

Each one of the circuits proposed in Figure 2.6 has a logical state of the voltage: high voltage level (H) or low voltage level (L) at its output in accordance with the state of the switch. Let's consider the logical H is equal to power supply voltage V_{CC}, and logical L has a potential of common ground (0 Volt).

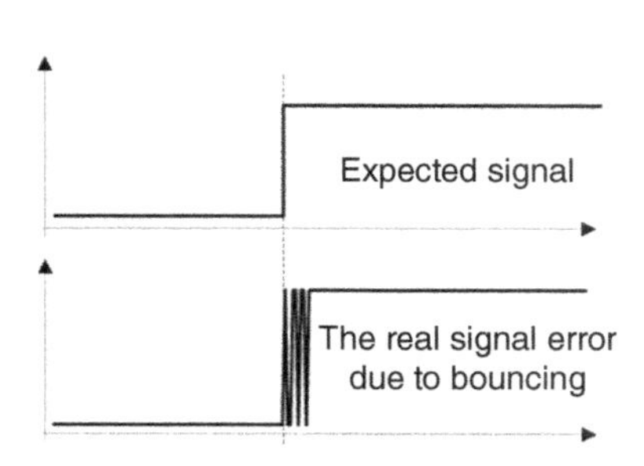

Figure 2.7 Typical switch bounce spikes.

Insofar as a switch is a mechanical device, it has its own dynamics due to built-in springs and masses of copper contacts. See, for example, Crisp (2000). The moving contact of the switch collides with a stationary contact at the commutation time. This tends to cause vibration known as a contact bounce. This causes a series of very fast ON and OFF states of the switch for a short time, until the contacts stop bouncing in the closed position. The time of bouncing is very short, as shown in Figure 2.7, but the digital circuit recognizes such spurious pulses as additional switching actions.

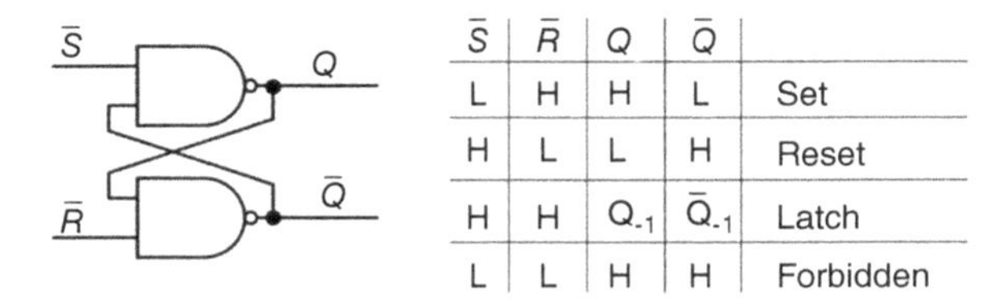

$\bar{S}$	$\bar{R}$	Q	$\bar{Q}$	
L	H	H	L	Set
H	L	L	H	Reset
H	H	Q_{-1}	$\bar{Q}_{-1}$	Latch
L	L	H	H	Forbidden

Figure 2.8 The $\bar{R}\bar{S}$-flip-flop (RSFF) topology and a table of states.

There are many ways to avoid the effect of contact bounce. The most elegant method involves the use of a latching element (also known as an asynchronous RS-flip-flop) to make the contact bounce irrelevant (Crisp 2000).

The simple flip-flop is composed of two NAND logic gates, as shown in Figure 2.8. The NAND gate has logic low (L) voltage when both of its inputs are set to high (H) logic value. Thus, when $\bar{S} \neq \bar{R}$, then $Q \neq \bar{Q}$ (see the table in Figure 2.8). These states are known as *set* and *reset*. The *latch* or *memory* state takes place when $\bar{S} = \bar{R} = H$. In this case, the values of Q and $\bar{Q}$ cannot be changed. The flip-flop "freezes" the values without change until the next set or reset action. A forbidden state is prohibited from the point of view of logic. In this state, $Q = \bar{Q} = H$. Such values cannot be stored in memory. The latch state, whenever it follows immediately after a forbidden one, puts the logical values of outputs at set or reset position arbitrarily and stores it.

The $\bar{R}\bar{S}$-flip-flop (RSFF) is very effective in avoiding the switch bounce effect. Figure 2.9 illustrates the way of using the latch together with a double-action switch in practice to provide clean pulses. The double-action, "break before make" switch in normal position, connects the reset ($\bar{R}$) junction to ground (L) and the ($\bar{S}$) junction to V_{CC} (H). This is a reset state of the RSFF. When the button is activated, the switching procedure happens in the following order: first, setting $\bar{R}$ junction voltage to logic H by disconnecting it from the ground, see (a–b) interval in time-domain diagram in Figure 2.9. This sets a "memory" state of RSFF. Neither Q nor $\bar{Q}$ can be changed during this period. Next, the switch is short-circuiting the $\bar{S}$ junction to the ground at time b in Figure 2.9. This is a set state and the output Q will be at logic H until the subsequent reset. The c–d interval is a latch again. Any further pulses due to switch bounce will be ignored at this state. At d, the

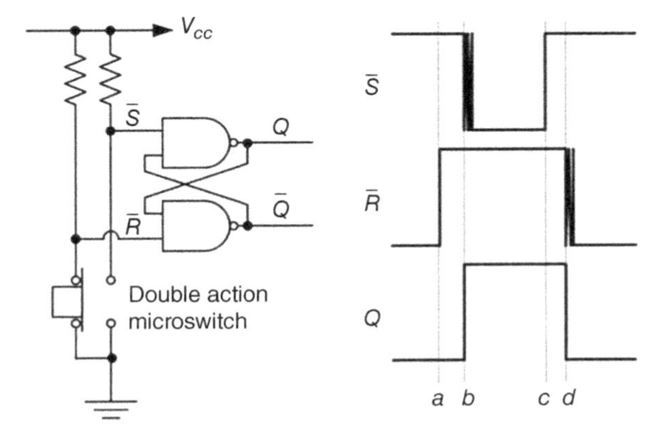

Figure 2.9 The basic scheme of bounceless switch with double-action "break before make" switch and double NAND logic latch circuit. The waveforms of the Q (output) demonstrate that this circuit cleans the contact bounce out.

latch turns back to reset state. The switching process is finished. The use of a "break before make" rather than a "make before break" switch is important, as it ensures that during the changeover period (time *a* to time *b* and time *c* to time *d* in Figure 2.9) both inputs are at logic H rather than in the forbidden state where both inputs would be logic L.

Using optocouplers instead of the microswitch also solves the contact bounce problem due to the lack of mechanical contacts. In addition, the lack of mechanical contacts greatly increases the reliability and lifetime of the system. Among others, the optocoupler does not constrain the movement of the bumper or resist it (Marston 1998 and 1999). The two main configurations of the optocouple switch (slotted and reflective) are depicted in Figure 2.10. Subfigures show the principle of their operation. The principle of operation of a reflective optocouple is shown in subfigure Figure 2.10a. Its possible realization is shown in Figure 2.10b. The Figure 2.10c subfigure depicts a slotted optocouple scheme. Both types include a pair of phototransistors and LED. The beam of light is emitted by LED. It activates the phototransistor. This happens when the beam is reflected from an external reflecting surface in the case of reflective type or by direct illumination in the case of a slotted type. The phototransistor plays a role of a switch that is activated by illuminance instead of a button. Thus, the topologies of connections are the same as in the case of microswitch; see Figure 2.6. In Figure 2.10, the pull-up resistor R_2 is connected to a collector (c) terminal of the phototransistor just as the emitter (e) terminal is grounded. The output voltage is the voltage taken at the collector. The LED is always on and its current is

$$I_{LED} = \frac{V_{CC} - V_D}{R_1} \tag{2.2}$$

Where V_{cc} is the power supply voltage and V_D is a built-in voltage of the LED that depends on emitted wavelength. For infrared LED, V_D is about 1.5–1.6 V.

When the pull-up (or pull-down) resistor is chosen correctly, the output analog voltage changes its value substantially with illumination variation, so the digital signal can be formed with a comparator only. The comparator is a type of amplifier that has a pair

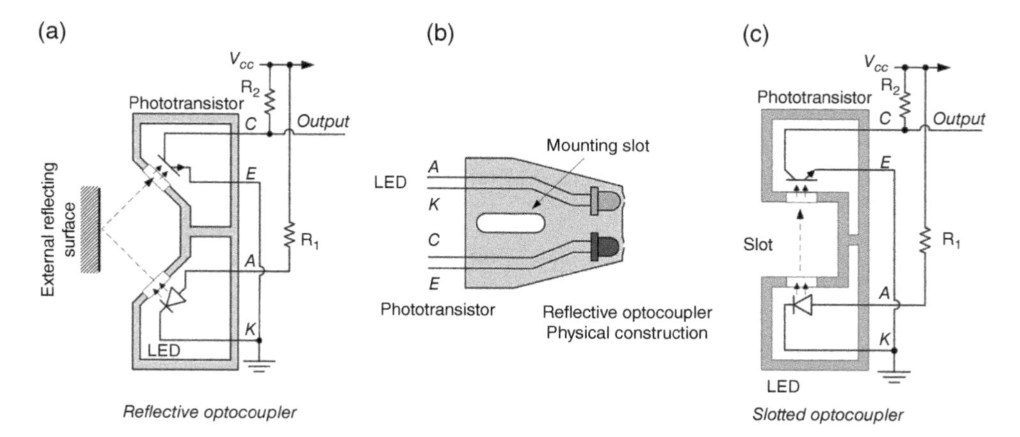

Figure 2.10 Common optocoupler topologies: (a) reflective optocoupler; (b) physical implementation of the reflective optocoupler; and (c) slotted optocoupler. *A, K, E,* and *C* means anode and cathode of the LED and emitter and collector of the phototransistor correspondently.

of input terminals and a single output terminal. The gain of the comparator is close to infinity. The comparator amplifies the voltage difference between input terminals. In practice, the values of voltages at the input terminals cannot ever be equal to each other. Even a very small difference between the input voltages causes the saturation of the comparator. Thus, the voltage at its output terminal is always equal to the voltage value of its positive or negative power source. In digital circuits, the upper saturation value of the comparator is determined by the power supply voltage V_{CC} and the lower saturation value is most often the potential of the ground (0 V). In practice the comparator often performs as an "open collector" or "open drain." In this case, an additional pull-up resistor must be used at the comparator's output.

Figure 2.11 shows four basic topologies of the photosensitive sensor connection: Figure 2.11a pull-up topology, with a resistor connected to a power source and a noninverting comparator; Figure 2.11b pull-up topology, with a resistor connected to the

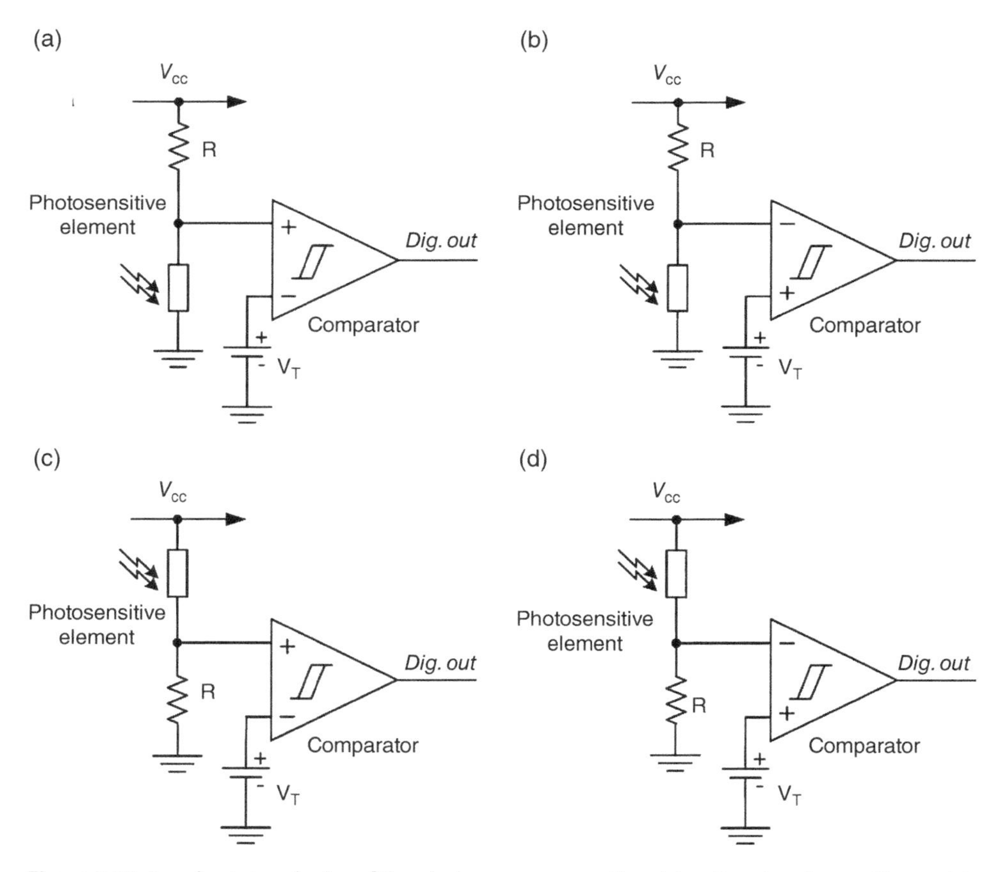

Figure 2.11 Four basic topologies of the photosensor connection: (a) pull-up topology, with a resistor connected to a power source with a noninverting comparator; (b) pull-up topology, with a resistor connected to the power supply and with an inverting comparator; (c) topology pull-down, with a grounded resistor and with a noninverting comparator; (d) pull-down topology with a grounded resistor and with an inverting comparator.

power supply and the inverting comparator; Figure 2.11c topology pull-down, with a grounded resistor and the noninverting comparator; Figure 2.11d pull-down topology with a grounded resistor and an inverting comparator. The comparator compares the values of voltage on a photosensitive element with the reference to voltage value. The reference voltage is normally selected as an average value between "lit" and "dark" values of voltage on the photosensitive element. An inverting comparator outputs a logical "one" in case the voltage on the photosensitive element is below the reference voltage value and logical "zero" otherwise. The noninverting comparator, conversely, outputs a logical "one" when the voltage on the photosensitive element is above the reference voltage value. A photoresistor, a photodiode in the conductive mode, or most often a phototransistor can be used as a photosensitive element in this system.

The photoelectric switch has no mechanical contacts, thus, it is free of contact bounce. Nevertheless, another serious problem arises in this case. This is a problem of optical noise and interference. The light-sensitive sensors are heavily influenced by solar radiation, the artificial lighting as well as the invisible spectrum radiation. Thermal noise and electronic interference also affect the output signal, albeit to a lesser degree. Figure 2.12a shows the response of the noninverting comparator on the noisy input signal. The noise causes that during the main signal level alteration (in our case, from low to high voltage) the noisy signal curve repeatedly traverses

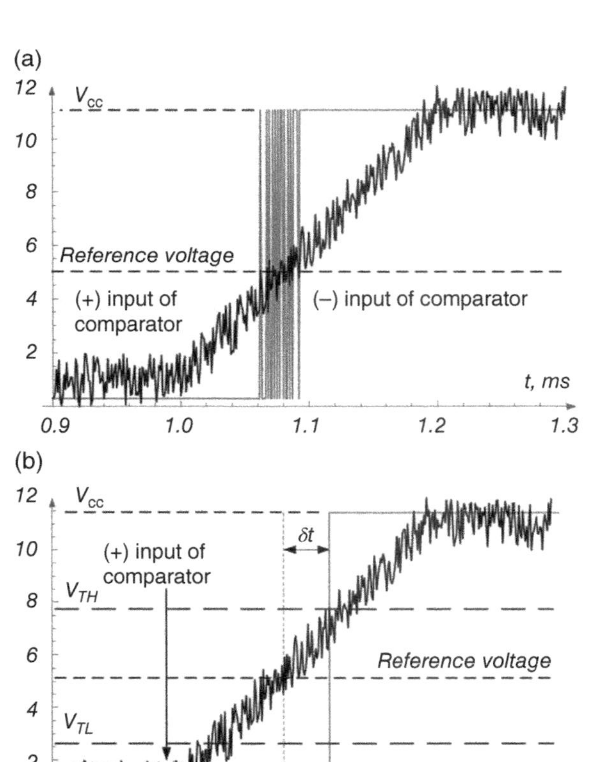

Figure 2.12 The digital output of noninverting comparator (gray curve) for noisy input signal (black curve): (a) without hysteresis; and (b) with hysteresis.

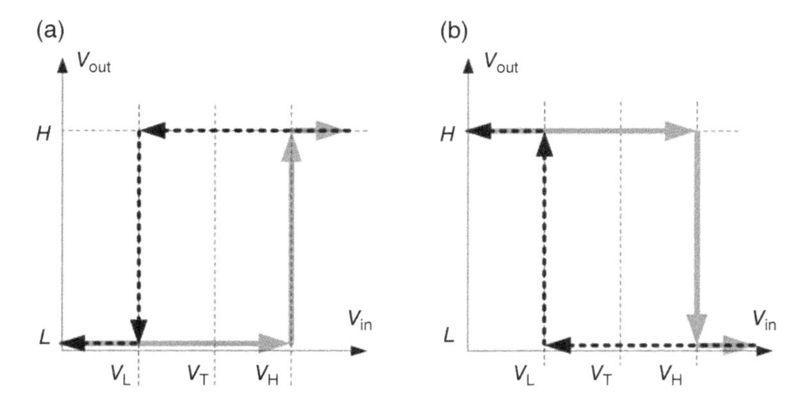

Figure 2.13 The input-output characteristics of (a) noninverting; and (b) inverting comparators with hysteresis. The gray line corresponds to an upward variation of the input signal from low to high value and the black dashed line corresponds to its downward variation.

the dashed line corresponding to the reference voltage value in both upward and downward directions. Thus, noise results in multiple pulses at comparator's output instead of the desired logic level change.

The common way to avoid pulsation at comparator output due to noise is usage of comparator with hysteresis also known as Schmitt trigger (Crisp 2000). The Schmitt trigger concept is that instead of a constant reference voltage V_{REF} the V_{TH} reference value used when the input signal varies from low to high value, and the second reference value V_{TL} used in the case where the input signal varies from high to low value. The V_{TL} has lower value than V_{TH}. The input-output characteristics of the inverting and noninverting comparator with hysteresis are shown schematically in Figure 2.13.

An example of operation of comparator with hysteresis is depicted in Figure 2.12b. The picture shows the main input signal that changes its value from low to high. The corresponding input noise is also shown. In this case, the V_{TH} reference voltage is used. The curve of the signal with the noise crosses the line of the reference voltage V_{TH}. At this point the comparator changes its state from low to high. From then on, the reference value V_{TL} is used. Thus, if the difference between V_{TH} and V_{TL} (noise margin) is higher than half of noise peak-to-peak value, the noise cannot result in a reverse switching of the comparator. The same happens when the main signal varies from high to low state. The reference value remains V_{TL} until the signal with noise reaches this value. Then the comparator toggles and the reference value changes back to V_{TH}.

V_{TL} and V_{TH} values must alternate automatically depending on the state of the comparator. Some digital devices include the Schmidt trigger. But often, the designer has to build a Schmidt trigger by himself, using a comparator chip and resistors. Standard schemes of the Schmidt trigger are shown in the following Figure 2.14.

The following analysis can be used to calculate the values of resistors. Consider the inverting and noninverting versions separately.

$$V_T = V_{CC}\frac{R_2}{R_2 + R_1} \tag{2.3}$$

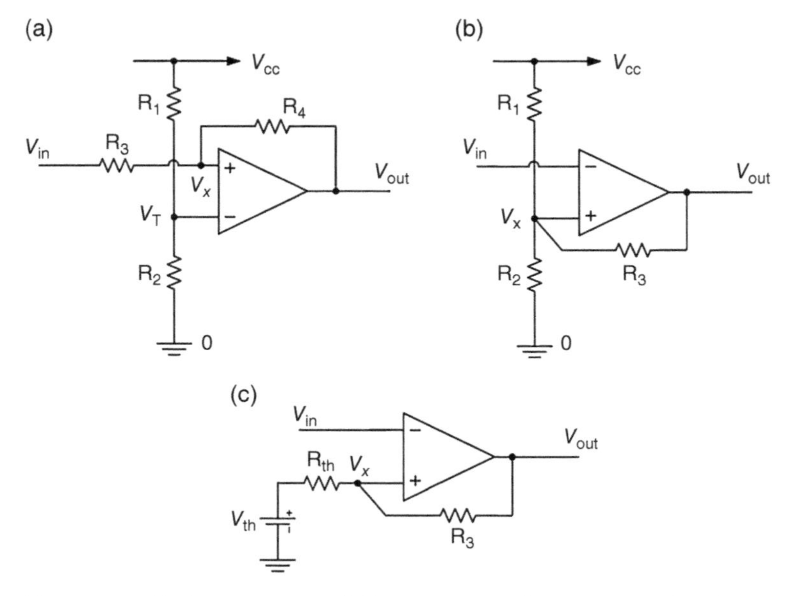

Figure 2.14 (a) Noninverting; and (b) inverting topologies of the Schmidt trigger comparator. Subfigure (c) shows Thevenin's equivalencies of subfigure (b). Pull-up resistor should be added to output in the case of open-collector or open drain comparator type.

1) Noninverting Schmidt trigger. The voltage dividers R_1 and R_2 set the threshold voltage V_T. Let's start with the case of V_{out} is Low.

$$V_x = V_{in}\frac{R_4}{R_4 + R_3} \tag{2.4}$$

The comparator toggles its state once $V_T = V_x$, thus $V_{in} = V_{TH}$ so

$$V_T = V_{TH}\frac{R_4}{R_4 + R_3} \tag{2.5}$$

$$V_{TH} = V_T\frac{R_4 + R_3}{R_4} = V_T\left(1 + \frac{R_3}{R_4}\right) \tag{2.6}$$

In the opposite case, the V_{out} is High. Thus V_{in} is also High or V_{cc}. Thus, from superposition of voltages,

$$V_x = V_{in}\frac{R_4}{R_4 + R_3} + V_{CC}\frac{R_3}{R_4 + R_3} \tag{2.7}$$

As in the previous case, $V_T = V_x$ is a critical value. In this case, $V_{in} = V_{TL}$.

$$V_T = V_{TL}\frac{R_4}{R_4 + R_3} + V_{CC}\frac{R_3}{R_4 + R_3} \tag{2.8}$$

$$V_{TL} = \left(V_T - V_{CC}\frac{R_3}{R_4 + R_3}\right)\frac{R_4 + R_3}{R_4} = V_T\left(1 + \frac{R_3}{R_4}\right) - V_{CC}\frac{R_3}{R_4} = V_T + \frac{R_3}{R_4}(V_T - V_{CC}) \tag{2.9}$$

In the most common case $V_T = V_{CC}/2$, thus

$$V_{TH} = \frac{V_{CC}}{2}\left(1 + \frac{R_3}{R_4}\right) \quad (2.10)$$

$$V_{TL} = \frac{V_{CC}}{2}\left(1 - \frac{R_3}{R_4}\right) \quad (2.11)$$

2) Inverting Schmidt trigger topology: let's change the voltage dividers R_1, R_2 by their Thevenin's equivalence V_{TH}, R_{TH} as shown in Figure 2.14c (see Boylestad 2010):

$$V_{TH} = \frac{R_2}{R_2 + R_1} \quad (2.12)$$

$$R_{th} = \frac{R_1 R_2}{R_1 + R_2} \quad (2.13)$$

We repeat now the analysis we did in the previous case, in the annex to the inverting comparator. First state: the input is low, thus the output is high:

$$V_x = V_{th}\frac{R_3}{R_{th} + R_3} + V_{CC}\frac{R_{th}}{R_{th} + R_3} \quad (2.14)$$

$$\begin{aligned} V_{TH} = V_{th}\frac{R_3}{R_{th} + R_3} + V_{CC}\frac{R_{th}}{R_{th} + R_3} &= V_{th}\left(1 - \frac{R_{th}}{R_{th} + R_3}\right) + V_{CC}\frac{R_{th}}{R_{th} + R_3} \\ &= V_{th} + \frac{R_{th}}{R_{th} + R_3}(V_{CC} - V_{th}) \end{aligned} \quad (2.15)$$

The second state: the input is high and the output is low:

$$V_x = V_{th}\frac{R_3}{R_{th} + R_3} \quad (2.16)$$

$$V_{TL} = V_{th}\frac{R_3}{R_{th} + R_3} = V_{th}\left(1 - \frac{R_{th}}{R_{th} + R_3}\right) \quad (2.17)$$

In the most common case $V_{th} = V_{CC}/2$, so $R_1 = R_2 = R$, $R_{th} = R/2$.

$$V_{TH} = \frac{V_{CC}}{2}\left(1 + \frac{R}{R + 2R_3}\right) \quad (2.18)$$

$$V_{TL} = \frac{V_{CC}}{2}\left(1 - \frac{R}{R + 2R_3}\right) \quad (2.19)$$

An additional example of a logic buffer with hysteresis assembled from two logic inverters and resistors that can perform the Schmidt-trigger function is shown in Figure 2.15.

As in the previous circuit, the auxiliary resistors set the level of hysteresis. As far the threshold voltage of the logic inverter is $V_{cc}/2$, we can derive the value of the V_{TH} and V_{TL} of the Schmidt trigger.

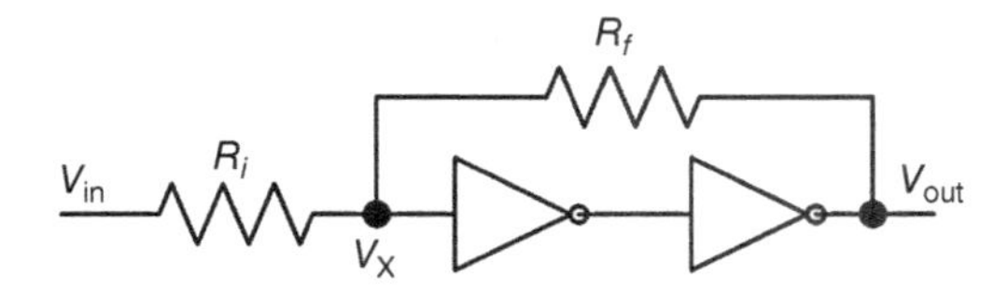

Figure 2.15 Example hysteresis logic buffer composed of two inverters and resistors.

In this case, the V_{in} is low, therefore V_{out} is low too (about 0 V). Thus,

$$V_x = V_{\text{in}} \frac{R_f}{R_i + R_f} \tag{2.20}$$

Like in the previous analysis, the V_{in} is equal to V_{TH} when V_x is equal to the $V_{\text{cc}}/2$.

$$\frac{V_{\text{CC}}}{2} = V_{\text{TH}} \frac{R_f}{R_i + R_f} \tag{2.21}$$

$$V_{\text{TH}} = \frac{V_{\text{CC}}}{2} \left(1 + \frac{R_i}{R_f}\right) \tag{2.22}$$

In the opposite case, the V_{in} is high, thus V_{out} is high too (about V_{cc}). Thus,

$$V_x = V_{\text{in}} \frac{R_f}{R_i + R_f} + V_{\text{CC}} \frac{R_i}{R_i + R_f} \tag{2.23}$$

In this case V_{in} is equal to V_{TL} when V_x is equal to the $V_{\text{cc}}/2$.

$$\frac{V_{\text{CC}}}{2} = V_{\text{TL}} \frac{R_f}{R_i + R_f} + V_{\text{CC}} \frac{R_i}{R_i + R_f} \tag{2.24}$$

$$V_{\text{TL}} = \frac{V_{\text{CC}}}{2} \left(1 - \frac{R_i}{R_f}\right) \tag{2.25}$$

This circuit is very useful, because the logic gates are commonly available and there is no need in comparator integrated circuit.

2.3 The Odometry Sensor

An odometry sensor registers the length of the way made by the robot from a specified starting point (Borenstein, Everett, and Feng 1996). In robotics the odometry sensor is an encoder of length of the route covered. It takes into account neither curvature of the route nor the absolute direction of the beginning of the route. It only registers the length of the path from the beginning of the movement to a particular point. The odometer's action is similar to the vehicle mileage counter.

The odometer converts the distance traveled into a binary digital code corresponding to the covered distance. The odometer is a mechano-electro-optical system. The most common type of an odometer is a wheel equipped with an optical encoder. One of the robot's own wheels or an extra wheel, serving exclusively for measuring the length of the route covered, can perform the function of the encoder. Sometimes it is more convenient to use an extra wheel to measure distance. It is important that the wheel does not slip. The wheel has to be as close as possible to the axis of motion. Otherwise, when driving in a curve, the wheels distanced from the axis of motion run different distances. The odometer for a robot differs from a vehicle odometer by its resolution. A typical resolution for a mobile robot is centimeters or even tenths of a centimeter.

The distance by which the center of the wheel will travel along the surface due a single revolution of the wheel will be equal to

$$L_1 = 2\pi R \tag{2.26}$$

where R is an external radius of the wheel.

To increase the resolution, the number of fractions of a wheel revolution should be counted rather than number of complete revolutions. Thus, the distance covered due to revolution of the wheel by Θ, the nth part of 2π angle will be

$$\Delta L = 2\frac{\pi}{n}R = \Theta_n R \tag{2.27}$$

The full distance covered due to the number N of nth parts of the wheel revolutions can be expressed by the relation

$$L = N\,\Delta L = N\Theta_n RL \tag{2.28}$$

Thus, an odometer must enroll each increment of the rotation angle by Θ, and count the total number of increments. The binary number corresponding to the number of such increments is the output of the odometer.

There are many different types of optical encoders used for odometry. The simplest optical encoder, chopper, is shown in Figure 2.16. The chopper consists of a punched disc mounted coaxially with the wheel. The number of punches defines the resolution of odometer. The number N of expression (2.28) is equal to the number of punches. The different choppers may contain from one to thousands of slots. The angular

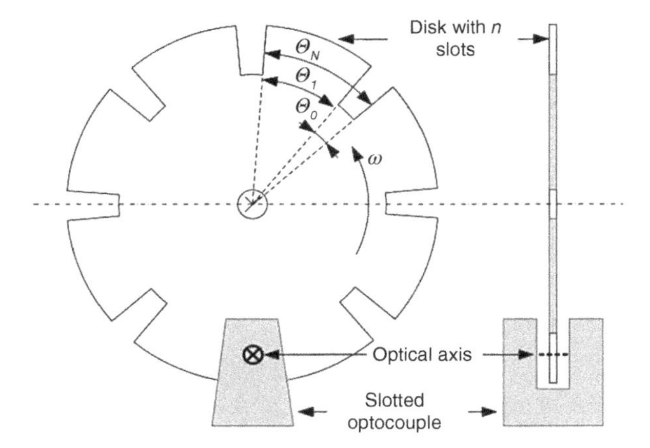

Figure 2.16 Front and side views of an optical chopper, the opto-mechanical part of the odometer.

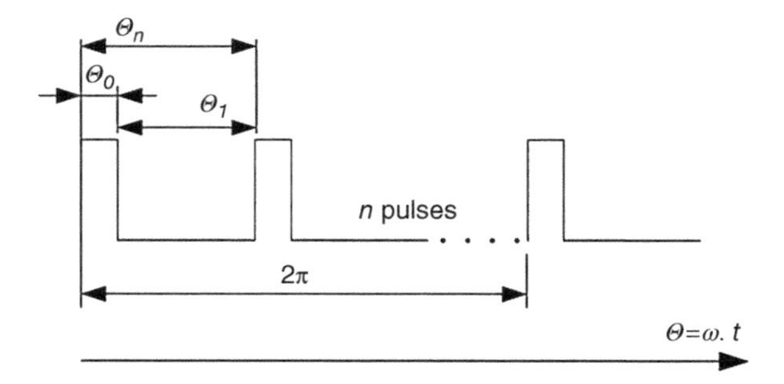

Figure 2.17 The digital pulse train at the output of odometer's optocouple.

displacement of the disc is equal to the displacement of the wheel. Once the slot of the disk is out of the optical axis of the optocouple, the beam of the optocouple is interrupted. The beam interruption leads to the pulse of voltage at the output of the optocoupler. The signal conditioning and pulse forming circuitry of the optocouple is similar to the one already explained.

As a result of continuous rotation of the wheel (and the punched disk together with it), pulse sequence is generated on the electrical output optocouplers. This pulse sequence is shown in Figure 2.17. Each subsequent pulse means that the wheel has turned an additional n^{th} portion of a full rotation cycle, by angle Θ_n.

To count the number of pulses generated since the beginning of the movement, a binary counter circuit may be used (Crisp 2000). Figure 2.18 shows the simple circuit of the up/down series binary counter.

The scheme utilizes the JK-Flip-flops (JKFFs) with both data inputs J and K set to a logical one. This type of connection converts the JKFF to a toggle-flip-flop (TFF). The TFF will "toggle" on the falling edge (high-to-low transition) of the clock signal input. In the case when the input up/down is set to a logical zero, the "clock" input of each one of the TFFs except the first one is equal to the value of the output Q of previous TFF. Otherwise, when the up/down input is set to logical one, the "clock" input of each one of the TFFs except the first one is equal to the value of the output $\bar{Q}$ of the previous TFF. The clock input of the first TFF is the pulse sequence to be counted. Obviously, this scheme can be extended to include more bits by adding more flip-flops.

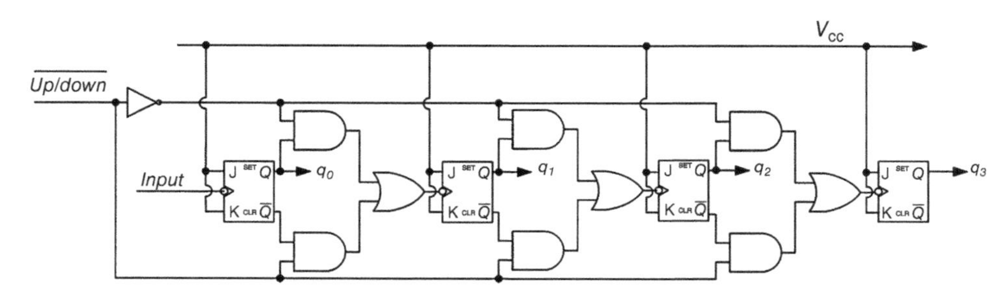

Figure 2.18 The 4-bit series binary counter using JK-flip-flops.

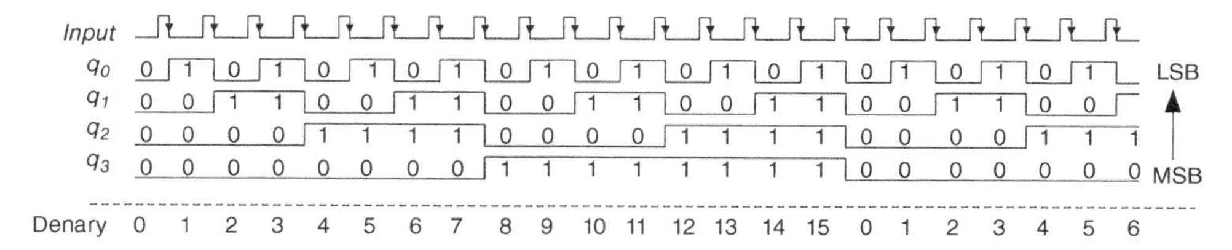

Figure 2.19 The typical waveforms of the circuit shown in Figure 2.18, in the case of up/down input is set to "up" (logical zero).

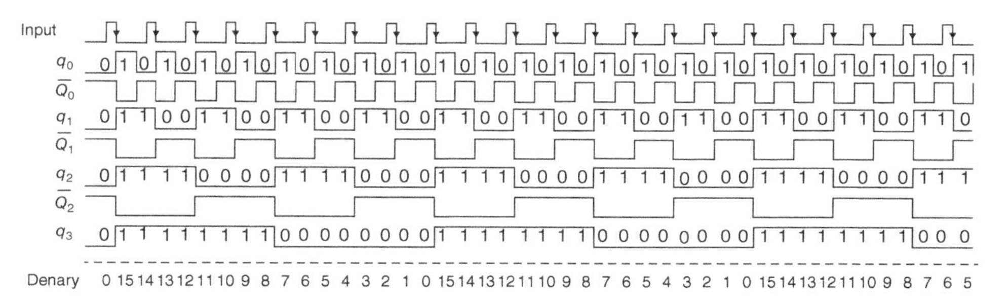

Figure 2.20 The typical waveforms of the circuit shown in Figure 2.17 in the case of up/down input is set to "down" (logical one).

Figures 2.19 and 2.20 shows the waveforms of signals of the scheme in Figure 2.18. Figure 2.19 corresponds to a case of "up" counting. In this case, the waveforms of $\bar{Q}$s are irrelevant. For reading the binary output, one has to reorganize the order of the outputs $q_0 - q_2$ so as to obtain the least significant bit (LSB) at q_0 and the most significant bit (MSB) at q_2. The bits are arranged so that they form a binary number. In the case of upcounting, each input pulse increases the binary output by 1 until it reaches the maximum possible value for the given bit-depth. After this value, the circuit starts counting again from zero.

Figure 2.20 shows the waveforms of the same circuit, but with up/down input set as "down" (logic one). The clock input of each flip-flop except the first is a value of output $\bar{Q}$ of the previous flip-flop. As one can see, the waveforms of output bits $q_0 - q_2$ form the binary numbers.

The value of the up/down bit can be set by a microcontroller that controls the robot. This bit must change its value when the wheel changes the direction of revolution. However, very often the encoder is provided with an additional device that allows it to automatically detect the wheel of revolution direction and to change the direction of the counter. This device involves the use of an auxiliary optical pair, which is mounted in such a way that when the optical axis of the main optocoupler (call it A) is open, the optical axis of the auxiliary optocoupler (call it B) is partially closed. Figure 2.21a depicts such a device. This arrangement ensures that the output signals of A and B optocouplers are 90° shifted in phase relative to one another. When the wheel rotates in the positive direction, the signal A leads, as shown in Figure 2.21b. As soon as the direction of rotation changes, the order of waveforms A and B changes, too.

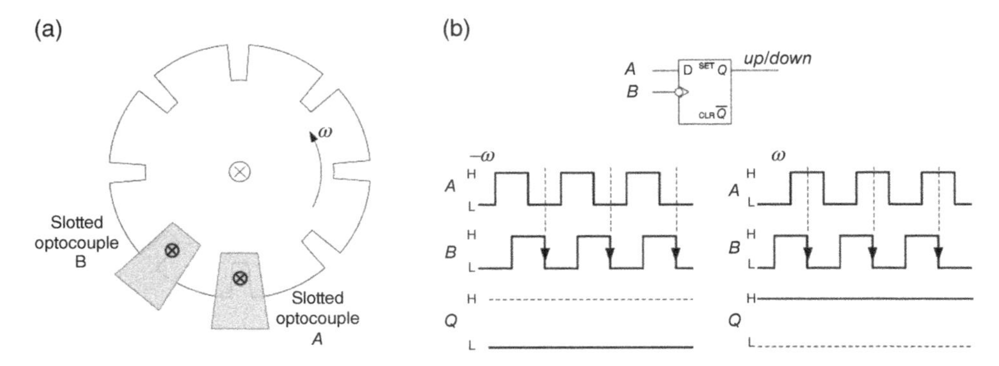

Figure 2.21 Using an auxiliary optocoupler and D-flip-flop for detecting the direction of the wheel rotation: (a) the scheme of mechanical and electrical connection; (b) waveforms for direct (ω) and opposite ($-\omega$) directions of rotation.

The D-flip-flop (DFF) can serve as a lead/lag detector in this case (see Figure 2.21a). Whenever the pulse sequence B on terminal $c\bar{l}k$ of the DFF leads an A sequence on its D input, the DFF remains in the "set" position. Its output Q stays at a logical 1. Otherwise, when the A-signal leads, the DFF stays in the position "reset" and its output Q is a logical 0. Thus, the DFF can set up/down direction of the counter when outputs of A and B optocouples are connected to D and $c\bar{l}k$ inputs of the DFF and its output Q is connected in its turn to "up/down" input of the counter.

2.4 Distance Sensors

A distance sensor, also known as a range finder, enables a robot to detect obstacles in its path (Borenstein, Everett, and Feng 1996). Unlike the bumper switches that alert you when they have been hit, the range finder sensor can alert you to an obstacle in the path of the robot prior to hitting it. This can allow you time to safely navigate around obstacles. The sensor can be used to determine distances to objects. It can be used as a tool to determine if any objects are in the robot's path at all. To increase the sensing range, the sensor can be mounted on a turntable to allow it to rotate. The most useful in robotics are laser and ultrasonic rangefinders.

The main types of laser rangefinders are time-of-flight (ToF), phase-shift, triangulation, and interference. Interference range finder is the highest-precision method of optical distance measuring: its precision is parts of micrometers, but its range of measurement is very short. This is why this method has not been used for robotics and is not described here.

Here we consider the principle of operation of these basic types of sensors.

2.4.1 The ToF Range Finders

The ToF range finder consists of an optical part, opto-electric sensor, electronic signal conditioning system, pulse-shaping circuitry, delay-to-voltage converter, and, finally, a

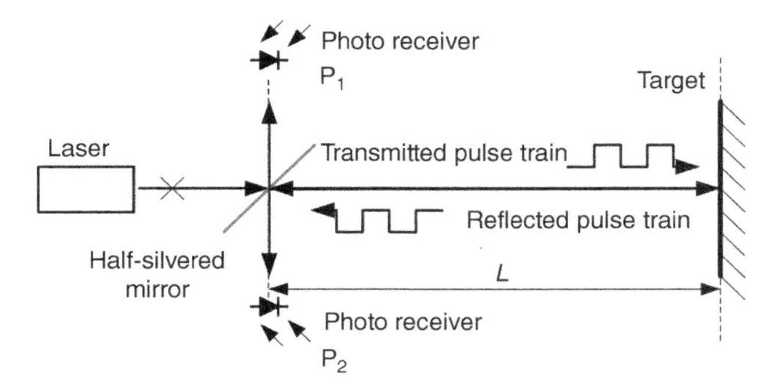

Figure 2.22 Bursts of light sent by the laser toward a target are reflected back. The half-silvered mirror splits the portion of forward going ray to a photo receiver P_1 and turns a backward light to photo receiver P_2.

high resolution analog-to-digital signal converter (Siciliano and Khatib 2016). The optical portion of the ToF range finder is shown in Figure 2.22. A laser sends busts of light toward the target. Bursts of light come back to the starting point being reflected from the target. Thus, light passes the doubled distance from the instrument to the target. The light travels through the air at a finite speed so that there is a time interval between the time when a burst of light is transmitted and the time when it is received.

The ToF is calculated as

$$\mathrm{ToF} = 2\frac{L}{c_a} \tag{2.29}$$

where c_a is the light propagation speed in air and L is a distance to the target. Thus, the distance L can be derived from (2.29) as follows:

$$L = \frac{TOF}{2}c_a \tag{2.30}$$

A photodiode operating in photovoltaic mode with a current amplifier is most commonly used for light registration; see Figure 2.23a. The photons of the laser beam lead

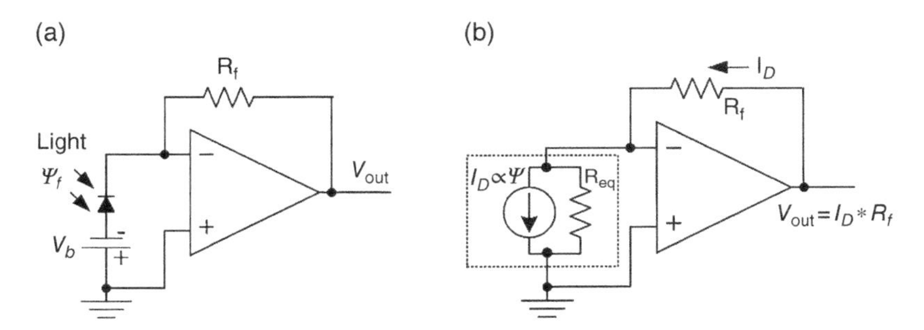

Figure 2.23 The photo receiver: (a) photodiode in photovoltaic operation mode with amplifier; and (b) equivalent electric circuit of the photodiode with amplifier. Ψ is the intensity of light. The diode's current I_D is proportional to Ψ. The R_{eq} is a parasitic shunt resistance of the photodiode equivalent circuit.

the electric current at the photodiode's junction. The current is to be amplified by a trans-impedance amplifier. This type of amplifier converts the input current to output voltage and can be implemented using fast operational amplifier, as it is shown in Figure 2.23.

The operational amplifier is a device that measures the voltage drop between its (+) and (−) input terminals, and amplifies the result by the open loop gain A_{ol}. The A_{ol} of ideal opamp is assumed to be infinite. This assumption is not far from the real situation. The amplification gain of hundreds of millions can be assumed as infinity in many cases. Thus, no opamp can operate in linear mode in open loop. Only negative feedback permits linear operation mode and finite amplification. To explain the principle of current amplifier as any other linear circuits with one or more opamps, the two "golden rules" describing the ideal amplifier principle can be used (Horowitz and Hill 1989).

The first rule is derived from the assumption that voltage measuring circuits at the opamp's input terminals (+) and (−) are almost-ideal ones with infinite resistance. Thus, the first rule claims that "no current flows into/out of the input terminals."

The second golden rule is the "rule of virtual zero." It states that "the voltage drop between its input terminals is zero when the scheme with operational amplifiers operates in linear mode." The input signal multiplied by an infinite gain gives a finite value. This means that the input signal is very small. It should be negligibly small compared with all the signals in the circuit – that is, practically zero. Alternatively, the virtual zero rule can be formulated as follows: "The operational amplifier with negative feedback in a linear regime has at its output a voltage that maintains equal voltage values at the input terminals."

Operating with both of these two rules, we can easily analyze the trans-impedance amplifier of Figure 2.23b where the photodiode is replaced by its equivalent circuit. The (+) terminal is grounded, so its potential is zero, so the potential of (−) terminal is zero too. The potential drop at R_{eq} is zero, so no current flows via this resistor. It also means keeping in mind the first golden rule that the entire photo-current I_D flows via feedback resistor R_f. Thus, the output voltage is

$$V_{out} = R_f I_D \tag{2.31}$$

Figure 2.24 depicts one of the possible electronic schemes of the ToF rangefinder. The two equal branches of photo detection circuits are able to register both transmitted and reflected light. The photodiodes' current is amplified by identical current amplifiers. The analog signals from amplifiers are shaped into pulses by Schmidt trigger circuits. The delay between the transmitted pulse and the reflected pulse can be detected by XOR logic gate. The output of the XOR is a train of short pulses of time T_{on}. The T_{on} is in fact a ToF. The longer the distance to the target, the greater the value of time T_{on} at XOR's output.

One should keep in mind that even for large distances the time of flight is a very small time, of the order of hundreds of nanoseconds. Registration of such intervals with high resolution is very difficult and costly when using digital electronics. In such cases, the help comes from analog electronics. The laser transmits the light bursts with constant period T. This means that the pulses having the width T_{on} that depends on the measured distance are repeated with a constant period T. Averaging of such a pulse train gives the DC voltage of value:

$$V_{DC} = \frac{T_{on}}{T} V_{CC} \tag{2.32}$$

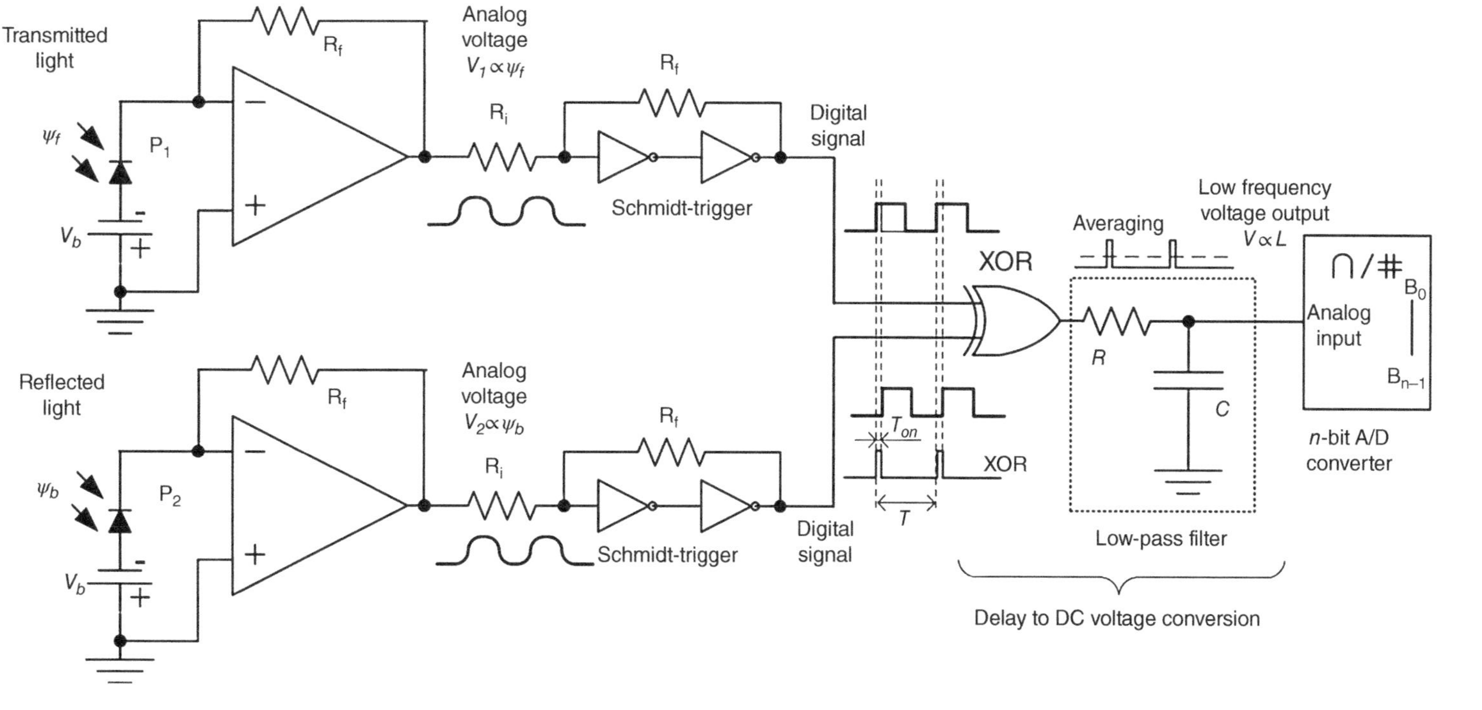

Figure 2.24 Possible electronic scheme of a ToF laser rangefinder.

Average voltage value over the period is easily obtained using low-pass filter of the first or higher order. If the bandwidth of LPF is at least two orders lower than frequency $1/T$, the ratio of ripple to the signal will be very low. The ADC with high resolution can digitize a resulting DC voltage.

Since the two photo detection, amplification, and processing branches are identical, a signal propagation delay (t_{pd}) of each element is compensated and does not matter. However, absolutely identical elements do not exist. The signal propagation time is slightly different from chip to chip. This difference determines the resolution of the method. Summary imbalance signal propagation time in the two branches is no less than a nano-second. Assuming the speed of light in the air as equal to the speed of light in the vacuum (approximately 3.00×10^8 m s^{-1}), we can estimate the resolution of the measured distance:

$$L_{1ns} = \frac{1 \times 10^{-9}}{2} \cdot 3 \times 10^8 \left[\frac{\text{sec m}}{\text{sec}}\right] = 0.15\,\text{m} = 15\,\text{cm} \tag{2.33}$$

Thus, it can be concluded that the laser rangefinder of TOF type has a relatively low resolution, and its circuitry is quite expensive. However, it is very good for long distances of hundreds of meters up to several kilometers and it is very fast in operation (Siciliano and Khatib 2016).

2.4.2 The Phase Shift Range Finder

The phase shift range finder has a better resolution than ToF rangefinder; it operates in the range from centimeters to tens of meters, but some slowly than the previous method. An optical scheme of the system for the laser rangefinder with the phase-shift measuring as a whole is similar to the scheme of the ToF range finder described above. Like in the previous case, the laser transmits the ray toward the target. But in this case, the ray's intensity is modulated by a modulating signal with frequency f_{m}. Note the carrier frequency of the laser stays constant. The modulating frequency is normally below 500 MHz. The spatial length of the modulating wave is

$$\lambda = \frac{c_a}{f_m} \tag{2.34}$$

Where c_a is a speed of light in the air. The principle of operation is shown in Figure 2.25.

The double distance between the source and the target contains an integer number N of wavelengths and a portion of wavelength of the modulating signal.

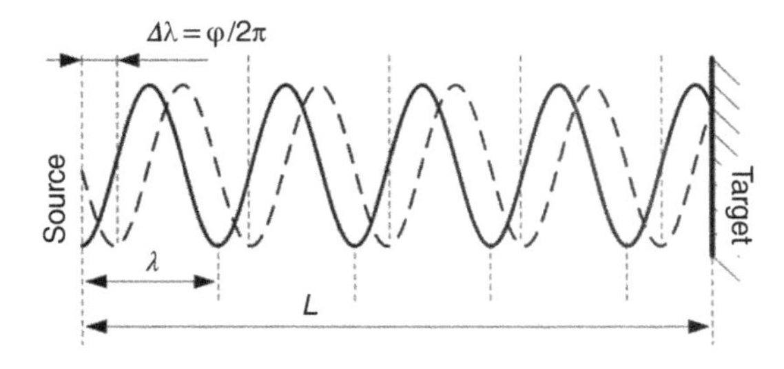

Figure 2.25 Schematic representation of the envelope of the modulated signal transmitted by a laser (solid curve) and the reflected signal. L is the distance from the source to the target, λ is a wavelength of the envelope, and $\Delta\lambda$ is a portion of a wavelength. φ is a phase shift of reflected wave relative to transmitted wave at the origin.

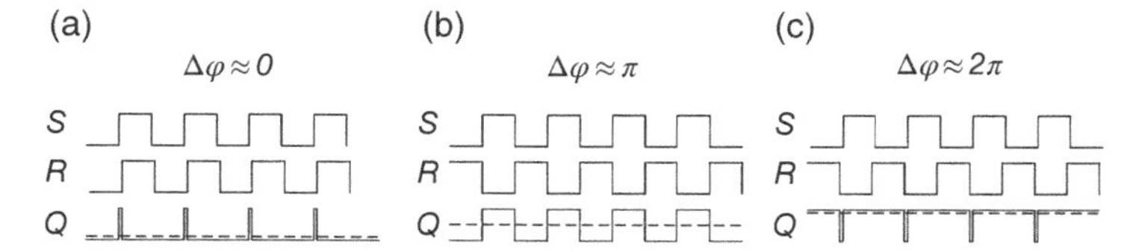

Figure 2.26 Phase-shift detection using asynchronous RS-flip-flop. *R* and *S* are reset and set inputs correspondently. *Q* is the flip-flop's output. The dashed line shows the average voltage level of different phase shifts between the input signals.

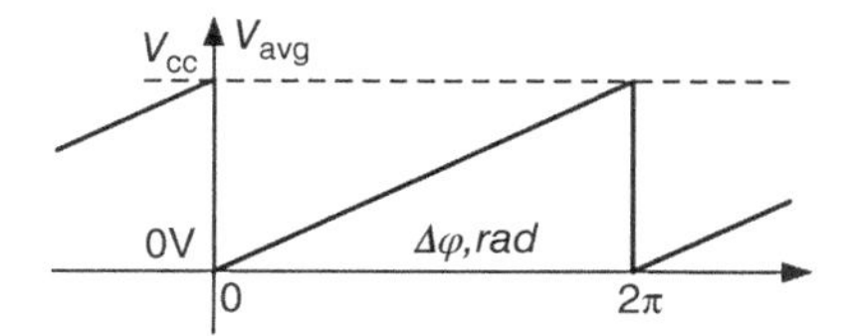

Figure 2.27 The average voltage of the RSFF output *Q* versus the phase-shift between its input signals.

$$L = \frac{\lambda \cdot N + \Delta\lambda}{2} = \frac{\lambda}{2}\left(N + \frac{\varphi}{2\pi}\right) = \frac{c_a}{2f_m}\left(N + \frac{\varphi}{2\pi}\right) \tag{2.35}$$

where φ is a phase shift of the reflected signal. The phase shift φ can be precisely measured in many ways. The number N in the expression cannot be measured directly. The system of equations should be solved to calculate it.

One of the popular methods of phase-shift detection is digital phase detection. The simple asynchronous RS flip-flop having on its R and S inputs some square waveforms of the same frequency, as shown in Figure 2.26, returns at its output Q the pulse train. The width of the pulses is proportional to a phase-shift between the two signals.

As it is shown in Figure 2.26, the high-frequency terms of the output signal can be filtered out by low-pass filter. Thus the low-frequency (or DC) average signal will be proportional to a phase-shift between the flip-flop's inputs. Figure 2.26 shows at (a), (b), and (c) subfigures some characteristic phase-shifts. The 0 rad leads the 0 V average voltage. The π rad phase shift leads to a half V_{CC} average voltage, and 2π phase shift leads to a V_{CC} average voltage. The full V_{avg} versus the phase-shift function is shown in Figure 2.27.

An electronic scheme of the phase-shift laser rangefinder is shown in Figure 2.28. The modulating signal (in our case of sine form) varies amplitude of the laser with modulating frequency f_m. The photo detector converts the amplitude of the reflected amplitude-modulated signal into a voltage. This process is known as envelope detection. The envelope of the reflected signal has the same frequency f_m as the modulating signal. The phase shift of the detected envelope signal relative to the modulating signal is measured by converting both signals into square waves by Schmidt triggers and by feeding the square waves to a SRFF input. Filtering out high-frequency components of the flip-flop's output Q returns us a DC voltage, proportional to a phase shift - φ. The A/D converter digitizes this DC voltage with a high resolution.

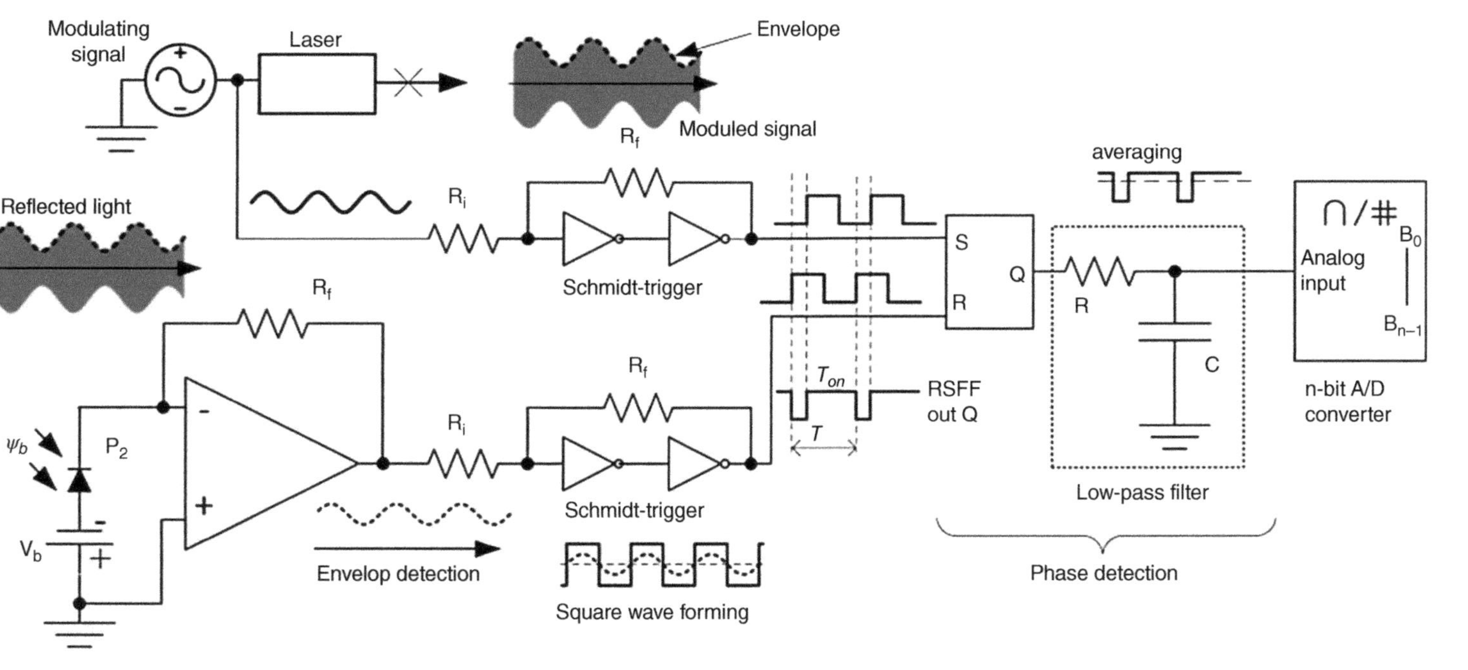

Figure 2.28 Simplified block-scheme of the phase-shift laser rangefinder.

Now, after phase shift is detected, we return to expression (2.35). There are two unknown variables in the expression: φ and N. It was shown above that the variable φ can be measured precisely. The number N cannot be measured but calculated only. There are several methods for approximate calculation of N. The most popular method is to use two modulation frequencies, f_{m1} and f_{m2}. Thus,

$$L = \frac{c_a}{2f_{m1}}\left(N_1 + \frac{\varphi_1}{2\pi}\right) = \frac{c_a}{2f_{m2}}\left(N_2 + \frac{\varphi_2}{2\pi}\right) \tag{2.36}$$

If f_{m1} and f_{m2} are close to one another, then

$$N_1 = N_2 = N \tag{2.37}$$

In this case, the N could be eliminated from (2.36):

$$L = \frac{c_a}{4\pi}\frac{\varphi_1 - \varphi_2}{f_{m1} - f_{m2}} \tag{2.38}$$

The N can be calculated as

$$N = \frac{f_{m2}\varphi_2 - f_{m1}\varphi_1}{2\pi\left(f_{m1} - f_{m2}\right)} \tag{2.39}$$

This method has good precision, but due to deep averaging integration time and the processing time it is quite slow and not very suitable for real-time measurement or for scanning.

2.4.3 Triangulation Range Finder

The triangulation type of rangefinder uses a linear image sensor. It is a one-dimensional array of photo sensors of the same type, normally charge-coupled devices (CCD) (Siciliano and Khatib 2016). The modern CCD arrays of pixels include up to tens of thousands of pixels of several micrometer size. A principal scheme of the triangulation-type range finder is shown in Figure 2.29. A pinhole is used instead of the lens to simplify the visualization. In a real devise, the lens is used to improve the light concentration and to reduce the mechanical size of the system. Nevertheless, the principle of operation of the pinhole and the lens are the same. Each pixel of the photodetecting device can see through the pinhole only one point located on the axis of the laser pointer's ray.

Thus, the pixel number 0 "knows" what happens in point L_{max} only. The pixel n registers the value of luminosity at point L_{min} only. All other pixels of the array can see the points between points L_{min} and L_{max}. This range is shown in Figure 2.29 as R. When some object, let's call it a target, interrupts the beam of the laser pointer within the R, the spot of the laser on the target's surface will be registered by a pixel i corresponding to a point on the distance L of the pointers' axis. Thus, the distance L should be calculated as:

$$L = L_{min} + \frac{L_{max} - L_{min}}{n} i \tag{2.40}$$

This method is very useful for short-range measurements. It can provide a fairly good accuracy, not very fast, but cheap.

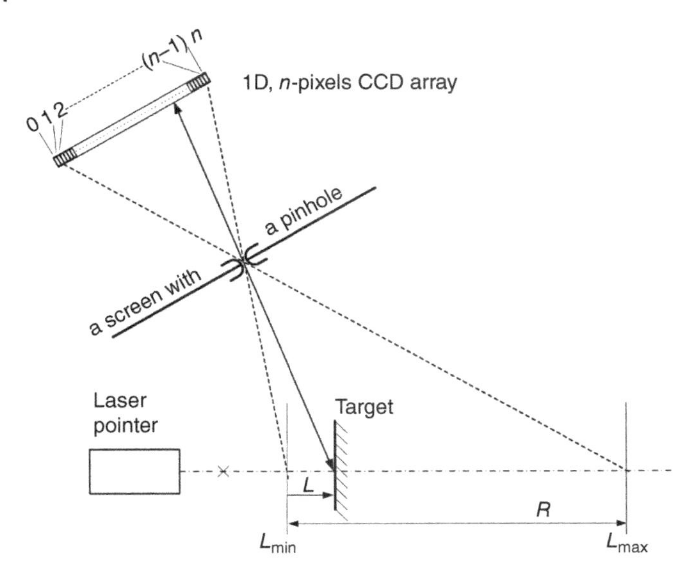

Figure 2.29 Explanation of the triangulation method for finding the distance L to the target. For the sake of simplicity, a pinhole is used here instead of a lens.

2.4.4 Ultrasonic Rangefinder

An ultrasonic rangefinder sensor enables a robot to detect obstacles in its path by utilizing the propagation of ultrasonic acoustic waves (Marston (1998) and Siciliano and Khatib (2016). The sensor's sound transmitter (speaker) emits a 40 kHz ultrasonic wave bursts, which bounces off a reflective surface and returns to the sound receiver (microphone) sensor. Then, using the amount of time it takes for the wave to return to the sensor, the distance to the object can be computed. Unlike the bumper switch that alerts you when it has been hit, the ultrasonic range finder alerts you about an obstacle in the path prior to hitting it. The ultrasonic rangefinder can also be used to determine a distance to objects within a range of a couple of meters.

The idea of the ultrasonic rangefinder is very close to a ToF laser rangefinder discussed above. In dry air at 20 °C, the speed of sound is 343.2 m s^{-1}. Increasing the temperature of the surrounding air increases the speed of sound by about 0.6m s^{-1} per centigrade. Humidity also has a small but measurable effect on the speed of sound (causing it to increase by about 0.1–0.6%). Thus, the time of flight of 1 ms corresponds to about of 0.34 cm of distance.

Figure 2.30 depicts a block-scheme of the ultrasonic rangefinder. The trigger pulse at input (a) enables the burst generator that produces the sequence of pulses of ultrasonic frequency (b) and sets the RSFF's output Q to a logic "one" value. The standard case is eight pulses of 40 kHz frequency. The power amplifier drives the ultrasonic transmitter with this signal, and the transmitter in its turn produces a short chirp of sound. The sound propagates in the air and returns from an object whenever it appears in the range of the rangefinder. The receiver converts the echo from the object into voltage and the amplifier increases the amplitude of the voltage (c). The voltage (c) is compared with a

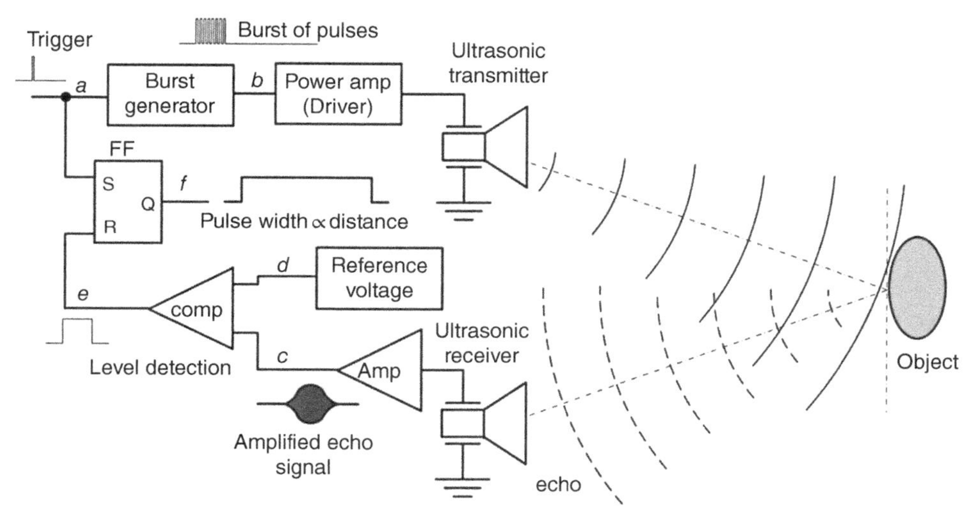

Figure 2.30 Block diagram of the ultrasonic rangefinder.

reference voltage (d) and if echo is received, the comparator switches its state to a logic "one" at point (e). This resets the RSFF (f) to a logic "zero." The time when the RSFF is ON corresponds to the time of propagation of the sound to a target and back (doubled distance to a target).

Unlike a laser beam, the sound propagates in the air by waves with an almost spherical front. That's why the transmitter and the receiver haven't been placed on the same axis. Some distance between transmitter and receiver is even desirable, to avoid the influence of spurious signal that the receiver may pick up directly from the transmitter via mechanical parts and the printed circuit board (PCB). The problem of sound propagation is that the amplitude of the acoustic signal reduces exponentially with the distance, unlike laser light intensity. Figure 2.31 shows the different forms of signals at the comparator's input (c) of Figure 2.30. The C_1 waveform is the result of a spurious signal. It happens simultaneously with the burst of ultrasonic signal C_1. The C_2 is a signal of a lower amplitude than the C_1. It is the result of the echo from the object at short distance. The C_3 is the voltage amplitude of the signal that is

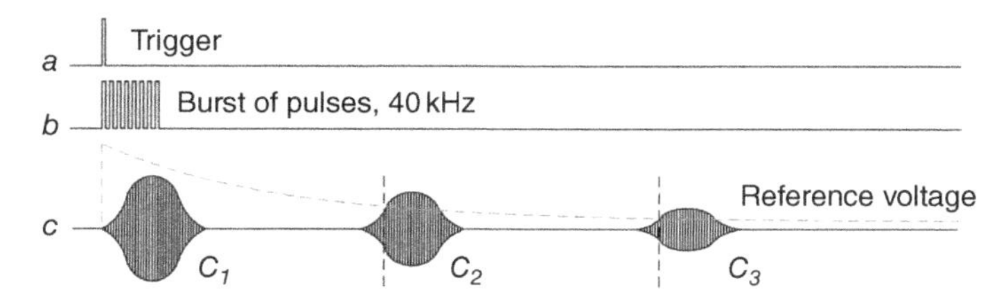

Figure 2.31 The main waveforms of the ultrasonic rangefinder: (a) trigger; (b) burst of pulses. This burst is converted into acoustic wave; (c) is the received echo converted into a voltage and amplified. The C_1 is a spurious signal passing through the mechanical contact, C_2 is the echo from the object at a short distance, C_3 is the echo from the object at long distances.

reflected from the object at long distance. This inequality of amplitudes with a distance (and correspondently with time of signal propagation) demonstrates the necessity of designing such a reference voltage generator (see the dashed line in the figure) that would provide a voltage at the second comparator input (*d*) of Figure 2.30.

The C-R circuit connected to a RSFF's output (*f*) solves this problem. The transient of the C-R circuit is a signal that is exponentially reducing from V_{cc} to zero with time constant τ = RC. The voltage *d* thus varies with time as

$$V_{\text{ref}} = V_{CC}e^{-\frac{t}{\tau}} \tag{2.41}$$

See Figure 2.32.

Thus, all waveforms in the diagram in Figure 2.30 are shown in Figure 2.33. T is the period between the triggers. T_{on} is the time of sound traveling toward the target and back. The ratio T_{on} over T is a duty cycle of the output pulse (*f*). The duty cycle is proportional to the time of propagation of the sound and, correspondently, proportional to the distance to the target. T_{on} relates to T as average voltage V_{f_avg} at output (*f*) relates to V_{cc}.

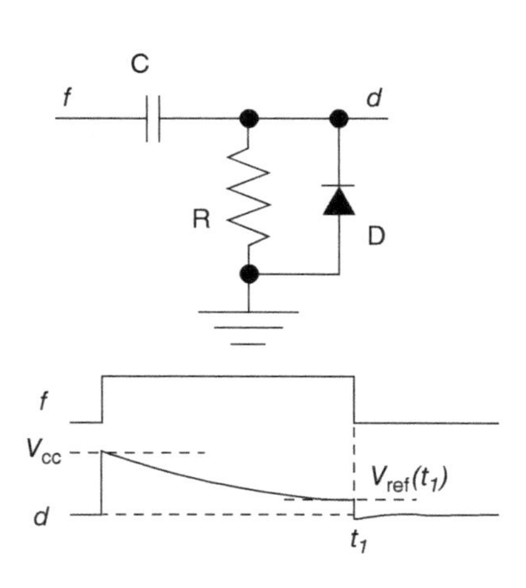

Figure 2.32 The C-R circuit and the waveforms at its input (f) and output (d). The diode D prevents the negative pulse of the voltage (d) at commutation time t_1 and discharges the capacitor C in a fast way.

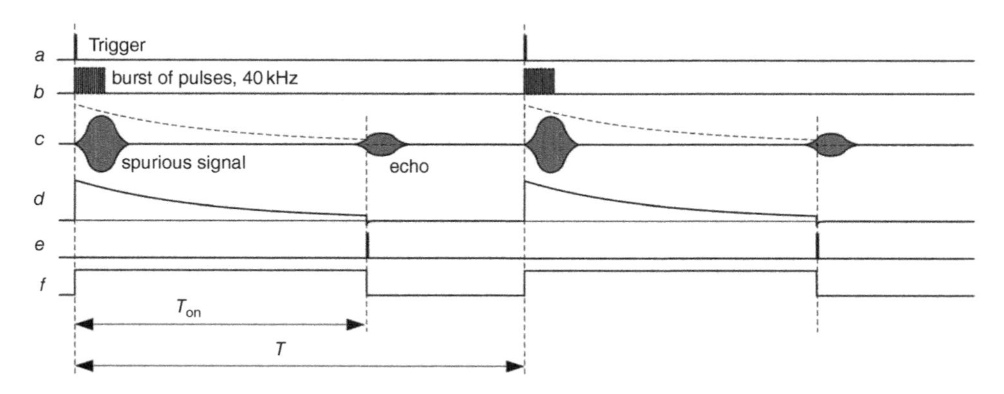

Figure 2.33 Waveforms diagram of the system shown in Figure 2.30.

$$T_{\text{on}} = T \cdot \frac{V_{f_{\text{avg}}}}{V_{\text{CC}}} \tag{2.42}$$

So the doubled length L in centimeters can be found from the T_{on} in microseconds as

$$L[\text{cm}] = T_{\text{on}}[\mu s] \cdot \frac{0.034\left[\frac{\text{cm}}{\mu s}\right]}{2} \tag{2.43}$$

By this sensor we finalize the consideration of the methods of basic perception. Certainly, the robot can be equipped by additional sensors, for example, by the cameras and corresponding programs for image and video processing or by the sensors that allow perception of specific data. Additional information about such sensors can be found in appropriate literature; for a detailed review of the sensors and literature sources see the book by Borenstein, Everett, and Feng (1996).

2.5 Summary

In the chapter, we considered the basic types of sensors that are used in the mobile robots. The chapter starts with the basic scheme of sensor (Section 2.1), and then presents its implementations for different types of data:

1) The bumper (Section 2.2) is the basic and commonly used obstacle. In its basic configuration, it is a simple switch that is pressed when the robot meets the obstacle. In more complicated setting, the sensor includes optical elements and appropriate electrical circuits for data analysis.
2) The odometry sensor (Section 2.3) measures the length of the robot's trajectory that is necessary, especially for mobile robots. In its simple configuration, the sensor includes an optical-mechanical part that encodes the length units into binary sequences and electrical circuits that transform these sequences into the formatted data.
3) Finally, the chapter considers different types of distance sensors (Section 2.4). It starts with optical range finders (time-of-light and phase shift laser rangefinders) that are based on the properties of the reflected light rays, and then considers the most popular ultrasonic rangefinder that uses the properties of reflected ultrasound waves.

Certainly, the indicated three basic sensors do not exhaust the variety of the sensors used in mobile robots. In particular, the robots can be equipped with common purpose microphones and cameras (with the appropriate software) or with specific sensors for activity in definite environments.

References

Borenstein, J., Everett, H.R., and Feng, L. (1996). Where am I? Sensors and methods for mobile robot positioning. *University of Michigan* 119 (120): 27.

Boylestad, R.L. (2010). *Introductory Circuit Analysis*. Upper Saddle River, NJ: Prentice Hall Press.

Crisp, J. (2000). *Introduction to Digital Systems*. Amsterdam: Elsevier.
Dudek, G. and Jenkin, M. (2010). *Computational Principles of Mobile Robotics*. Cambridge: Cambridge University Press.
Hasan, K.M. and Reza, K.J. (2014). Path planning algorithm development for autonomous vacuum cleaner robots. In: *2014 International Conference on Informatics, Electronics & Vision (ICIEV)*, 1–6. IEEE.
Horowitz, P. and Hill, W. (1989). *The Art of Electronics*. Cambridge: Cambridge University Press.
Jahns, R., Greve, H., Woltermann, E. et al. (2012). Sensitivity enhancement of magnetoelectric sensors through frequency-conversion. *Sensors and Actuators A: Physical* 183: 16–21.
Marston, R.M. (1998). *Security Electronics Circuits Manual*. Oxford and Boston: Newnes.
Marston, R.M. (1999). *Optoelectronics Circuits Manual*. Oxford: Butterworth-Heinemann.
Siciliano, B. and Khatib, O. (eds.) (2016). *Springer Handbook of Robotics*. Berlin: Springer.
Siegwart, R., Nourbakhsh, I.R., Scaramuzza, D., and Arkin, R.C. (2011). *Introduction to Autonomous Mobile Robots*. Cambridge, MA: MIT Press.
Tönshoff, H.K., Inasaki, I., Gassmann, O. et al. (2008). *Sensors Applications*. Wiley-VCH Verlag GmbH & Co. KGaA.
Zumbahlen, H. (2007). *Basic Linear Design*, 8–21. Norwood, MA: Analog Devices.

3

Motion in the Global Coordinates

Nir Shvalb and Shlomi Hacohen

This chapter considers the models of motion in global coordinates system (in two and three dimensions) such that the robot has complete information about the robot's location and velocity with respect to the origin. Such information is provided by permanent access to global positioning system or to hovering cameras. In particular, the chapter introduces the concept of configuration space and considers transformations of vehicle coordinates; turning radius, instantaneous velocity center, rotation, and translation matrices.

3.1 Models of Mobile Robots

3.1.1 Wheeled Mobile Robots

Numerous types of ground wheeled robots exist; we can roughly categorize them as follows, and shown in Figure 3.1a–f:

1) Car-like vehicles
2) Bi-wheeled vehicles
3) Tracked vehicles
4) Omni-wheeled vehicles
5) Hilare-type vehicles

Car-like vehicles (or robots) possess three or more wheels and are constructed with at least one steered axle. The motorized wheels are chosen to be standard (*a standard wheel* is one that is fixed to roll along a straight line path). The various car-like designs differ from one another by the choice of the steered axle, usually the front axle (see Figure 3.1a) and the choice of the motorized wheels (*front drive* or *rear drive*).

Consider a car-like vehicle with front steering, as depicted in Figure 3.2, observe that in order for the vehicle to turn without slip, the angle of the front inner turning wheel (right in the figure) steering angle should be greater than the one for the outer wheel – this is termed *Ackermann steering* in literature. Such a geometry may be achieved applying different designs (some of which will end up with a limited turning angles). Since the control scheme for such vehicles is strongly dependent on the specific steering design we shall not discuss it here.

Autonomous Mobile Robots and Multi-Robot Systems: Motion-Planning, Communication, and Swarming, First Edition. Edited by Eugene Kagan, Nir Shvalb and Irad Ben-Gal.

Companion website: www.wiley.com/go/kagan/robotsystems

Remark When designing a car-like vehicle, the set of considerations when choosing between the front or rear drive includes space efficiency (and therefore energy efficiency); traction considerations (e.g., while accelerating, the front wheels experience less traction and therefore, it is preferable with this regard to place the motorized wheels at the rear); the need for a drive shaft. This is out of the scope of this short introduction. For a detailed discussion, the reader may turn to Happian-Smith (2001).

Bi-wheeled vehicles (Figure 3.1b) like bikes or bicycles are attractive for some applications due to their narrow cross section and their improved maneuverability. However, they require a continuous control process for stabilization, which is expensive in the computational and energy consumption senses (e.g., consider Murayama and Yamakita (2007), which incorporate an inverted pendulum as a balancer). For these reasons, bi-wheeled robot vehicles are rare and therefore will not be discussed here.

Track mechanisms (Figure 3.1c) introduce an exact straight motion and suit uneven terrains, which is common when dealing with off-road scenarios. On the other hand, they are large (Ahmad, Polotski, and Hurteau 2000) and are characterized by low energy efficiency for rotation compared with other driving mechanism types. Skid steering is commonly used for these vehicles. But the reader should note that such maneuvers involve a complicated ground-track interaction due to slippage phenomena, which is still a subject

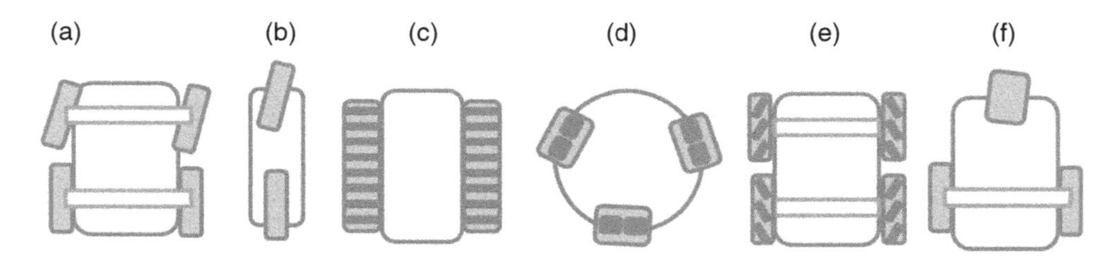

Figure 3.1 Types of vehicles.

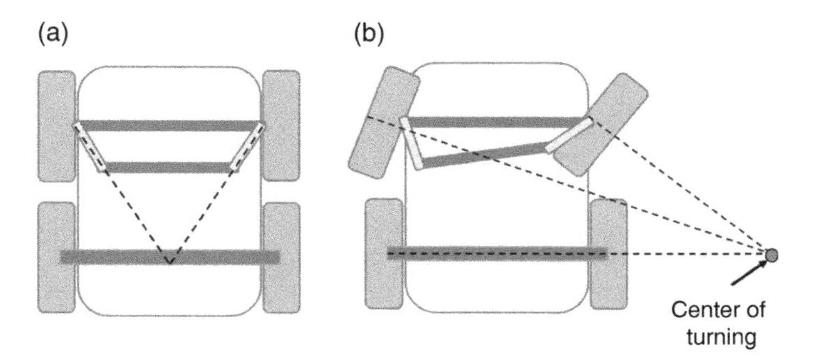

Figure 3.2 Ackermann steering geometry.

for research in the field of ground mechanics. So, in order to correctly model and control such robots, a great deal of experimentation needs to be handled prior to the actual control scheme planning.

Omni-direction vehicles can translate in all planar directions at any given configuration. A side translation is achieved by using specially designed wheels that do not introduce slip while moving in the direction of their axis. These are:

- *Omni-wheels* or *poly-wheels*, which are wheels with rollers on their circumference perpendicular to the rotation axis of the wheel. Two wheels situated in a nonparallel manner (see, e.g., Figure 3.1d) can translate to any given direction by rotating them in different rotational velocities ratios. Moreover, such wheels can also be used as an enhanced Hilare (see below) type by exploiting their nonslip character (Griffin and Allardo 1994).
- *Mecanum wheels* or *Swedish wheels* (Figure 3.1e) possess a set of rollers on their circumference perpendicular to the rotational axis of the wheel, with their axis of rotation oriented at 45° to the plane of the wheel and at 45° to the axis of rotation of the wheel. Two such wheels that are situated in a parallel manner can translate sideways by rotating them opposite rotational velocities.

3.1.2 Aerial Mobile Robots

In recent years, aerial mobile robots have become more and more attractive for researchers as well as for military and civil applications. The change is mostly accredited to the improvement, minimization, and low cost of the electronic devices. This is attributed to the fact that sensors such as inertial measurement units (IMUs) or cameras have become smaller and cheap; high-performance microprocessors and microcomputers are readily available; and so are batteries, communication devices, and more. Aircrafts can be categorized into those that are heavier than air and those that are lighter (literature also distinguishes between motorized aircrafts and nonmotorized aircrafts, but this is out of the scope of our discussion). We shall now briefly discuss the following types of aircrafts:

- Fixed-wings aircrafts are motorized or unmotorized and are heavy.
- Rotors-based aircrafts are motorized and are heavy.
- Zeppelins, blimps, and balloons are motorized and are light.

We can categorize them as follows:

1) *Fixed-wings aircrafts are most widely used for unmanned aerial missions*. This is due to their simple mechanism and their low energy consumption. These can carry heavy loads for the mission, and their control and kinematics are comparatively simple. Moreover, this type of aircraft can be motorized or nonmotorized, since the wing shape and area enable gliding, which makes it very attractive in the sense of energy efficiency.

   On the other hand, fixed-wings aircraft require long runways, and their maneuver performances requires a wide area, so this type of robot is quite limited in urban and indoor environments. For example, a U-turn for an airplane require about 10 times its wingspan.

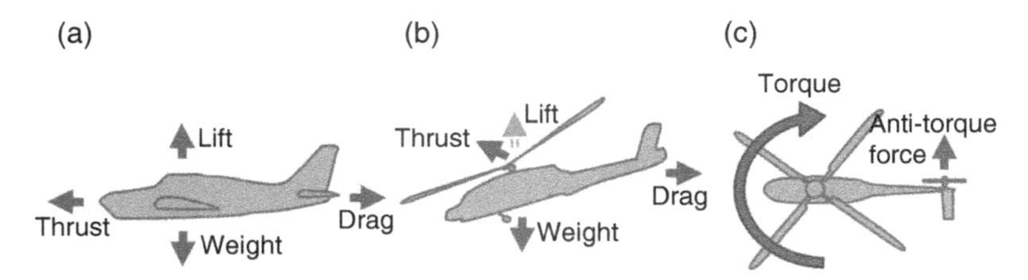

Figure 3.3 The forces acting on aircraffts.

The forces that act on an airplane include the drag and thrust in the horizontal axis and the lift force and weight in the vertical axis (see Figure 3.3a).

The airplane is controlled by two inputs: (i) the power of the main motor, which sets the horizontal speed by changing the thrust force magnitude; and (ii) the airplane's attitude, which is determined by changing the orientations of the rudder, elevator, and ailerons (i.e., changing the wing's and tail's profiles). Changing the aircraft's attitude changes the thrust vector and therefore the linear acceleration is modified.

2) *Rotors-based aircrafts are always motorized, making them expensive energy wise.* However, the maneuvers that a rotorcraft can perform are much complicated than that of a fixed-wings aircraft. Along with vertical takeoffs; hovering capabilities and flying backward make them attractive for indoor and urban environments.

   Let us consider, for simplicity, a regular helicopter with a main rotor and an anti-torque rotor located at its tail. Here, the forces are the same as in the fixed-wings aircraft except for the direction of the thrust, which is determined by the main rotor attitude (see Figure 3.3b) In addition, the main rotor creates a torque, which is perpendicular (!) to the rotor surface direction. In order to control the helicopter given this torque, one uses the tail rotor. Its orientation is perpendicular to the main rotor, such that the force it creates multiplied by the arm length exerts an opposite torque (see Figure 3.3c). Other configurations introduce a *coaxial rotor* in such a way that one rotor is on top of the other rotating in opposite directions.

   Finally, note that to control the helicopter, three inputs are available – the attitude and speed of the main-rotor, which sets the direction and the magnitude of the thrust, and the speed of the tail-rotor, which sets the rotational acceleration of the aircraft.

3) *Zeppelins, blimps, and balloons differ from one another in their structure.* The zeppelin possesses a rigid structure. Its framework is composed of solid spars enveloped by flexible casing. In these airships, the lifting gas is contained in multiple gas cells along the airship. Such a structure enables the aircraft to carry heavy loads. In blimps, non-rigid structures, the shape is maintained due to the lifting-gas pressure (c.f. semi-rigid airships).

   The altitude of such airships is determined by changing the air density (via hitting or compression). The vertical maneuvers are generated by motorized propellers in the bottom of the airship, and attitude changes are accomplished using vertical and horizontal fins.

   A balloon uses hot air for lift while for horizontal movements the pilot descends/ascends to "ride" wind currents.

   This section presented a short review of wheeled and aerial robots. We haven't mentioned anything about vessel robots as discussed in Fahimi (2008). Other robot structures like in Salomon (2012) and Shvalb (2013) or Schroer, Robert T., et al. (2004) are outside the scope of our discussion.

3.2 Kinematic and Control of Hilare-Type Mobile Robots

As already mentioned, the most common configuration for mobile robots is the Hilare-type robot (known also as differential steering) (see Figure 3.1f). Such a robot has two independent motors. The mobile platform is governed by the rotational velocity (and direction) of each wheel. Commonly, one assumes a no-slip between the wheels and the floor during the whole motion.

3.2.1 Forward Kinematics of Hilare-Type Mobile Robots

Consider a wheel of radius r. Assume the wheel is rotating from an initial 0° angle to an angle α. Imagines the wheel roles over a flat plane without slip. It is easy to notice that the wheel advances$\times\alpha$. According to the same rationale, if the wheel is rotating at a constant angular velocity ω, then its linear velocity is $r \times \omega$.

Now, we are interested in the path of the robot when its wheels are turning. Let us suppose for now that the left wheel is not turning ($\omega_l = 0$), and the right wheel is having a constant rotational velocity ω_r. Since the robot is a rigid body it easy to understand that the robot body will change its orientation relatively to the center of the left wheel, e.g., all the robot points change their location apart from the pivot (see Figure 3.4a, and compare to Figure 3.4b where non of robot's points is fixed). This pivot is the center of the left wheel, and since the distance between the wheels is constant the right wheel will make a motion along a circular arc. The new orientation of the robot is proportional to the arc length. To find the angle between the initial orientation and the final angle, we can write this differential equation:

$$\frac{d\phi}{dt} = \frac{r\omega_r}{L}$$

where r is the wheels radius. Suppose the movement starts at $t = 0$ and $\phi(0) = 0$. Integrating the above yields:

$$\phi(t) = \frac{r\omega_r}{L}t \tag{3.1}$$

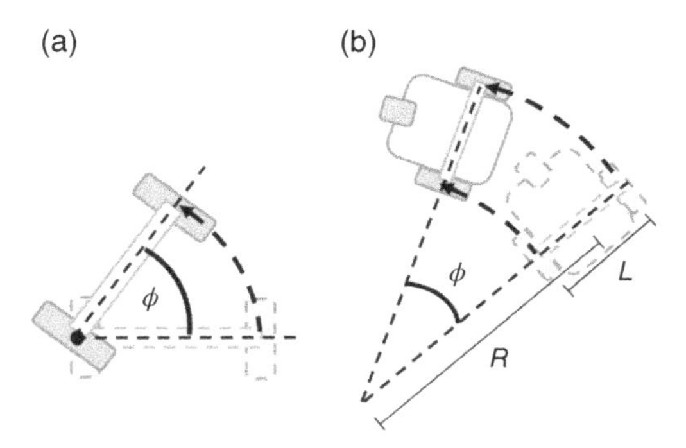

Figure 3.4 (a) A turn around the center of the left wheel. (b) A turn around some point.

Adding the initial orientation, the new orientation is

$$\theta(t) = \frac{\pi}{2} - \phi(t)$$

Now suppose that the left wheel has a small rotational velocity so the pivot is moving, too. But, in the same manner as in the previous description, the orientation is changing relatively to the rotational velocity of the right wheel. Now, when the left wheel is moving as well, the only option for the two wheels to be at the same orientation and to keep the distance between the wheels constant is to have a common pivot for the two wheels, e.g. the movement is along a circular arc with radius R. The arc angle is then similar to Eq. (3.1):

$$\phi(t) = \frac{r}{L}(\omega_r - \omega_l)\, t$$

Usually one is interested in the location of the robot in global coordinates. It is convenient to consider the reference point as the middle of the axle rather than the central mass (see Figure 3.5).

Having said that, when the velocities of both wheels are constant, the robot will travel in a circular curve. If the velocities are given, one can determine its location in global coordinates. This is called the *forward kinematics*. Let us now refer to Figure 3.5 and construct the forward kinematics:

Given the angular velocities of the robot wheels ω_r and ω_l, we can write:

$$\omega_l = \dot{\phi}\left(R - \frac{L}{2}\right); \omega_r = \dot{\phi}\left(R + \frac{L}{2}\right) \tag{3.2}$$

Subtracting yields:

$$\omega_r - \omega_l = \phi\frac{r}{L} \Rightarrow \dot{\phi} = \frac{r}{L}(\omega_r - \omega_l) \tag{3.3}$$

Substituting back into Eq. (3.2), we find:

$$R = L\frac{(\omega_r + \omega_l)}{2(\omega_r - \omega_l)} \tag{3.4}$$

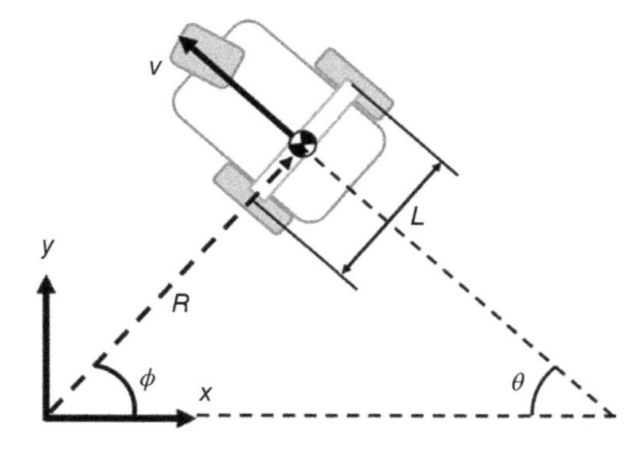

Figure 3.5 Demonstration of the parameters of the forward kinematics of Hilare type mobile robot.

The velocity of the robot in body coordinates is marked as v (see Figure 3.5) is the tangent velocity:

$$v = \dot{\phi} R = \frac{r}{2}(\omega_r + \omega_l)$$

Finally, back to the global coordinates, the velocity of the robot is as follows:

$$\dot{x} = -vS_{\phi};\ \dot{y} = vC_{\phi}$$

To find the location, assuming constant velocities, the equations should be integrated over time. In practice, these are computed in a discrete manner:

$$\begin{bmatrix} x_{k+1} \\ y_{k+1} \end{bmatrix} = \begin{bmatrix} x_k \\ y_k \end{bmatrix} + v\Delta t \begin{bmatrix} -\sin(\phi_k) \\ \cos(\phi_k) \end{bmatrix}$$

where: Δt is the sampling time, and the orientation angle of the velocity ϕ_k is

$$\phi_{k+1} = \phi_k + \frac{r\Delta t}{L}(\omega_r - \omega_l)$$

In an arbitrary scenario of nonconstant velocities, one can use small time increments so that the assumption of constant velocities will hold.

3.2.2 Velocity Control of Hilare-Type Mobile Robots

This subsection will introduce some strategies for finding the controlled velocities that are essential for tracking a path.

Let us suppose the robot is located at a given location and a given orientation, and a single target is located in a given distance from the current location in an area free of obstacles. The simple and naive way to reach the target is twofold. First, change the orientation without translating, and then transverse in a straight line to the target, so $\omega_r = \omega_l = \frac{1}{r} v_d$ where v_d is the desired velocity. So the first step is done by rotating the wheels in the same velocity magnitude and with opposite signs. Substituting into Eq. (3.4) yields

$$\omega_l = -\omega_r = \frac{L}{2r}\dot{\phi}_d$$

where $\dot{\phi}_d$ is the desired rate for the orientation change.

Note that, though simple, in this motion scheme the robot must stop each time that change of direction is needed. Moreover, the method requires sharp changes in the direction and velocities.

A smoother motion can be obtained by changing the orientation while traveling:

$$\omega_r(t) = \frac{1}{r}\left(u_d(t) + \frac{L}{2}\dot{\phi}_d(t)\right); \omega_l(t) = \frac{1}{r}\left(u_d(t) - \frac{L}{2}\dot{\phi}_d(t)\right)$$

The resulting path in the *turn-then-travel* scheme is obviously shorter, but the maneuver is not smooth. In the *turn-while-traveling* method, the path is along a smooth curve.

The simplest example for a curve is a circular arc (see Figure 3.6) with a radius that can be computed based on the angel ϕ and the distance d to the target:

$$R = \frac{d}{2\sin\beta}$$

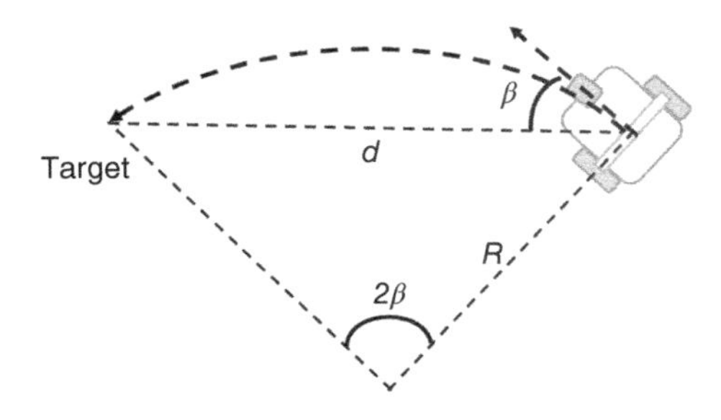

Figure 3.6 A velocity control in circular arc.

3.2.3 Trajectory Tracking

In many applications, one must track a given path. There are numerous algorithms in which the path may be computed, but this is out of the scope of this chapter. The challenge here is to end up with a path for which its deviation from the computed path is as small as possible. Deviations are due to the fact that the robot cannot make all maneuvers needed for perfect tracking. Generally, Hilare-type mobile robots have an advantage in this sense, since they can change their orientation without changing their location (see "turn-then-travel" in Section 3.2.2). Since the path is given in global coordinates, it is more convenient to present the kinematic equations with respect to the orientation θ than the arc angle ϕ. The kinematics equations we shall use are, then:

$$\dot{x} = v(t)\cos(\theta(t));\dot{y} = v(t)\sin(\theta(t))$$

Now, suppose the desired path is given by: $x_d(t)$, $y_d(t)$, and the time derivatives are $\dot{x}_d(t), \dot{y}_d(t)$, respectively. Since we assume the wheels do not slip, the desired orientation is given by the desired velocities:

$$\theta_d(t) = \arctan\left(\frac{\dot{y}_d}{\dot{x}_d}\right)$$

In order to simplify the kinematic model, one can change the control inputs and the states of the robot configuration. The vector that describes the robot's configuration is referred to as *states- vector* (or states, for short). The *new states* are given by

$$z_1 = x$$

$$z_2 = T_\theta$$

$$z_3 = y$$

The desired trajectory is given in the new states coordinate system:

$$z_1^d = x_d$$

$$z_2^d = \tan\theta_d = \frac{\dot{y}_d}{\dot{x}_d}$$

$$z_3^d = y_d$$

We denote the new control inputs u_1, u_2 to be the derivatives of z_1 and z_2, so

$$\begin{aligned} u_1 &= \dot{z}_1 v\cos\theta \\ u_2 &= \dot{z}_2 = \dot{\theta}(1 + \tan^2\theta) \end{aligned} \tag{3.5}$$

The derivative of z_3,which we shall need as well, is $\dot{z}_3 = \dot{y} = v\sin\theta = v\cos\theta\tan\theta = u_1 u_2$. Thus, the desired control inputs are

$$u_1^d = \dot{x}_d$$
$$u_2^d = \dot{z}_2^d = \frac{d}{dt\left(\frac{\dot{y}_d}{\dot{x}_d}\right)}$$

We define the *tracking error* $\tilde{z}_i = z_i^d - z_i$ and the *inputs error* $\tilde{u}_i = u_i^d - u_i$ as the subtraction of the measured value from the desired value. We use these in order to close a control loop (see any basic control theory textbook; c.f. Ogata 2001). The derivatives of the states-error functions are given in Eq. (3.6):

$$\begin{aligned}\dot{\tilde{z}}_1 &= \dot{z}_1^d - \dot{z}_1 = \tilde{u}_1\\ \dot{\tilde{z}}_2 &= \dot{z}_2^d - \dot{z}_2 = \tilde{u}_2\\ \dot{\tilde{z}}_3 &= \dot{z}_3^d - \dot{z}_3 = \tilde{u}_1^d z_2^d - u_1 z_2 = -\tilde{u}_1\tilde{z}_2 + u_1^d\tilde{z}_2 + \tilde{u}_1 z_2^d\end{aligned} \tag{3.6}$$

If we assume small tracking errors, all quantities that are marked with $\sim$ are small, but since the value of $\tilde{u}_1\tilde{z}_2$ is of order two, we may neglect it altogether.

We shall now write down the (linear) control law equation (Eq. (3.7)), which connects the current configuration (the states) with the desired control signals by means of matrix multiplication. Similarly, the *transition equation* Eq. (3.8) defines the evolution of the states vector at each time step by means of multiplication with a transition matrix. Recall that such a system would remain close to the desired path (*finite tracking error*) if all eigenvalues of such a transition matrix would be of negative real part. Thus, in order to obtain a stable system, we choose the control law to be

$$\begin{bmatrix}\tilde{u}_1\\ \tilde{u}_2\end{bmatrix} = \begin{bmatrix}k_1 & 0 & 0\\ 0 & k_2 & \frac{k_3}{u_1^d}\end{bmatrix}\begin{bmatrix}\tilde{z}_1\\ \tilde{z}_2\\ \tilde{z}_3\end{bmatrix} \tag{3.7}$$

where, k_1, k_2, and k_3 are the (real valued) controller gains. Substituting the control law in Eq. (3.6), we get the linear closed loop system:

$$\begin{bmatrix}\dot{\tilde{z}}_1\\ \dot{\tilde{z}}_2\\ \dot{\tilde{z}}_3\end{bmatrix} = \begin{bmatrix}k_1 & 0 & 0\\ 0 & k_2 & \frac{k_3}{u_1^d}\\ k_1 z_1^d & u_1^d & 0\end{bmatrix}\begin{bmatrix}\tilde{z}_1\\ \tilde{z}_2\\ \tilde{z}_3\end{bmatrix} \tag{3.8}$$

For the system to be stable, the eigenvalues of the matrix in Eq. (3.8) must have negative real part. Here the eigenvalues are

$$\lambda_{1,2,3} = \mathrm{k}_1, \frac{k_2 \pm \sqrt{k_2^2 - 4k_3}}{2}$$

Exploring all solutions, we can come to a conclusion that for stability, the controller gains must satisfy $k_1 < 0$ and $k_2 < 0$. Note that although the system is time dependent (since u_1^d changes in time), the stability does not depend on u_1^d so the linear approximation is asymptotically stable.

Using the fact that $u_i = u_i^d - \tilde{u}_i$ and Eq. (3.5), one can write:

$$v = \frac{r}{2}(\omega_r - \omega_l)$$
$$v = \frac{r}{2}(\omega_R - \omega_L)$$
$$\dot{\theta} = \frac{r}{2L}(\omega_R - \omega_L)$$

and:

$$\omega_R = \frac{1}{r}\left(v + L\dot{\theta}\right) = \left(u_1^d - \tilde{u}_1\right)\left(\frac{1}{C_\theta} + \frac{L}{1+\tan^2\theta}\right)$$
$$\omega_L = \frac{1}{r}\left(v - L\dot{\theta}\right) = \left(u_1^d - \tilde{u}_1\right)\left(\frac{1}{C_\theta} - \frac{L}{1+\tan^2\theta}\right)$$

3.3 Kinematic and Control of Quadrotor Mobile Robots

The main advantages of quadrotors are their simple structure and their impressive performance. One of the most important capabilities of the quadrotor is the ability to move in any direction from almost every configuration. However, these are accompanied with challenges in the controlling process. The fact that the quadrotor has six degrees of freedom (DoF) but only four controlled inputs empowers this challenge.

This section is organized as follows: First, we shall introduce a short review of the quadrotor's dynamics. Next, a simple proportional-derivative (PD) controller for stabilizing the quadrotor will be presented. Finally, we introduce a trajectory tracking control scheme.

3.3.1 Dynamics of Quadrotor-Type Mobile Robots

Let us assume a symmetric structure of the quadrotor, which means that the four motors are located at the same distance from the quad's center of mass. We also assume that the arms are perpendicular. We fix two coordinate systems, as depicted in Figure 3.7:

- Body coordinates' system is located at the quad's center of mass with the Z_q axis pointing up and the X_q and Y_q axis along the arms.
- World coordinate system is designated by X_W, Y_W, Z_W.

The assumption of symmetry and the selection of the quad's coordinate system in this way implies a diagonal form of the inertia matrix:

$$\mathrm{I}_q = \begin{bmatrix} I_{xx} & 0 & 0 \\ 0 & I_{yy} & 0 \\ 0 & 0 & I_{zz} \end{bmatrix}$$

The configuration of the quadrotor is defined by six parameters: three for the central mass location (X_W, Y_W, Z_W) and three for the quad's orientation measured in the world

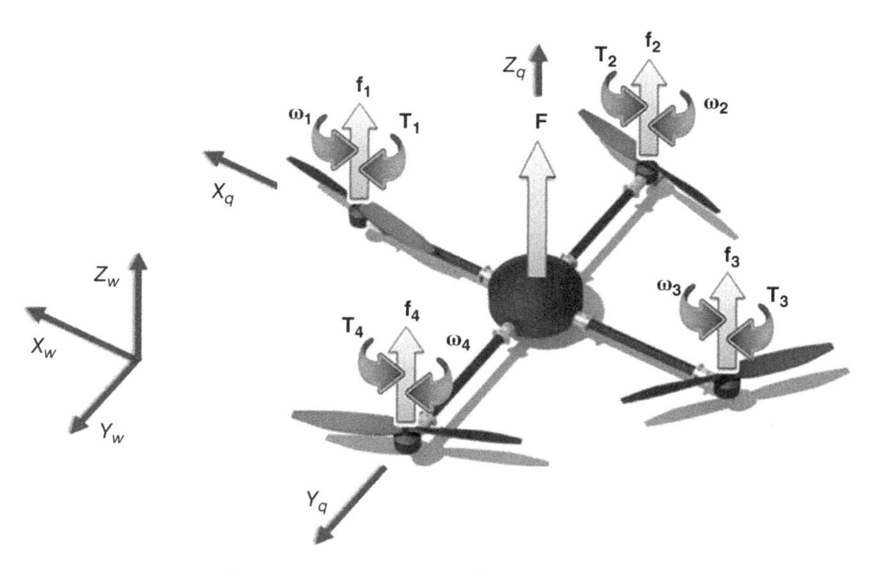

Figure 3.7 Forces and torques acting on a quadrotor.

coordinate system. The *roll* ϕ determines the rotation about the X_W axis, the *pitch* θ is the angle about the Y_W axis, and the *yaw* ψ is the angle about the Z_W axis.

Note that the intuitive notion of these may not hold when the quad orientation is "far" from (X_W, Y_W, Z_W). The configuration vector is therefore defined:

$$q = [x, y, z, \theta, \phi, \psi]^T$$

and the velocity is defined as the vector's derivative with respect to time: the center of mass linear velocity and the quad's angular velocity $\dot{q} = [\dot{x}, \dot{y}, \dot{z}, \dot{\theta}, \dot{\phi}, \dot{\psi}]^T$.

The controlled inputs are the angular velocities of the rotors, denoted by $\{\omega_i\}_{i=1..4}$, and torque exerted by the *i*th rotor to the quadrotor body is denoted by T_i. Note that said torque is counter-directed to the angular velocity of the rotor ω_i. In order to better understand why this is so, imagine a swimmer immersed in water with both of his arms aligned, and note that when sweeping back his left arm while sweeping his right arm forward, he will experience an opposite directed torque.

The lift force at the center of the *i*th rotor is marked by f_i. Finally, we denote the total lift force, which is located at the center of mass by F.

3.3.2 Forces and Torques Generated by the Propellers

As already mentioned, the forces and torques are due to the angular velocity of the rotors. We assume a simple model where the resulting lift force is proportional to the squared angular velocity. The lift force is therefore:

$$f_i = \rho d^4 C_T \omega_i^2 \triangleq C_1 \omega_i^2$$

where ρ is the air density, d is the propeller's diameter, and the thrust coefficient C_T is a function of the propeller's advanced ratio: $J = \frac{V}{\omega_d}$ where V is the airplane velocity. An elaborated discussion is available in Chapter 6 in Kimberlin (2003).

The thrust is directed to Z_q and given by the sum of all four rotors' forces as

$$\mathrm{F} = \sum_{i=1}^{4} f_i$$

The torque that each rotor exerts around its own axis is proportional to the square of the angular velocity. The proportion coefficient b, is a function of the air density (a tenth power of the) radius of the propellers and the geometry of the propeller (c.f. Kimberlin 2003). On top of that, one should also add a second term that expresses the moment needed to rotate the propeller itself. This sums up to

$$T_i = \rho d^5 C_Q \omega_i^2 + I_r \dot{\omega}_i \tag{3.9}$$

with I_r the propeller polar moment of inertia. Here, C_Q is the torque coefficient that depends on J, the air properties, and the geometry of the propeller. So, the total torque in the X_q and Y_q axis are created by the lift forces multiplied by the arm's length L. The total torque in the Z_q direction is created by the summation of all T_i's. For simplicity of the model, we can assume that I_r is small enough so the second term in Eq. (3.9) is negligible. So, after neglecting, it became

$$T_i = C_2 \omega_i^2$$

For summary the thrust and torques for the quad is given by:

$$\begin{bmatrix} F \\ M_p \\ M_q \\ M_r \end{bmatrix} = \begin{bmatrix} C_1 & C_1 & C_1 & C_1 \\ 0 & -LC_1 & 0 & LC_1 \\ -LC_1 & 0 & LC_1 & 0 \\ C_2 & -C_2 & C_2 & -C_2 \end{bmatrix} \begin{bmatrix} \omega_1^2 \\ \omega_2^2 \\ \omega_3^2 \\ \omega_4^2 \end{bmatrix} \tag{3.10}$$

where, F is the total thrust, M_p is the torque around X_q, M_q is the torque around Y_q, and M_r is the torque around Z_q.

3.3.3 Relative End Global Coordinates

Since some computations are more simple in the body frame and some are needed in the inertial frame, a transformation is required. The logic for having this transformation is to rotate the coordinate system three times, once for each axis. The matrix associated with rotation around the Z_W axis is

$$R_Z(\psi) = \begin{bmatrix} C_\psi & S_\psi & 0 \\ -S_\psi & C_\psi & 0 \\ 0 & 0 & 1 \end{bmatrix}$$

the rotations around the Y_W and X_W axes are, respectively:

$$R_Y(\theta) = \begin{bmatrix} C_\theta & 0 & -S_\theta \\ 0 & 1 & 0 \\ S_\theta & 0 & C_\theta \end{bmatrix}; R_X(\phi) = \begin{bmatrix} 1 & 0 & 0 \\ 0 & C_\phi & S_\phi \\ 0 & -S_\phi & C_\phi \end{bmatrix}$$

The rotation matrix that describes the quad's orientation in the inertial frame is obtained by consequentially multiplying these matrices:

$$R = R_{Z(\psi)R_Y(\theta)R_X(\phi)} = \begin{bmatrix} C_\theta C_\Psi & S_\theta S_\phi C_\psi - C_\phi S_\psi & S_\theta C_\phi C_\Psi + S_\phi S_\Psi \\ C_\theta S_\Psi & S_\theta S_\phi S_\Psi + C_\phi C_\Psi & S_\theta C_\phi S_\Psi - S_\phi C_\Psi \\ -S_\theta & C_\theta S_\phi & C_\theta C_\phi \end{bmatrix}$$

For what comes next, note that the time derivative of the rotation matrix R with Euler angles as above is

$$\dot{R} = \Omega_{[\dot{\phi},\dot{\theta},\dot{\psi}]} R$$

where

$$\Omega_{[\dot{\phi},\dot{\theta},\dot{\psi}]} \triangleq \begin{bmatrix} 0 & -\dot{\psi} & \dot{\theta} \\ \dot{\psi} & 0 & \dot{\phi} \\ -\dot{\theta} & \dot{\phi} & 0 \end{bmatrix}$$

We would like to establish the kinematics, so we would like to write down the angular velocity transformation from the quad's system to the inertial system (world system). We can differentiate with respect to time as follows:

$$\dot{R} = \Omega_{[\dot{\phi},0,0]} R_Z(\psi) R_Y(\theta) R_X(\phi) + R_Z(\psi)\{\Omega\}_{[0,\dot{\theta},0]} R_Y(\theta) R_X(\phi) + R_Z(\psi) R_Y(\theta) \Omega_{[0,0,\dot{\psi}]} R_X(\phi)$$

On the other hand, R can also be written in the quad's frame of reference with $[p, q, r]$ angle velocities. So if we think of it as such, we can differentiate as follows:

$$\dot{R} = \Omega_{[p,q,r]} R$$

Equating and simplifying, we establish the desired relation:

$$\begin{bmatrix} p \\ q \\ r \end{bmatrix} = R(\phi)R(\theta) \begin{bmatrix} 0 \\ 0 \\ \dot{\psi} \end{bmatrix} + R(\phi) \begin{bmatrix} 0 \\ \dot{\theta} \\ 0 \end{bmatrix} + \begin{bmatrix} \dot{\phi} \\ 0 \\ 0 \end{bmatrix}$$

The reader should note that the change in the orientation of the quad in the inertial frame is described by a nonorthogonal vector, so the transformation is nontrivial.

Intuitively, we can think of this result by considering the consecutive rotations that constitute R: We think of a small angle change (i.e., we think of $\delta\theta$ rather than $\dot{\phi}$) and try to estimate the effects on the resulting rotated vector. Note that $\delta\psi$ undergoes two additional rotations, $\delta\theta$ is followed by an additional (one) rotation, and the final Euler angle $\delta\phi$ is not followed by any additional rotations.

The transformations are then as follows:

$$\begin{bmatrix} p \\ q \\ r \end{bmatrix} = \begin{bmatrix} 1 & 0 & -S_\theta \\ 0 & C_\phi & C_\theta S_\phi \\ 0 & -S_\phi & C_\theta C_\phi \end{bmatrix} \begin{bmatrix} \dot{\phi} \\ \dot{\theta} \\ \dot{\psi} \end{bmatrix}$$

and

$$\begin{bmatrix} \dot{\phi} \\ \dot{\theta} \\ \dot{\psi} \end{bmatrix} = \begin{bmatrix} 1 & S_\phi T_\theta & C_\phi T_\theta \\ 0 & C_\phi & -S_\phi \\ 0 & S_\phi / C_\theta & C_\phi / C_\theta \end{bmatrix} \begin{bmatrix} p \\ q \\ r \end{bmatrix}$$

For what comes next, we shall use these notations:

$$\Theta = \begin{bmatrix} p \\ q \\ r \end{bmatrix}, \dot{\Phi} = \begin{bmatrix} \dot{\phi} \\ \dot{\theta} \\ \dot{\psi} \end{bmatrix}, W = \begin{bmatrix} 1 & 0 & -S_\theta \\ 0 & C_\phi & C_\theta S_\phi \\ 0 & -S_\phi & C_\theta C_\phi \end{bmatrix}$$

One can write:

$$\Theta = W\dot{\Phi} \tag{3.11}$$

3.3.4 The Quadrotor Dynamic Model

In order to design a controller, one needs to formulate the dynamic mathematical model, namely, the force and torque differential equations, which we shall pursue now. The total *force equation* can be written in inertial coordinates, which includes the forces exerted by the rotors, the drag force, and the gravity:

$$\begin{bmatrix} \ddot{x} \\ \ddot{y} \\ \ddot{z} \end{bmatrix} = \frac{1}{m} R \begin{bmatrix} 0 \\ 0 \\ F \end{bmatrix} - \frac{1}{m} \begin{bmatrix} c_x \dot{x} \\ c_y \dot{y} \\ c_z \dot{z} \end{bmatrix} - \begin{bmatrix} 0 \\ 0 \\ g \end{bmatrix} = \frac{F}{m} \begin{bmatrix} S_\theta C_\phi C_\psi + S_\phi S_\psi \\ S_\theta C_\phi S_\psi - S_\phi C_\psi \\ C_\theta C_\phi \end{bmatrix} - \frac{1}{m} \begin{bmatrix} c_x \dot{x} \\ c_y \dot{y} \\ c_z \dot{z} \end{bmatrix} - \begin{bmatrix} 0 \\ 0 \\ g \end{bmatrix} \tag{3.12}$$

where c_x, c_y, c_z are the drag coefficient. Other aerodynamical effects are commonly neglected, and so we shall also do here.

The differential equation that describes the orientation change of the quad's frame is the Euler equation (which results from a simple differentiation of both Θ and the quad's coordinates X_q, Y_q, Z_q, which change due to rotation):

$$\mathrm{I_q}\ddot{\Theta} + \Theta \times (\mathrm{I_q}\dot{\Theta}) = \mathrm{M} - \mathrm{M_{gyr}}$$

where M is the resulting torque (moment) vector exerted by the rotors, I_q denotes the quadrotor's moment of inertia matrix:

$$I_q = \begin{bmatrix} I_{xx} & 0 & 0 \\ 0 & I_{yy} & 0 \\ 0 & 0 & I_{zz} \end{bmatrix}$$

and M_{gyr} is the gyroscopic moment (in literature, it is misleadingly named the *gyroscopic force*) that acts to maintain each of the blades in the Z_q direction (as the case of a rotating top):

$$\mathrm{M_{gyr}} = \mathrm{I_r}\,\Theta \times [0,0,1]^{\mathrm{T}} (\omega_1 - \omega_2 + \omega_3 - \omega_4)$$

Here, I_r is the rotors' moment of inertia, so by substituting, one get the *torque equation*:

$$\ddot{\Theta} = I_q \begin{bmatrix} -qr \\ pr \\ 0 \end{bmatrix} + I_q^{-1} \begin{bmatrix} M_p \\ M_q \\ \frac{1}{2} M_r \end{bmatrix} - I_r I_q^{-1} \begin{bmatrix} q \\ -p \\ 0 \end{bmatrix} (\omega_1 - \omega_2 + \omega_3 - \omega_4) \tag{3.13}$$

Note that by differentiating Eq. (3.11) the angular acceleration written in the inertial coordinates is

$$\ddot{\Phi} = \frac{d}{dt}\left(W^{-1}\dot{\Theta}\right) = \frac{d}{dt}\left(W^{-1}\right)\dot{\Theta} + W^{-1}\ddot{\Theta} \tag{3.14}$$

Note that the movement equation Eq. (3.12) neglects the influence of the aerodynamic effects, while the directed thrust sets the second derivatives (accelerations) of Γ.

3.3.5 A Simplified Dynamic Model

The dynamic model given in Eqs. (3.12) and (3.13) is a bit cumbersome.

Nevertheless, for most tasks a simplified model will do. In Eq. (3.12) one can assume that the velocities (second term) are small with respect to the thrust (first term). So, the drag force can be neglected without losing much. The approximation for Eq. (3.12) yields an approximated *force equation*:

$$\begin{bmatrix} \ddot{x} \\ \ddot{y} \\ \ddot{z} \end{bmatrix} = \frac{1}{m} R \begin{bmatrix} 0 \\ 0 \\ F \end{bmatrix} - \begin{bmatrix} 0 \\ 0 \\ g \end{bmatrix} \tag{3.15}$$

Furthermore, under the assumption of small angular velocities, the quadratic terms of Eq. (3.13) (qr and pr) may be omitted, and since the $I_r \ll I\_xx$, Eq. (3.13) approximately becomes:

$$\ddot{\Theta} = \mathrm{I_q^{-1}}\,\mathrm{M} \tag{3.16}$$

substituting Eq. (3.14) into Eq. (3.16) and neglecting the angular velocities once again yields Eq. (3.17):

$$M = I_q\, W\, \ddot{\Phi} \tag{3.17}$$

which is our approximated *torque equation*.

Remark Note, this approximation can be used only for the controller development (i.e., when applying a controller to a real quad). In order to write a controlling/dynamics simulation, one should use the complete version given in Eq. (3.12) for the linear acceleration and Eqs. (3.13) and (3.14) for the angular acceleration.

3.3.6 Trajectory Tracking Control of Quadrotors

For many aerial application, the quadrotor mission involves tracking a predefined trajectory. As written previously, one of the major advantages of the quadrotor is the ability to make almost any maneuver with almost any configuration.

The most commonly used controller is the proportional-integral-derivative (*PID*) due to its simple implementation and tuning. The control signal generated by this controller is

$$u(t) = k_P e(t) + k_I \int_0^t e(\tau) d\tau + k_D \dot{e}(t)$$

where k_P, k_I, and k_D are the *controller gains* that have to be determined somehow. Generally, the proportional part with the gain k_P is used to reduce the tracking errors and tune the convergence time; increasing k_P decreases the convergence time. The integral gain k_I of the controller is mainly used to decrease the steady-state error. The derivative gain k_D accounts for the error's rate of change.

The purpose is to reach a sequence of way-points on a predefined trajectory. Note that it is not necessary to precisely cross every way-point in order to do so. Also note that the integral gain will "notice" a constant error and will try to minimize it, i.e. a robot following a parallel trajectory with a constant error to the desired one. The k_D and k_P, on the other hand, will not "mind" such an error. Since we do not care about such errors, a PD controller is preferable.

In the dynamic model previously discussed, the quadrotor configuration is defined by the vector $q = [x, y, z, \phi, \theta, \psi]^T$. We define the path to traverse by a set of desired way-points $\Gamma_i \in R^3$ for $I \in \{0, 1, \dots, N\}$. In addition, we define a unit vector normal to the segment connecting Γ_i and Γ_{i+1}, and think of the vector's tail as connected to the quad's center of mass. This unit vector is marked as n. We also denote the current quad's location by $\Gamma(t) = [x, y, z]^T$. Figure 3.8 depicts these notations.

The controller objective is to decrease the tracking error between the quad's position and the way-points. The next way-point to follow is selected using a simple criteria:

Promote i to pursue $\Gamma_d = \Gamma_{i+1}$ when n pass through the Γ_{i+1} way-point.

We denote the tracking error as follows:

$$e_\Gamma = \Gamma_d - \Gamma(t) \tag{3.18}$$

On top of orienting the quad's maneuver to a desired direction, we have to change the quad's attitude to suit the thrust force corresponding to that direction. Thus, we design

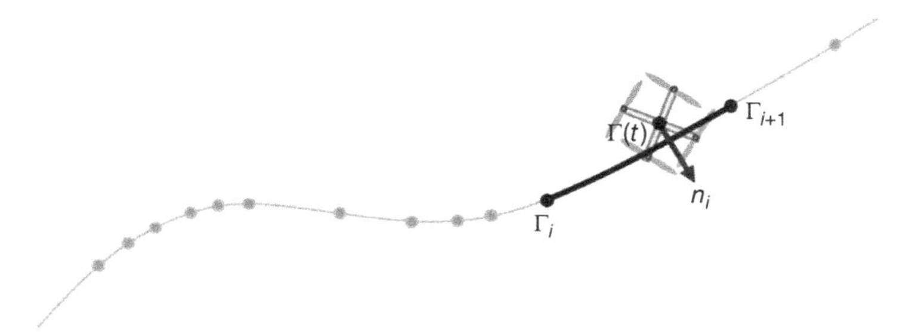

Figure 3.8 Quadrotor configuration where current quad's location is $\Gamma(t) = [x, y, z]^T$.

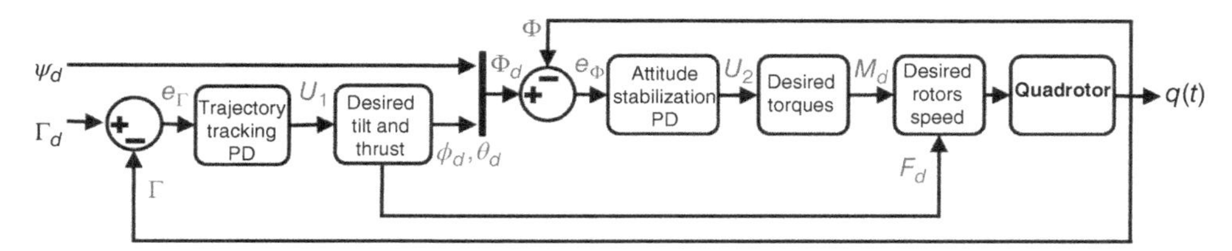

Figure 3.9 Trajectory tracking control double loop. The implementation of this diagram is described in the last section of this chapter.

the controller structure having two loops. The outer loop controls the quadrotor's location while the inner loop adjusts the quad's attitude (see Figure 3.9).

The outer loop inputs are the quad's linear accelerations. Following Zuo (2010), the PD control law with control gains is given by the positive definite matrices K_P^Γ and K_D^Γ:

$$\ddot{e}_\Gamma + K_D^\Gamma \dot{e}_\Gamma + K_P^\Gamma e_\Gamma = 0$$

defining the virtual control inputs as $U_1 = \ddot{\Gamma} = [u_1, u_2, u_3]^T$. So, Eq. (3.15) can be rewritten:

$$U_1 = \frac{1}{m} R \begin{bmatrix} 0 \\ 0 \\ F \end{bmatrix} - \begin{bmatrix} 0 \\ 0 \\ g \end{bmatrix} \tag{3.19}$$

which takes the explicit form (and since $R^{-1} = R^T$ for all rotation matrices):

$$R^T \begin{bmatrix} u_1 \\ u_2 \\ u_3 + g \end{bmatrix} = \frac{1}{m} \begin{bmatrix} 0 \\ 0 \\ F \end{bmatrix}$$

and for finding U_1 (assuming $\Gamma_d = 0$):

$$U_1 = K_D^\Gamma (\dot{\Gamma}_d - \dot{\Gamma}) + K_P^\Gamma (\Gamma_d - \Gamma) \tag{3.20}$$

The dynamic model certainly depends on the quad's attitude, which sets the rotation matrix R in Eq. (3.19), which is the purpose of the inner loop. Expanding Eq. (3.19) yields the following:

$$C_\theta C_\psi u_1 + C_\theta S_\psi u_2 - (u_3 + g)S_\theta = 0 \tag{3.21}$$

$$(S_\theta C_\psi S_\phi - S_\psi C_\phi)u_1 + (S_\theta S_\psi S_\phi + C_\psi C_\phi)u_2 + (u_3 + g)C_\theta S_\phi = 0 \tag{3.22}$$

$$(S_\theta C_\psi C_\phi + S_\psi S_\phi)u_1 + (S_\theta S_\psi C_\phi - C_\psi S_\phi)u_2 + (u_3 + g)C_\theta C_\phi = \frac{F}{m} \tag{3.23}$$

We may assume that $C_\theta \neq 0$ and then divide both sides of Eq. (3.21) by C_θ:

$$\theta = \arctan\left(\frac{u_1 C_\psi + u_2 S_\psi}{u_3 + g}\right)$$

multiplying Eq. (3.22) with C_ϕ and Eq. (3.23) with S_ϕ and subtracting, we get:

$$S_\psi u_1 - C_\psi u_2 = \frac{F}{m} S_\phi$$

since we haven't found F yet, the term $\frac{F}{m}$ must be found some other way. Rearranging and squaring Eq. (3.19) gives the following:

$$\begin{bmatrix} 0 \\ 0 \\ \frac{F}{m} \end{bmatrix}^T \begin{bmatrix} 0 \\ 0 \\ \frac{F}{m} \end{bmatrix} = \left(U + \begin{bmatrix} 0 \\ 0 \\ g \end{bmatrix}\right)^T \left(U + \begin{bmatrix} 0 \\ 0 \\ g \end{bmatrix}\right)$$

Note that the above equation is right since R is an orthogonal matrix that sustains $R^T = R^{-1}$ and thus, $\frac{F}{m} = \sqrt{u_1^2 + u_2^2 + (u_3 + g)^2}$, which yields:

$$\phi = \arcsin\left(\frac{u_1 S_\psi - u_2 C_\psi}{\sqrt{u_1^2 + u_2^2 + (u_3 + g)^2}}\right)$$

The desired roll and pitch given U and ψ_d are

$$\theta_d = \arctan\left(\frac{u_1 C_{\psi_d} + u_2 S_{\psi_d}}{u_3 + g}\right); \phi_d = \arcsin\left(\frac{u_1 S_{\psi_d} - u_2 C_{\psi_d}}{\sqrt{u_1^2 + u_2^2 + (u_3 + g)^2}}\right) \tag{3.24}$$

Then, the desired thrust can be calculated by Eq. (3.23) as

$$F_d = m\left[u_1 \left(S_{\theta_d} C_{\phi_d} C_{\psi_d} + S_{\phi_d} S_{\psi_d}\right) + u_2\left(S_{\theta_d} C_{\phi_d} S_{\psi_d} - S_{\phi_d} C_{\psi_d}\right) + + (u_3 + g)C_{\theta_d} C_{\phi_d}\right] \tag{3.25}$$

For interim summary, the outer loop is proposed to control the quad's position. The maneuvers in the directions of x and y are made by setting roll and pitch and by having the correct thrust, which may be found in Eqs. (3.24) and (3.25). As for the yaw, it does not influence the actual maneuver of the quad, so it can be selected to be any required value. The desired roll, pitch, yaw, and z are controlled by the inner loop.

As for the inner closed loop, the error can be defined with the same logic of Eq. (3.18):

$$e_\Phi = \Phi_d - \Phi$$

where $\Phi_d = [\phi_d, \theta_d, \psi_d]$ and Φ is the current orientation. Here, the PD control low is:

$$U_2 = K_P^{\Phi} e_{\Phi} + K_D^{\Phi} e\Phi \tag{3.26}$$

Assume the new virtual control inputs to be $U_2 = \ddot{\Phi}$. Thus, based on Eq. (3.17), one can get the desired torques:

$$M_d = I_q W U_2 \tag{3.27}$$

Finally, having the desired thrust (Eq. (3.25)) and torques (Eq. (3.27)), the motors rotational speeds can be computed by Eq. (3.10). This attitude control, on the one hand, is very simple but on the other it has some shortcomings: The most important is the need to calculate the attitude error derivatives, which cost much in the sense of computation efforts. For a more sophisticated attitude control scheme, see Zuo (2010) or Altuğ (2005).

Table 3.1 Quadrotor controller parameters.

Notation	Quantity	Value	Units
m	Quad's mass	0.5	Kg
I_q	Quad's moment of inertia	$diag[5, 5, 9] \cdot 10^{-3}$	$Kg \cdot m^2$
C_i	Drag force in the i-th direction	1	$N \cdot sec/m$
I_r	Rotor polar moment of inertia	$3.4 \cdot 10^{-5}$	$Kg \cdot m^2$
C_1	Lift force coefficient	$3.4 \cdot 10^{-5}$	$N \cdot sec^2/rad^2$
C_2	Total torque coefficient	$1.2 \cdot 10^{-7}$	$N \cdot sec^2/rad^2$
l	Arm length	0.2	m

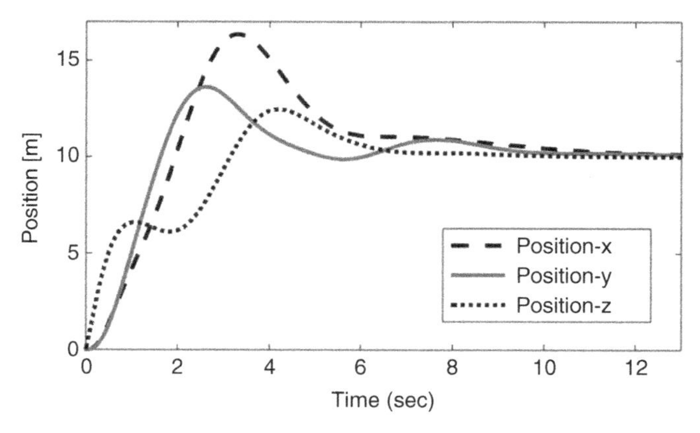

Figure 3.10 The controller performance for $K_P^{\Gamma} = diag[2.5,2.5,2.5]$ and $K_D^{\Phi} = diag[2.5,2.5,2.5]$. The different lines indicate the x, y, and z location axes.

3.3.7 Simulations

In this section, we shall present some experimental results for the quadrotor control and path tracking. The quad's physical characteristics taken here match regular toy aircrafts, and the controller parameters are listed in Table 3.1.

The diagram blocks in Figure 3.9 are implemented (from left to right): Eq. (3.20) is for trajectory tracking parameters U_1; Eqs. (3.24) and (3.25) find the desired orientation and thrust force; Eq. (3.26) finds the virtual control inputs U_2; Eq. (3.27) estimates the desired torque; Eq. (3.10) translates U_2 to rotors' angular velocities; and finally Eqs. (3.12)–(3.14) are used to calculate the quadrotor dynamics (required for simulation alone) (Figures 3.10–3.12).

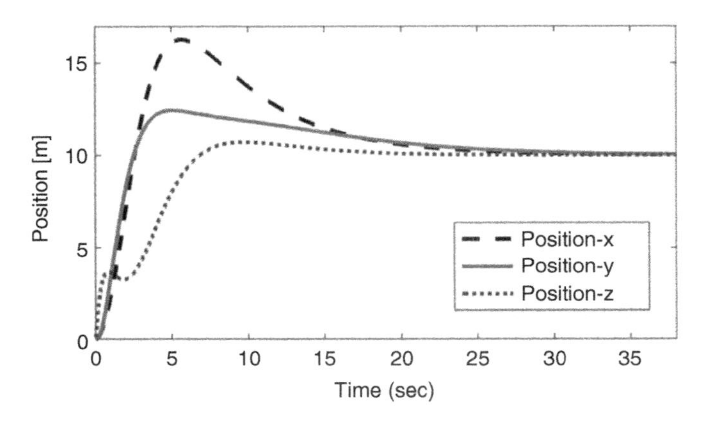

Figure 3.11 The controller performance for $K_P^\Gamma = K_D^\Gamma = K_P^\Phi = \text{diag}[1,1,1]$ and $K_D^\Phi = \text{diag}[6,6,6]$. The different lines indicate the *x*, *y*, and *z* location axes.

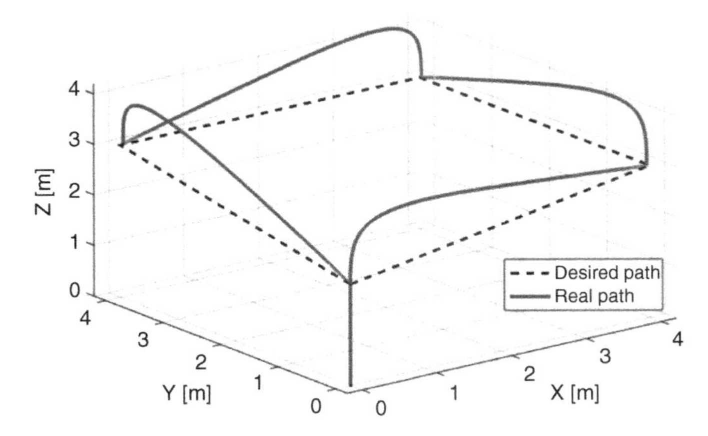

Figure 3.12 The controller performance for controller gains as the same as in Figure 3.11. The dashed line indicates the desired path, gray solid line the actual controlled path that the quad assume

References

Ahmad, M., Polotski, V., and Hurteau, R. (2000). Path tracking control of tracked vehicles. In: *ICRA'00. IEEE International Conference on Robotics and Automation*, 2938–2943. IEEE.
Altuğ, E.J. (2005). Control of a quadrotor helicopter using dual camera visual feedback. *The International Journal of Robotics Research* 24 (5): 329–341.
Fahimi, F. (2008). *Autonomous Robots: Modeling, Path Planning, and Control*, vol. 107. Springer Science & Business Media.
Griffin, R., and Allardo, E. J. (1994). "Robot transport platform with multi-directional wheels." U.S. Patent No. 5,323,867. 28 June.
Happian-Smith, J. (2001). *An Introduction to Modern Vehicle Design*. Elsevier.
Kimberlin, R.D. (2003). *Flight Testing of Fixed-Wing Aircraft*. American Institute of Aeronautics and Astronautics.
Murayama, A. and Yamakita, M. (2007). Development of autonomous bike robot with balancer. In: *SICE, 2007 Annual Conference*, 1048–1052. IEEE.
Ogata, K. (2001). *Modern Control Engineering*. Prentice Hall PTR.
Salomon, O. A. (2012). Nir Shvalb, and Moshe Shoham. "Vibrating robotic crawler." U.S. Patent No. 8,294,333. 23 October.
Shvalb, N.A. (2013). A real-time motion planning algorithm for a hyper-redundant set of mechanisms. *Robotica* 31: 1327–1335.
Zuo, Z. (2010). Trajectory tracking control design with command-filtered compensation for a quadrotor. *IET Control Theory and Applications* 4: 2343–2355.
Schroer, R.T., Boggess, M.J., Bachmann, R.J. et al. (2004). Comparing cockroach and Whegs robot body motions. In: *IEEE International Conference on Robotics and Automation, 2004. Proceedings. ICRA'04*, vol. 4, 3288–3293. Institute of Electrical and Electronics Engineers (IEEE).

4

Motion in Potential Field and Navigation Function

Nir Shvalb and Shlomi Hacohen

This chapter introduces motion-planning methods that do not assume knowledge of global coordinates, but, instead, are based on the artificial potential field specified over the domain and implement the navigation function scheme. Similar to general gradient descent techniques, the main idea of potential field and navigation function (NF) methods is to construct a "surface" over which the robot "slides" from initial configuration to the goal configuration. In addition, the chapter presents the novel method of path planning in uncertain environment with probabilistic potential.

4.1 Problem Statement

In general, the problem of motion planning for a mobile robot requires determination of an optimal path in two- or three-dimensional domain without colliding with static or dynamic obstacles, where optimality refers to minimal length of the path, minimal energy consumption, or similar parameters that are interpreted as a cost of the motion. Obstacles in this framework depend on the considered mission of the robot and may be walls, furniture, other players, or agents that move in the robot's environment, or even the regions where the robot cannot communicate with the base station or cannot maneuver.

The methods and techniques for solving the problem of motion planning are usually determined as the motion-planning scheme, which formally is aimed to provide a path of the robot and minimizes a definite *cost function* and maintains a set of requirements that cannot be violated; these requirements are referred to as the *constrains*. Notice that the constraints may be both physical constraints such as limited velocity and real objects in the workspace.

From now on, we will use the term *configuration* to designate a system's state $\mathfrak{S}$. For an arm manipulator, configuration is the set of its actuator's values, while for a mobile robot configuration may represent the robot's location $\vec{x}$, which is a vector $\vec{x} = (x,y)$ or

Autonomous Mobile Robots and Multi-Robot Systems: Motion-Planning, Communication, and Swarming, First Edition. Edited by Eugene Kagan, Nir Shvalb and Irad Ben-Gal.

Companion website: www.wiley.com/go/kagan/robotsystems

$\vec{x} = (x,y,z)$ of the robot's Cartesian coordinates in two- or three-dimensional (2D or 3D) domain. If it is needed, the configuration $\mathfrak{s}$ may also include the robot's orientation, velocity, and similar parameters. Consequently, the configuration or state space $\mathfrak{S}$ of the system is defined as follows.

Definition 4.1 *A configuration space $\mathfrak{S}$ is the set of all configurations $\mathfrak{s}$, that can be acquired by the system if no obstacles are at reach.*

Definition 4.2 *The free configuration space $\mathfrak{S}_{free} \subset \mathfrak{S}$ is a set of the configurations $\mathfrak{s}$, that avoid physical obstacles within the robot's workspace.*

Using a set of real-value functions $g_i : \mathfrak{S} \rightarrow \mathbb{R}$, $i = 0, 1, 2, \ldots$, it is convenient to define the free configuration space as a set of configurations $\mathfrak{S}_{\text{free}} = \{\mathfrak{s} \mid g_i(\mathfrak{s}) > 0, \forall i\}$. This implies that $\mathfrak{S}$ is considered as a perfectly known configuration space; in the other words, it is assumed that the environment can be sensed by the robot in a very precise way, which, certainly, is not obvious.

The problem considered in the chapter is formulated as follows:

Problem 4.1 *Generate an optimal discrete path $\langle \mathfrak{s}_0, \mathfrak{s}_1, \mathfrak{s}_2, \ldots, \mathfrak{s}_n \rangle$ from an initial position $\mathfrak{s}_0 \in \mathfrak{S}_{free}$ to a destination position $\mathfrak{s}_n = \mathfrak{s}_d \in \mathfrak{S}_{free}$, where optimality refers to the minimum of a cost function $c : \mathfrak{S} \rightarrow \mathbb{R}$.*

There are many methods and algorithms for solving this problem, based on the following common assumptions:

1) Perfectly known environment represented by the environmental map
2) Perfect knowledge about positions of the robot and the obstacles
3) Simplistic geometries of the robot and obstacles
4) Simplistic dynamic model of the robot

The widely known algorithms for generating optimal or near-optimal paths using these assumptions are rapidly exploring random tree planner (RRT), visibility graph, and roadmaps, which implement different search techniques (Choset et al. 2005; Siegwart et al. 2011). In this chapter, we consider two of the most widely used approaches for motion planning called *artificial potential field* method and *navigation function* scheme.

The artificial potential field method was introduced in 1986 by Khatib (1986) for arm manipulators and in 1989 by Borenstein and Koren (1989) for mobile robots. The main idea behind the potential field scheme is constructing a function that serves as a "surface" over which the robot "slides" from the initial configuration to the goal configuration. The method of navigation function (NF), which was introduced in 1990 by Koditschek and Rimon (1990), follows the approach of the potential field. However, a robot that follows a navigation function will not fall prey to local minima as often happens when following a potential field. Both methods are extensions of gradient descent method of optimization.

4.2 Gradient Descent Method of Optimization

Gradient descent refers to the widely spread local optimization method, where it is needed to find the policy that minimizes (or maximizes) certain costs $c(\mathfrak{s})$ (or rewards) of the configurations. In each time step, this policy has to provide the motion direction such that it minimizes the cost function c in the configuration $\mathfrak{s}$ that is – the gradient $-\nabla c(\mathfrak{s})$. For clarity, we will first ignore any constraints and assume the configuration space is empty.

4.2.1 Gradient Descent Without Constraints

Let us consider the motion of mobile robot; in this case, the configuration of the system $\mathfrak{s}$ represents the location vector $\vec{x}$ of the robot. Consequently, in this chapter, the terms *configuration, location,* and *position* will be used in the same meaning as points of the configuration space $\mathfrak{S}$, which is represented by two- or three-dimensional Cartesian domain X.

Denote an initial location of the robot at time t_0 by $\vec{x}_0 = \vec{x}\,(t_0)$ and a goal position by $\vec{x}_g$. A simple example for the cost of the location $\vec{x} \in X$ is:

$$c(\vec{x}) = \left(\vec{x} - \vec{x}_g\right)^2. \tag{4.1}$$

Intuitively, if the cost function c is differentiable at the current location $\vec{x}_t = (x_t, y_t, z_t)$, then the next step that decreases the distance to a local minimum should be chosen in the opposite direction to the gradient of the cost function:

$$\nabla c\left(\vec{x}_t\right) = \left(\frac{\partial c(x_t, y_t, z_t)}{\partial x}, \frac{\partial c(x_t, y_t, z_t)}{\partial y}, \frac{\partial c(x_t, y_t, z_t)}{\partial z}\right)^T. \tag{4.2}$$

This direction is denoted by

$$p_t = -\frac{1}{\|\nabla c\left(\vec{x}_t\right)\|}\nabla c\left(\vec{x}_t\right); \tag{4.3}$$

Notice that the vector p_t is normalized to unit.

Algorithmically, the extent d_t, by which the robot moves in this direction, is better to be as large as possible. However, the robot certainly should not continue moving in the direction, which no longer agrees with the gradient. So, at each time the optimal size d_t^* of the step should be specified such that

$$d_t^* = \arg\min_{d_t}\left\{c\left(\vec{x}_t + d_t p_t\right)\right\}, \tag{4.4}$$

which can be found by the line search or similar methods. Notice that this equation implements the motion equation:

$$\vec{x}_{t+1} = \vec{x}_t + d_t p_t; \tag{4.5}$$

In Algorithm 4.1, we will use this equation repeatedly until the termination condition is achieved.

The termination condition is defined over the size of the step proportionally to the values $\|\nabla c\left(\vec{x}_t\right)\|$ or $\frac{\|\vec{x}_t - \vec{x}_{t-1}\|}{\|\vec{x}_t\|}$. Since for each location $\vec{x}_t$ the gradient of the cost function c defined by Eq. (4.1) is $\nabla c\left(\vec{x}_t\right) = 2\left(\vec{x}_t - \vec{x}_g\right)$, that is, a straight line between $\vec{x}_t$ and $\vec{x}_g$, the optimal path of the robot from initial to goal position is a line between $\vec{x}_0$ and $\vec{x}_g$. The algorithm that implements the gradient descent method without constraints is outlined as follows.

Algorithm 4.1 (Gradient descent). Given the domain X, the cost function c and termination criteria $\varepsilon > 0$, do:

1) Initialize $t = t_0$ and define initial location $\vec{x}_0 = \vec{x}\,(t_0)$.
2) Do:
3) Set $t = t + 1$.
4) Calculate direction p_t according to Eq. (4.3).
5) Set optimal size d_t^* of the step according to Eq. (4.4).
6) Update location $\vec{x}_t = \vec{x}_{t-1} - d_t^* p_t$ with respect to Eq. (4.5).
7) While $\|\vec{x}_t - \vec{x}_{t-1}\| \geq \varepsilon$ do:

The presented algorithm implements the one-step optimization defined by Eq. (4.4). Another widely known optimization method that can be applied for "good enough" cost functions is the Newton method.

For simplicity, let us start with the one-dimensional motion; in this case, the domain is $X \subset \mathbb{R}$ and the position $\vec{x}_t$ of the robot is represented by its coordinate x_t. As above, assume that it is required to find a path from initial location x_0 to the goal location x_g such that the derivative of the cost function c in this point is $c'\,(x_g) = 0$. For any point $x_t \in X$, the Tailor series for the function c in the neighborhood of x_t is

$$c(x_t) = c(x_t + \Delta x) + c'(x_t)\Delta x + \frac{1}{2}c''(x_t)\Delta x^2 + O\left(\Delta x^3\right). \tag{4.6}$$

Let us consider the second-order approximation of the function c. Recall that it is required to find Δx that leads the robot to move closer to the point, where the cost reaches its minimum (or maximum for the reward functions). The equalizing of the derivative of the series relative to Δx to zero then results in the equation

$$c'(x_t) + c''(x_t)\Delta x = 0, \tag{4.7}$$

which yields

$$\Delta x = -\frac{c'(x_t)}{c''(x_t)}. \tag{4.8}$$

Thus, for governing the robot toward the goal location x_g, instead of Eq. (4.5), the following equation is applied:

$$x_{t+1} = x_t - \frac{c'(x_t)}{c''(x_t)}. \tag{4.9}$$

For the problems in 2D and 3D domains, the method follows the same reasons with the multidimensional version of Eq. (4.9):

$$\vec{x}_{t+1} = x_t - \left(H\left(c\left(\vec{x}_t\right)\right)\right)^{-1} \nabla c\left(\vec{x}_t\right), \tag{4.10}$$

where $H\left(c\left(\vec{x}_t\right)\right)$ is the Hessian matrix of the function c in the point $\vec{x}_t$.

Notice that usually the Newton method converges faster than gradient descent method, but since the gradient or the inverse of the Hessian matrix vanish in the point with zero derivative of the cost function, it does not necessarily converge to maximum or minimum. As a result, the Newton method leads the robot either to local minimum, to local maximum, or to a saddle point. In contrast, the gradient descent method converges to local minimum or to local maximum with respect to the chosen sign of the direction vector p_t.

The difference between the methods executed over 2D domain with the points $\vec{x} = (x,y)$ is illustrated in Figure 4.1. The origin of the coordinate system is in the center of the domain. The methods evaluate the cost function

$$c\left(\vec{x}\right) = 0.05x^3 + 0.01(y-2)^3 + (x-15)^2 + y^2 + xy + 2y,$$

and compute the paths starting from two different initial points, $\vec{x}_0 = (-15,25)$ and $\vec{x}_0 = (10,28)$. The path created by the gradient descent method is depicted by the dashed lines, and the path created by the Newton method is depicted by the solid lines.

It is seen that starting from the initial point $\vec{x}_0 = (10,28)$, both gradient descent method and the Newton method create the paths that lead to the global minimum, and the path created by the Newton method is shorter than the path created by the gradient descent method. However, if the path starts from the initial point is $\vec{x}_0 = (-15,25)$, then the gradient descent method provides the path that leads to the global minimum, while the path created by the Newton method terminates in the saddle point of the cost function.

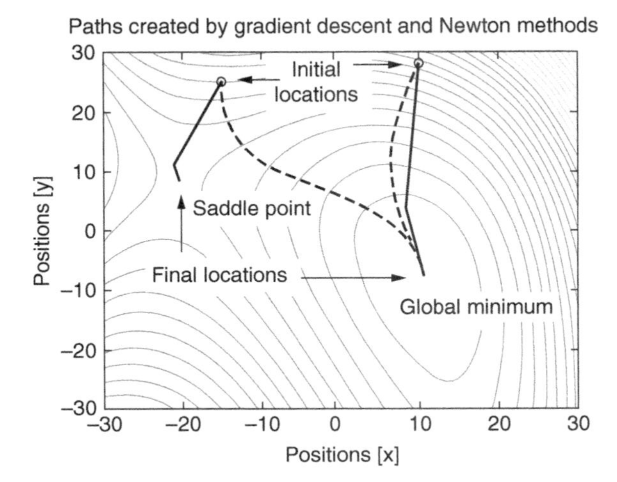

Figure 4.1 Paths of the robot trajectories created by the gradient descent method (dashed lines) and by the Newton method (solid lines).

4.2.2 Gradient Descent with Constraints

Now let us consider the gradient descent method with constraints as they were introduced above. Assuming that the configuration space represents a set of possible locations of the mobile robot, the constraint can be defined as a set of algebraic functions $g_i : X \to \mathbb{R}^+$, $i = 0, 1, 2, \ldots$ Then, the problem is to minimize the costs $c(\vec{x}) > 0$ subject to the inequalities $g_i(\vec{x}) > 0$.

In order to provide the algorithm of gradient descent with constraints, functions c and g must be merged into one function, which can be used repeatedly in the procedure of gradient descent. The usual straightforward definition of such function is the following:

$$\bar{c}(\vec{x}) = \begin{cases} f(\vec{x}) \text{ if } g_i(\vec{x}) > 0 \forall i, \\ \infty \quad \text{otherwise,} \end{cases} \tag{4.11}$$

where $\bar{c}$ is a new cost function and the function f defines the cost of the robot's location while the requirements regarding the constraints hold. In this formulation, the cost function has a sharp cliff around the constraint area and guarantees that the points with negative gradient are out of the constraint. However, notice that the cost function $\bar{c}$ is not smooth; hence, the methods that implement its derivatives cannot be applied.

The most popular method for solving the problem with such cost function is the *penalty method* (Siciliano and Khatib 2016, 879). The main idea of this method is to replace the cost function and the constraints by the function in Eq. (4.12):

$$\bar{f}(\vec{x}) = f(\vec{x}) + \alpha \sum_{i \in I} \left(\bar{g}_i(\vec{x})\right)^2, \tag{4.12}$$

where $\bar{g}_i\left(\vec{x}\right) = \min\left\{0, g_i\left(\vec{x}\right)\right\}$, $\alpha > 0$ is a penalty factor and I is a set of indices. The functions f and $\bar{f}$ have the same critical point in which the derivatives of the functions are zero; however, the points in which the gradient of $\bar{f}$ is $\nabla\bar{f}\left(\vec{x}\right) < 0$ are always out of the areas is restricted by the constraints. Nevertheless, notice again that this function is not differentiable everywhere.

For example, consider a mobile robot that acts in the circle and executes the following mission. Start from the initial position $\vec{x}_0$ and move to the goal position $\vec{x}_g$, avoiding collisions with circle obstacles or the boundary of the circle environment. The obstacles' and robot's radii are r_0 and r_i, $i = 1, 2, \ldots, n$, respectively, where n is a number of obstacles. The origin of the coordinates system is located at the center of the circle environment. The centers of the obstacles denoted by $\vec{x}_i$.

Define the cost function f and the constraints g as follows:

$$f(\vec{x}) = \|\vec{x} - \vec{x}_g\|^2, \quad g_i\left(\vec{x}\right) = \|\vec{x} - \vec{x}_i\| - r_i.$$

Recall that while for the obstacles the constraint's meaning is to be outside of the obstacles' areas, for the environment's boundary the meaning of the constraint is to be in it. Thus, the constraint that related to the environment's boundary is

$$g_0(\vec{x}) = r_0 - \|\vec{x} - \vec{x}_0\|.$$

Applying these functions to new cost function defined by Eq. (4.12), one obtains

$$\bar{f}\left(\vec{x}\right) = \|\vec{x} - \vec{x}_g\|^2 + \left(\min\left\{0, g_0\left(\vec{x}\right)\right\}\right)^2 + \alpha\sum_{i=1}^{n}\left(\min\left\{0, g_i\left(\vec{x}\right)\right\}\right)^2.$$

The considered penalty method provides the simple but rather efficient techniques for navigating the mobile robot in the environment with obstacles. Certainly, since in this setup, initial and goal positions as well as positions and the areas of the obstacles are known, the problem of creating an optimal path of the robot can be solved using more complicated optimization methods.

In particular, one of the widely accepted methods is the use of *Lagrange multipliers* (for the description of this method in the framework of navigation problems see, e.g., Stone (1975) and Kagan and Ben-Gal (2013, 2015). For simplicity, assume that there is a single constraint and the problem is to minimize the cost $c(\vec{x})$ subject to the constraint $g(\vec{x}) = 0$. The naïve intuition is based on the knowledge that the solution lies on the line $g(\vec{x}) = 0$, so the solution prescribes to walk along this line as the cost $c(\vec{x})$ decreases. However, such a solution is less effective. The gradient of $g(\vec{x})$ in the points where $g(\vec{x}) = 0$ is parallel to the plane spread by $\vec{x}$ and its direction is perpendicular to $g(\vec{x})$. If the walk is conducted along the spline $g(\vec{x}) = 0$ and is terminated when parallel to $g(\vec{x})$ component of the gradient $\nabla f(\vec{x})$ vanishes, then the termination point provides an optimal solution. In this point, the gradient $\nabla f(\vec{x})$ is perpendicular to $g(\vec{x})$; in the other words, the gradients of $g(\vec{x})$ and $f(\vec{x})$ are parallel and agree with the following equation:

$$\nabla f\left(\vec{x}\right) = -\lambda \nabla g\left(\vec{x}\right). \tag{4.13}$$

The constant λ is called the *Lagrange multiplier* and is needed since the magnitudes of the gradients on the left and the right sides of the equation are not necessarily equivalent.

Finally, let us replace the cost function and the constraint with the demands $g\left(\vec{x}\right) = 0$ and $\nabla f\left(\vec{x}\right) = -\lambda \nabla g\left(\vec{x}\right)$ and define the Lagrange equation:

$$\mathcal{L}\left(\vec{x},\lambda\right) = f\left(\vec{x}\right) + -\lambda g\left(\vec{x}\right). \tag{4.14}$$

Then, Eq. (4.13) obtains the following form:

$$\nabla_{\vec{x},\lambda} \mathcal{L}\left(\vec{x},\lambda\right) = \begin{pmatrix} \nabla f\left(\vec{x}\right) \\ 0 \end{pmatrix} + \begin{pmatrix} \lambda \nabla g\left(\vec{x}\right) \\ g\left(\vec{x}\right) \end{pmatrix} = 0. \tag{4.15}$$

Solution of this equation provides an optimal solution of the problem and, as a byproduct, the value of the Lagrange multiplier λ.

By the same reasons, in the case of multiple constraints the method of Lagrange multipliers results in this equation:

$$\mathcal{L}\left(\vec{x},\lambda_1,\lambda_2,\ldots\right) = f\left(\vec{x}\right) + \sum_{i \in I} \lambda_i g_i\left(\vec{x}\right). \tag{4.16}$$

The other usual methods known as *value iteration* and *policy iteration* algorithms (see, e.g., Kaelbling et al. 1996, 1998; Sutton and Barto 1998) implement dynamic programming techniques, and can be extended for navigation of the robots under uncertainties. In Section 7.2.2, we consider these algorithms in detail and apply them for navigation of the robot in belief space.

4.3 Minkowski Sum

The gradient descent approach just described provides a general framework for consideration of the navigation problem using the methods based on the potential and navigation functions. As indicated, a common technique used in the motion planning problems requires definition of free configuration space $\mathfrak{S}_{free}$, which in the case of navigation of mobile robot is associated with the domain $X_{free} \subset X \subset \mathbb{R}^n$, $n \geq 1$, where the robot can travel without colliding with obstacles, excluding the boundary. To this end, a prevalent method is to define X_{free} as the complement of the set X_{obs} – that is, the union of the *Minkowski sums* of the area occupied by the robot with the areas occupied by the obstacles.

Intuitively, the obstacles in the domain X are expended by the robot's volume, while the robot is considered as a point mass – the center of gravity, in which the robot's mass is concentrated (for simple intuition regarding this process, see the piano mover problem described in Section I.4). Explicitly, denote by A the set of vectors that specify the robot's geometry; these vectors are measured from the robot's center of gravity to each point on the robot body. Similarly, denote by B the set of vectors that specify geometry of the

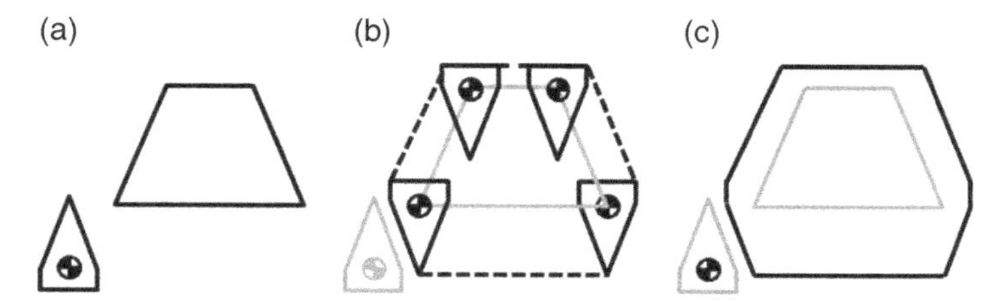

Figure 4.2 The Minkowski sum of the pentagon shaped robot and the trapezoid-shaped obstacle: (a) an initial setup; and (b) the robot rotated by 180°and placed in the corners of the trapezoid; perimeter of the spanning figure is depicted by the dashed line. The resulting configuration with the robot represented by its center of gravity and the obstacle enlarged up to the spanning figure is shown in (c).

obstacles; these vectors are measured from the origins of the obstacles to their body points. Then the set X_{obs} is defined as follows:

$$X_{obs} = B*(-A) = \{b-a \mid a \in A, b \in B\}, \tag{4.17}$$

where the asterisk $*$ stands for the Minkowski sum. Notice that in order to measure the distance between the point inside the robot and the point inside the obstacle, one should first rotate the robot at 180° , represented by the minus sign in Eq. (4.17). The sets A and B are the subspaces of X, and in general the set $B*(-A)$ is rather large, making computations difficult. One possible way to overcome this problem is to confine calculations to an intermediate time step; Figure 4.2 illustrates this process. In the figure, the obstacle is depicted by the trapeze and the robot by the pentagon with the center of gravity denoted by the sign ⊕.

Notice that in Figure 4.2b, the shortest distance from the lower-left corner of the obstacle to the robot's edge is equivalent to the shortest distance from the center of the robot to the edge of the obstacle after Minkowski summation. In the further considerations in this chapter (as well as in the next chapters if not otherwise indicated), the robot will be represented as point mass (center of gravity) and the obstacles by their Minkowski sum.

4.4 Potential Field

Obstacle avoidance can be treated in a somewhat more "holistic" manner: instead of separately taking care of the goal configuration and the obstacles, one may incorporate both in a single function $\mathcal{U}: \mathbb{R}^n \to \mathbb{R}$ called *artificial potential field* (Khatib 1986; Borenstein and Koren 1989). Potential fields naturally emerge when considering electrostatics. Such scenarios typically include several electrical charges (see Figure 4.3), and the question is: What is the vector force field that they induce?

Recall that in the case of electrostatics, the force induced on a point charge p located at the origin by a point charge q located at point with radius-vector $\vec{r}$ is

$$\|\vec{F}\| = k_e \frac{qp}{\|\vec{r}\|},$$

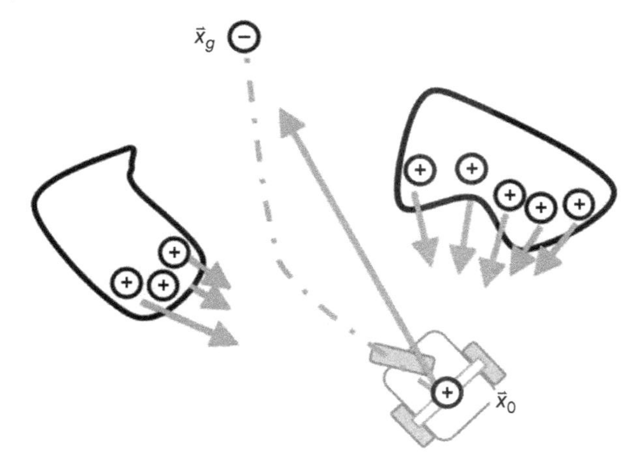

Figure 4.3 Analogy between navigation of mobile robot and potential field in electrostatics. The robot and the obstacles are "charged" positively and the goal position $\vec{x}_g$ is "charged" negatively. The robot starts in the point $\vec{x}_0$. Following electrostatics, the positively charged robot is attracted by the negatively charged goal point and is repulsed by the positively charged obstacles with respect to their areas. The resulting trajectory of the robot is depicted by the dashed-dotted curve.

in the direction from q to p, where k_e is the Coulomb constant, which for our purposes may be specified as $k_e = 1$. Thus, if q and p are the charges of the same sign, then $\vec{F}$ is a repulsive force, while for different signs of q and p it is an attracting force.

In the case where multiple point charges q_i, $i = 1, 2, \ldots$ acting on p, the forces are summed as depicted in Figure 4.3, and the force is

$$\vec{F} = \sum_i \vec{F}_i = \sum_i \frac{pq_i}{r_i^2}\hat{r}_i, \tag{4.18}$$

where $\hat{r}_i$ is the direction vector from p to q_i and $r_i = \|\vec{r}_i\|$.

For a motion-planning scheme, it is assumed that the robot has a positive charge, the goal has a negative charge, while the obstacles are positively charged as well. In addition, with respect to the consideration in Section 4.3, it is assumed that the robot has a point mass geometry and the obstacles can have any desired shape.

Without loss for generality, assume that the obstacle is specified by the binary function V such that $V(\vec{x}) = 1$ for each point $\vec{x}$ within the obstacle's area and $V(\vec{x}) = 0$ otherwise. Denote the charge density by ρ. Then, the artificial potential field for the ith obstacle, $i = 1, 2, \ldots$, with the center located in the point $\vec{x}_i$ is defined as follows:

$$\mathcal{U}(\vec{x}_i) = -\int \rho(\vec{x}) \frac{V(\vec{x})}{\|\vec{x}_i - \vec{x}\|^2} d\vec{x}, \tag{4.19}$$

For practical needs, the repulsive charges function $\mathcal{U}^{rep}$ and attractive charges function $\mathcal{U}^{atr}$ can be distinguished; then the total potential field $\mathcal{U}(\vec{x})$ in the point $\vec{x}$ is defined as a sum all repulsion and attraction fields $\mathcal{U}^{rep}(\vec{x})$ and $\mathcal{U}^{atr}(\vec{x})$ in this point. Attractive, repulsive, and total potential fields are illustrated in Figure 4.4.

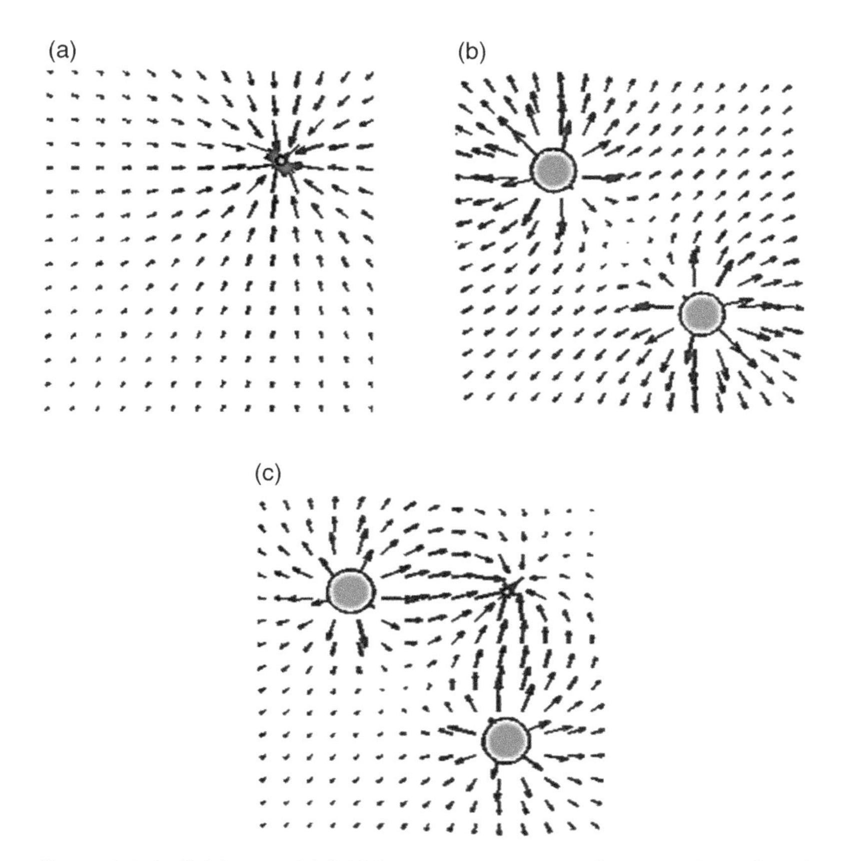

Figure 4.4 Artificial potential field for attractive target and two repulsive obstacles: (a) attractive potential field $\mathcal{U}^{atr}$ that corresponds to a single target; (b) repulsive potential field $\mathcal{U}^{rep}$ of two obstacles. (c) total potential field $\mathcal{U}$.

The gradient $\nabla\mathcal{U}$ of the potential field is the vector field that points away from the obstacles and toward the goal position. Now, the law of motion control can be defined by the gradient descent method in Section 4.2 and Algorithm 4.2, which summarizes this law as follows.

Algorithm 4.2 (motion in artificial potential field). Given the domain X, potential function $\mathcal{U}$, and termination criteria $\varepsilon > 0$, do:

1) Initialize $t = t_0$ and define initial location $\vec{x}_t = \vec{x}\,(t_0)$.
2) While $\|\nabla\mathcal{U}(\vec{x}_t)\| \geq \varepsilon$ do:
3) Set $t = t + 1$.
4) Update location $\vec{x}_t = \underset{t-1}{\vec{x}} + \nabla\mathcal{U}\left(\underset{t-1}{\vec{x}}\right)$.
5) End while.

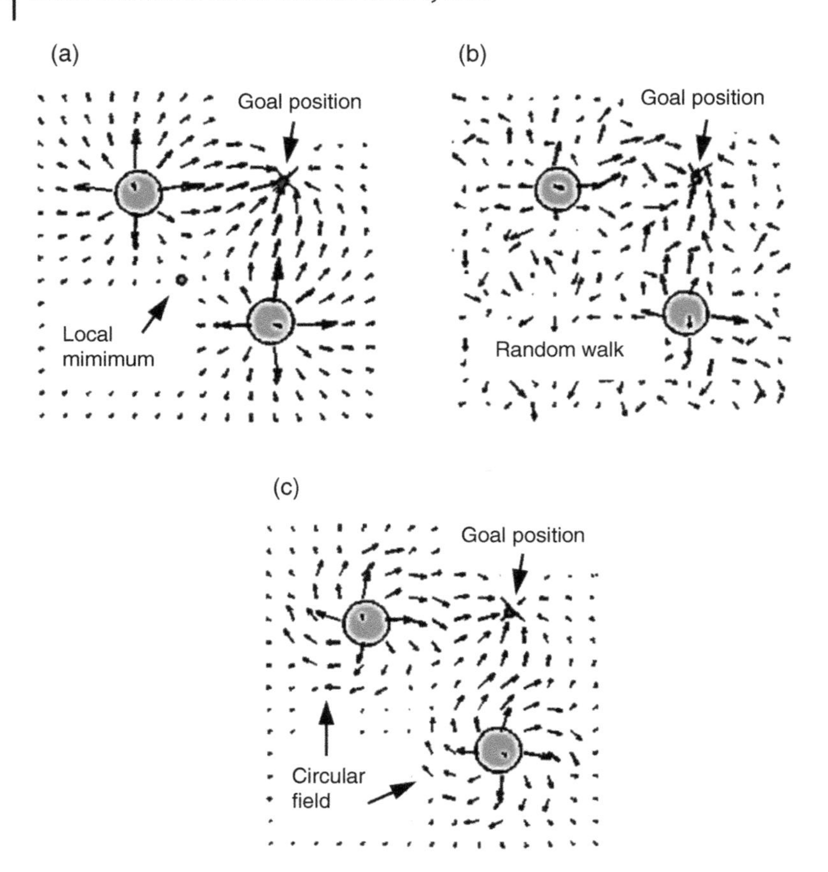

Figure 4.5 Artificial potential field for attractive target and two repulsive obstacles: (a) existence of the local minimum (cf. Figure 4.1). (b) potential field with additional random walk; and (c) potential field with additional circular field around the obstacles.

It is seen that this algorithm follows the same approach as Algorithm 4.1, but instead of the cost function c, it implements the potential function $\mathcal{U}$.

As indicated above, one of the main problems with this approach is that sometimes it get stuck in a local minimum, as shown in Figure 4.5a.

One advisable step to handle this problem is to add random walk to the vector field (see Figure 4.5b). Then the robot will leave the local minimum by random; however, it is not guaranteed that it will not fall in the local minimum again. In the worst case, it is also recommended to add circular vector field around the obstacles. Such fields prevent the robot from stopping in the areas close to the obstacles (see Figure 4.5c).

In order to prevent the undesired scenarios, in which the mobile robot terminates in the local minima or follows the goal position by nonoptimal paths as it occurs in the scenarios with random walk and circular fields around the obstacles, the potential field approach was modified and additional requirements to the potential function were implemented. The resulting function is known as the navigation function. In the next section, the NF is considered in detail.

4.5 Navigation Function

As indicated at the beginning of the chapter, the NF method was first introduced in 1990 by Koditschek and Rimon (1990) and completely developed (for deterministic setup) in 1992 (Rimon and Koditschek 1992). Following the approach of navigation using potential field, the NF is defined as a particular continuous smooth potential function that takes zero value at the target point and unity value on the boundaries of the environment and the obstacles. Moreover, in order to ensure a solution, it is required that all critical points of the navigation function are nondegenerated in such a sense that there are no such "plateau" areas where the gradient of the navigation function vanishes. This section presents three NF variants using three different scenarios: navigation in static deterministic environment, navigation in static uncertain environment, and finally, navigation in dynamic environment.

4.5.1 Navigation Function in Static Deterministic Environment

Following the original work by Koditschek and Rimon (1990), let us start with the navigation function that is defined over a domain X and provides a safe trajectory of the robot from initial position $\vec{x}_0 \in X$ to the goal position $\vec{x}_g \in X$. The discourse below extends the original consideration and is aimed to find such a function for motion planning that is coupled with a feedback control law calculation. As a result, the obtained path avoids obstacles and is confined to a given region called *work space* with guaranteed convergence from any initial position (subject to certain parameter).

Let us start with general definition. Let $X \subset \mathbb{R}^n$, $n \geq 1$, be a domain and, as above, denote by $X_{\text{free}} \subset X$ a free domain, where the robot can move without collisions.

Definition 4.3 *A function* $\varphi : X \to \mathbb{R}$ *is called a* navigation function *if it satisfies the following requirements:*

1) Function φ is analytic in X_{free}; more precisely, it is assumed that φ has at least the second derivatives in all points $\vec{x} \in X_{\text{free}}$.
2) Function φ is the Morse function in X_{free}; recall that the function is called the Morse function if its Hessian matrix does not vanish in any point of the considered domain.
3) Function φ is polar in X; that is, it has a unique minimum in X; it is assumed that the minimum is reached in the goal position $\vec{x}_g$.
4) Function φ is admissible in such a sense that it reaches its maximum on the boundary points of X_{free}.

It is clear that the first two requirements are rather general and are needed for analytical reasons. In contrast, the last two requirements make sense especially for navigation tasks: requirement 3 governs the robot to move toward the goal position, and requirement 4 leads the robot to stay in the domain without crossing the boundaries.

The most popular definition of the NF $\varphi_\kappa : X \to \mathbb{R}$ (with parameter $\kappa > 0$) is the following (Koditschek and Rimon 1990; Rimon and Koditschek 1992). Let $\vec{x} \in X$ be a point of the domain and $\vec{x}_g \in X$ be a goal position. Then:

$$\varphi_\kappa\left(\vec{x}\right) = \frac{\mathrm{r}_g\left(\vec{x}\right)}{\left[\mathrm{r}_g^\kappa\left(\vec{x}\right) + \beta\left(\vec{x}\right)\right]^{1/\kappa}}, \tag{4.20}$$

where $\mathrm{r}_g(\vec{x}) = \|\vec{x} - \vec{x}_g\|^2$ is a distance between the points $\vec{x}$ and $\vec{x}_g$ (c.f. definition of the cost function used in the penalty method in Section 4.2.2), β is the *obstacles function*

$$\beta\left(\vec{x}\right) = \prod_i \beta_i(\vec{x}), \tag{4.21}$$

where $i = 0, 1, 2, \ldots, N_{obs}$ are the indices of the permitted area ($i = 0$) and the obstacles ($i > 0$), and

$$\beta_i\left(\vec{x}\right) = \begin{cases} -\|\vec{x} - \vec{x}_0\|^2 + r_0^2 \text{ if } i = 0, \\ \|\vec{x} - \vec{x}_i\|^2 - r_i^2 \text{ if } i > 0, \end{cases} \tag{4.22}$$

where $\vec{x}_0$ stands for the center of the permitted area (that is usually considered as the origin of the coordinates system), $\vec{x}_i$ stands for the center of the ith obstacle, and r_0 and r_i are the radii of the permitted area and the ith obstacle. Notice that while considering the penalty method in Section 4.2.2 symbol r_0 was used for denoting the radius of the robot; however, as was indicated in Section 4.3, after application of the Minkowski sum, the robot is considered as a point mass and the obstacles are enlarged respectively to the robot's geometry. It is clear that, with respect to this remark, the definition of the functions β_i follows the same approach as definition of the constraint functions g_i in the gradient descent with constraints (see Section 4.2.2).

It is clear that in Eq. (4.20), the nominator is defined in such a manner that the robot is attracted to the goal position, while the denominator specifies repulsion by the obstacles and corresponding collision avoidance with respect to the functions β_i. Function β_0 for permitted area and β_i for the obstacles are illustrated in Figure 4.6.

Let us demonstrate that the function φ_κ defined by Eq. (4.20) meets the requirements of the navigation function listed in Definition 4.1. Formally, the function φ_κ is a composition of three functions, each of which provides a certain property of φ_κ.

Let us start with the function $\hat{\varphi}_\kappa$ that provides attraction of the robot by the goal position and repulsion of robot by the obstacles. In the other words, the purpose of the function is to decrease the distance between the robot and the goal position and to increase the distance between the robot and the areas occupied by the obstacles. Thus, if $\hat{\varphi}_\kappa$ is defined as

$$\hat{\varphi}_\kappa\left(\vec{x}\right) = \frac{\left(\mathrm{r}_g\left(\vec{x}\right)\right)^\kappa}{\beta\left(\vec{x}\right)}, \tag{4.23}$$

then since the distance r_g reaches its zero value only in the destination point, and β reaches its zero values on the boundaries of X_{free}, the function $\hat{\varphi}_\kappa$ has global minimum at the destination point and infinite value on the boundaries of X_{free}. Furthermore, by

(a)

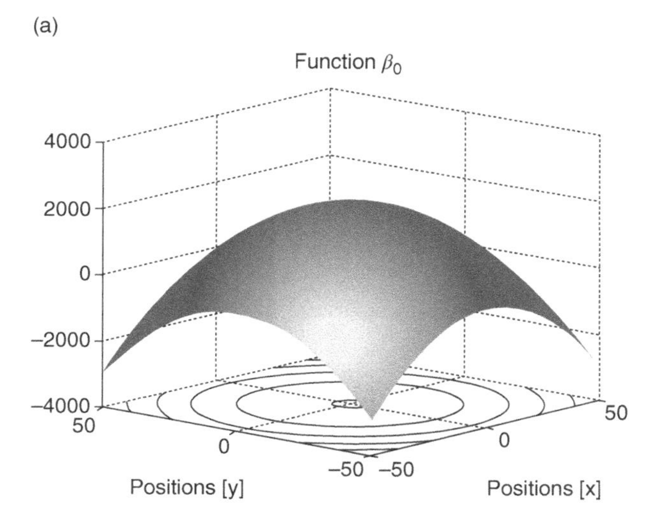

(b)

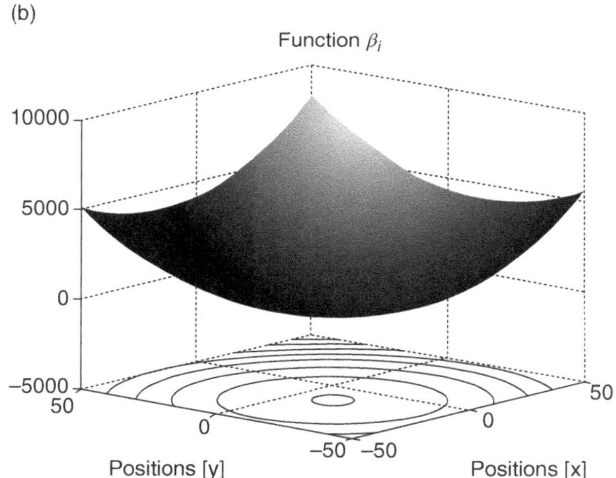

Figure 4.6 Components of the obstacle function: (a) function β_0 for permitted area; and (b) function β_i for the obstacle.

choosing the value the parameter κ sufficiently large, we guarantee that the minimum of $\hat{\varphi}_\kappa$ is unique. It is true since the value $\left\|\frac{\partial \beta}{\partial \vec{x}}\right\|$ does not depend on κ, while the value $\left\|\frac{\partial\, r_g^\kappa}{\partial \vec{x}}\right\|$ grows with κ; hence, the gradient $\nabla\hat{\varphi}_\kappa$ is always directed toward the goal position. As a result, function $\hat{\varphi}_\kappa$ provides the main aim of the NF to lead the robot toward the goal position and away from the obstacles.

Now, let us provide the bounds the function such that its maximal value in the domain is unit. For this purpose, the function $\hat{\varphi}_\kappa$ is composed with the sigmoid switch function $\sigma_\lambda(y) = \dfrac{y}{y+\lambda}$ that for positive λ and y maps $[0, \infty] \to [0, 1]$. As a result, the following function is obtained:

$$(\hat{\varphi}_\kappa{}^\circ\sigma_\lambda)\left(\vec{x}\right) = \sigma_\lambda\left(\hat{\varphi}_\kappa\left(\vec{x}\right)\right) = \frac{\left(\mathrm{r}_g\left(\vec{x}\right)\right)^\kappa/\beta\left(\vec{x}\right)}{\left(\left(\mathrm{r}_g\left(\vec{x}\right)\right)^\kappa/\beta\left(\vec{x}\right)\right)+\lambda}$$
$$= \frac{\left(\mathrm{r}_g\left(\vec{x}\right)\right)^\kappa/\beta\left(\vec{x}\right)}{\left(1/\beta\left(\vec{x}\right)\right)\left(\left(\mathrm{r}_g\left(\vec{x}\right)\right)^\kappa+\lambda\beta\left(\vec{x}\right)\right)} = \frac{\left(\mathrm{r}_g\left(\vec{x}\right)\right)^\kappa}{\left(\mathrm{r}_g\left(\vec{x}\right)\right)^\kappa+\lambda\beta\left(\vec{x}\right)}. \tag{4.24}$$

Finally, let us provide the Morse property. Notice that the obtained mapping $(\hat{\varphi}_\kappa{}^\circ\sigma_\lambda)\left(\vec{x}\right)$ has a unique minimum at the destination position, is bounded, and is analytical. Thus, it remains to guarantee that the destination point is a degenerate critical point. In order to obtain this property, the mapping $(\hat{\varphi}_\kappa{}^\circ\sigma_\lambda)\left(\vec{x}\right)$ is composed with the "distortion" function $\rho_\kappa(y) = y^{1/\kappa}$. As a result, the navigation function φ_κ defined by Eq. (4.20) is obtained; namely, for $\lambda = 1$:

$$(\hat{\varphi}_\kappa{}^\circ\sigma_1{}^\circ\rho_\kappa)\left(\vec{x}\right) = \rho_\kappa\left(\sigma_1\left(\hat{\varphi}_\kappa\left(\vec{x}\right)\right)\right) = \left(\frac{\left(\mathrm{r}_g\left(\vec{x}\right)\right)^\kappa}{\left(\left(\mathrm{r}_g\left(\vec{x}\right)\right)^\kappa+\beta\left(\vec{x}\right)\right)}\right)^{1/\kappa}$$
$$= \frac{\mathrm{r}_g\left(\vec{x}\right)}{\left[\mathrm{r}_g^\kappa\left(\vec{x}\right)+\beta\left(\vec{x}\right)\right]^{1/\kappa}}. \tag{4.25}$$

Unfortunately, there is no analytic way to choose the value of the parameter κ, and usually, it is defined with respect to the number of obstacle as $\kappa =$ # obstacles $\pm$ 1. Then, for the distortion function it follows that the function is sharper as κ increases that leads to the path that is shorter and closer to the boundaries of the free domain X_{free}.

Notice that for simplicity, we considered the circle domain X. The similar considerations for other shapes of the domain can be found in different literary sources and here are not addressed.

4.5.2 Navigation Function in Static Uncertain Environment

In the previous section, we assumed that the robots act in the deterministic domain with the known coordinates of the obstacles and their boundaries. In this section, we present an extension of the NF techniques to the uncertain environments, which implies that the robot should follow such a path that on one hand is the shortest one from its initial position $\vec{x}_0$ to the goal position $\vec{x}_g$, and on the other hand, minimizes the probability for collisions with the obstacles. For simplicity, we restrict the discussion by considering only Gaussian distributions (though this could be numerically extended to a more-general case as in Hacohen et al. 2017).

Recall that in the deterministic scenario, the definition of the NF is based on the functions $\beta_i(x)$ that specify the distances between the robot's position $\vec{x}$ and the boundary of the ith obstacle, $i = 1, 2, \ldots, N_{obs}$. An extension of the NF techniques to stochastic scenarios is based on the substitution of the distances $\beta_i(\vec{x})$ by the collision probabilities (but we shall use the same notation). For this end we define a certain threshold value Δ, which is obtained by replacing the geometric edges of the obstacles by the isoclines Ψ such that over Ψ the probabilities are equal.

In order to define the collision probabilities, at first let us implement the Minkowski sum concept to the probability density functions (PDFs) of the robot's position $\vec{x}_{rob}$ and locations $\vec{x}_i$ of the obstacles. In the stochastic case, the conceptual idea is to add (in some sense) the robot's position uncertainty to the location uncertainties of the obstacles. Note that this makes the location of the robot a deterministic value and the obstacle's uncertainty is "inflated." We do so by applying the convolution operator. Denote by $p_{rob}(\vec{x})$ and $p_i(\vec{x})$ the PDFs of the robot and the ith obstacle, $i = 1, 2, \dots, N_{obs}$, respectively. Then, using the convolution operator marked by "$*$" a new PDF of the ith obstacle given certain location of the robot is defined as follows:

$$p_i^*\left(\vec{x}\right) = p_{rob}\left(\vec{x}\right) * p_i\left(\vec{x}\right). \tag{4.26}$$

Assuming that the robot's location and the obstacles' locations are normally distributed:

$$\vec{x}_{rob} \sim \mathcal{N}(\bar{x}_{rob}, \Sigma_{rob}), \quad \vec{x}_i \sim \mathcal{N}(\bar{x}_i, \Sigma_i),$$

where $\bar{x}_{rob}$ and $\bar{x}_i$ are expected positions of the robot and the ith obstacle, respectively, and Σ_{rob} and Σ_i are their covariance matrices. Then, for a definite robot position and uncertain locations of the obstacles, it follows that

$$\vec{x}_{rob} \sim \mathcal{N}(\bar{x}_{rob}, 0), \quad \vec{x}_i \sim \mathcal{N}(\bar{x}_i, \Sigma_{rob} + \Sigma_i),$$

and the PDF of the inflated ith obstacle is (Hacohen et al. 2017):

$$\tilde{\mathfrak{f}}_i\left(\vec{x}\right) = p_i^*\left(\vec{x}, \bar{x}_i, \Sigma_{rob}, \Sigma_i\right) = \frac{(2\pi)^{-n/2}}{(\Sigma_{rob} + \Sigma_i)^{1/2}} \exp\left(-\frac{1}{2}\left(\bar{x}_i - \vec{x}\right)^T (\Sigma_{rob} + \Sigma_i)^{-1}\left(\bar{x}_i - \vec{x}\right)\right), \tag{4.27}$$

where n is a dimension of the domain X. Since the robot is represented by its center of gravity – a point mass – a collision between the robot and the obstacle occurs if the robot is in some point that is occupied by the obstacle (think of a concave shaped robot). So for circle-shaped obstacles with the radii r_i, ($i = 1, 2, \dots, N_{obs}$) defined as

$$D_i\left(\vec{x}\right) = \begin{cases} 1 \text{ if } \|\bar{x}_i - \vec{x}\| \le r_i, \\ 0 \text{ otherwise.} \end{cases} \tag{4.28}$$

the collision probability is defined by the convolution (i.e., integration of the PDF over the disc domain):

$$p_i\left(\vec{x}\right) = D_i\left(\vec{x}\right) * \tilde{\mathfrak{f}}_i\left(\vec{x}\right). \tag{4.29}$$

The convolution defined by Eq. (4.29) can be calculated numerically or analytically by approximating the Gaussian function in the circle area of the ith obstacle, $i = 1, 2, \dots, N_{obs}$ using the diagonal covariance matrix

$$\Sigma_i^* = I_n \sigma, \tag{4.30}$$

where σ is the maximal eigenvalue of the $\Sigma_{rob} + \Sigma_i$ and $I_n$is the identity matrix.

Using the matrices Σ_i^*, $i = 1, 2, \ldots$, the collision probabilities $p_i(\vec{x})$, $\vec{x} \in X$, are evaluated as follows:

$$p_i\left(\vec{x}, r_i, \sigma\right) = e^{-\frac{\|\vec{x}\|^2}{2\sigma}} \sum_{m=0}^{\infty} \left(\frac{\|\vec{x}\|^2}{2\sigma}\right)^m \frac{1}{m!} \bar{\gamma}\left(m + \frac{n}{2}, \frac{r_i^2}{2\sigma}\right), \tag{4.31}$$

where $\bar{\gamma}(a,b) = \frac{1}{\Gamma(a)} \int_0^b e^{-\xi} \xi^{a-1} d\xi$ is the normalized lower incomplete gamma function.

In order to guarantee a reasonably safe movement in the sense of probabilities $p_i(\vec{x}, r_i, \sigma)$ of collisions with the obstacles, these probabilities should be bounded by a predefined threshold value Δ (according to the risk one is willing to take). As already indicated, in the deterministic scenario, the function φ_κ defined by Eq. (4.20) decreases the distance to the goal position $\vec{x}_g$ that keeps the robot away from the obstacles (up to sliding along their boundaries). In order to obtain the same property in the stochastic scenario, the distances $\beta_i(\vec{x})$ are specified using the probabilities $p_i(\vec{x})$ provided by the corresponding probabilities p_i as follows:

$$\beta_i\left(\vec{x}\right) = \begin{cases} \Delta_i - p_i\left(\vec{x}\right) & \text{if } p_i\left(\vec{x}\right) \le \Delta_i, \\ 0 & \text{otherwise.} \end{cases} \tag{4.32}$$

Following this equation, the function β_i (and so – the function β) vanishes in the points where the collision probabilities are more than Δ. Similarly, the boundary of the domain of the function β_0 is defined as:

$$\beta_0\left(\vec{x}\right) = \begin{cases} -\Delta_0 + p_0\left(\vec{x}\right) & \text{if } p_0\left(\vec{x}\right) \le \Delta_0, \\ 0 & \text{otherwise,} \end{cases} \tag{4.33}$$

where the probability functions p_0 and Δ_0 refer to the external boundary computed based on the PDF of the robot alone. So, assigning Eq. (4.33) into Eq. (4.20) yields the desired probabilistic navigation function (PNF).

Following $\hat{\varphi}_\kappa(\vec{x})$ gradient will converge to the target (analytically proven in (Hacohen et al. 2019) while avoiding obstacles as long as they are not mutually tangent (as in such cases there might be no way out).

4.5.3 Navigation Function and Potential Fields in Dynamic Environment

In the previous sections, we considered an application of artificial potential field and navigation function for navigation of the mobile robot in static deterministic and stochastic environment. This section presents a simple modification of the considered techniques and their extension to navigation of the robots in the dynamic both deterministic and stochastic environment.

In the suggested solution, during the robot's motion its navigation function is updated with respect to the predicted locations of the obstacles and the robot's position. A possible block diagram of the robot's navigation using PNFs is shown in Figure 4.7.

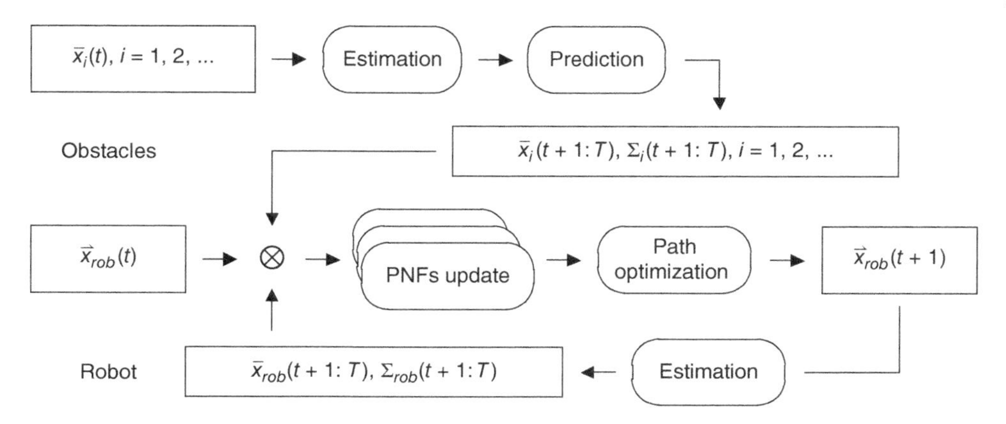

Figure 4.7 The scheme of the robot's navigation in dynamic environment using probabilistic navigation functions (PNFs).

As above, Figure 4.7 $\vec{x}_{rob}(t)$ stands for the robot's position at time t and $\bar{x}_i(t), i = 1,2,...N_{obs}$, marking the location of the ith obstacle at this time. Estimation and prediction up to time T results in the expected locations of the robot $\bar{x}_{rob}(t+1:T)$ and of the obstacles $\bar{x}_i(t+1:T)$ together with their covariance matrices $\Sigma_{rob}(t+1:T)$ and $\Sigma_i(t+1:T)$, $i = 1, 2, \dots, N_{obs}$.

The indicated solution is threefold: estimation, which provides the PDFs of the robot's and obstacles' positions in the case of a stochastic scenario; prediction, which results in the obstacles' positions in the next steps; and path optimization, which constructs the best path of the robot given the predicted position of the obstacles.

4.5.3.1 Estimation

The suggested NF-based algorithm requires the estimated PDFs of the robot's and the obstacles' positions with the statistical moments as realistic as possible. Notice that since the algorithm uses both the covariance and the expected value in the same extent, for most applications covariance is used for initial estimations. For our purposes, estimation may be conducted using a particle-filter algorithm (Doucet et al. 2000; Septier et al. 2009), which does not assume linearity of the processes and does not depend on the shape of the probability distribution. The Bayesian estimation methods are considered in Chapter 7 (for general information regarding the particle-filter algorithms, see, e.g., Choset et al. 2005; Siegwart et al. 2011).

4.5.3.2 Prediction

The prediction method depends on general assumptions regarding the system's dynamics. The motion of both the robot and the obstacles are governed by

$$\vec{x}_i(t+1) = f\left(\vec{x}_i(t), u(t), w(t)\right), i = 0, 1..., N_{obs} \quad (4.34)$$

where $\vec{x}_0(t) = \vec{x}_{rob}(t)$ is a position of the robot at time t and $\bar{x}_i(t)$, $i = 1,2,...,N_{obs}$ are the locations of the obstacles, $u(t)$ is a control signal, and $w(t)$ is a system noise (for many practical cases, this equation may be approximated by a linear equation). Obviously,

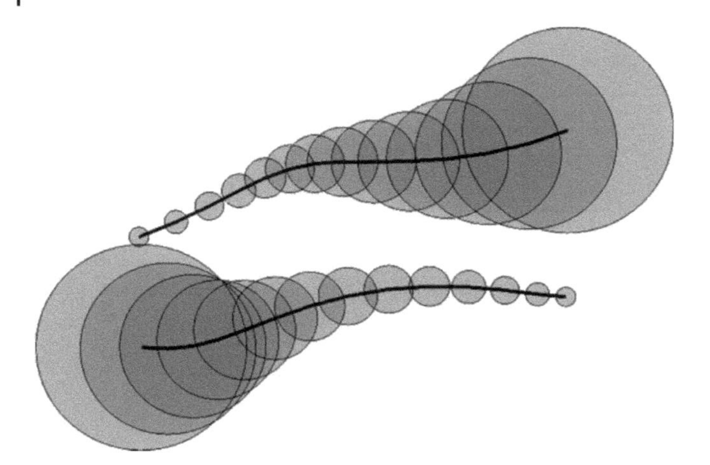

Figure 4.8 Two obstacles prediction in uncertain scenario – the uncertainty grows over time.

predicted obstacles' positions need to be taken into account when considering PNF (see Blackmore et al. 2006).

The NF is static by nature, so, in order to consider the dynamics of our model, we need to recalculate the PNF in each time step in order to evolve the PNF. So, the path will be represented by a sequence of static PNFs. In order to make the most out of the available knowledge (i.e., to use the predictions as well), the PNF will be calculated at the current time step k, as well as for a predetermined set at time steps $+1$, $k + 2, \ldots, k + N_{fwd}$. Since the uncertainty increases with the time, after a finite time steps of prediction the probability to find an obstacle in a specified location is uniformly in the whole environment, so the number of effective steps for prediction should be limited (Figure 4.8).

4.5.3.3 Optimization

Given the set of N_{fwd} PNFs, we would like to determine an optimal path. Consider the point λ_k generated at time step k. The PNF $\hat{\varphi}_\kappa(\vec{x})$ should be obviously updated to time step k and calculated for λ_k, and so forth. Since the PNF encapsulates the risk taken by the robot and the distance to the target, we may think of the PNF as a cost function. So optimizing the path $\Lambda = (\lambda_k, \ldots, \lambda_{k+N_{fwd}})$ simply means minimizing:

$$E = \min_{\Lambda} \left\{ \sum_{i=1}^{N_{fwd}} \varphi^{k+i}(\lambda_{k+1}) \right\}$$

Such an optimization can be achieved by applying various techniques such as simulated annealing or genetic algorithms (Davis 1987).

4.6 Summary

In this chapter, we introduced several motion-planning methods that do not assume knowledge of global coordinates, but, instead, are based on the artificial potential field idea specified over the domain. We introduced the navigation function scheme. Similar to general gradient descent techniques, the main idea of potential field and navigation function methods is in constructing a "surface" over which the robot "slides" from initial configuration to the goal configuration. These ideas can be extended to cases where uncertainty is involved. We introduced the PNF as a simple extension of the navigation function concept and showed how this could also be applied for dynamic scenarios.

References

Blackmore, L., Li, H., and Williams, B. (2006). A probabilistic approach to optimal robust path planning with obstacles. In: *American Control Conference*, vol. 7, 2831–2837. IEEE.

Borenstein, J. and Koren, Y. (1989). Real-time obstacle avoidance for fast mobile robots. *IEEE Transactions on Systems, Man, and Cybernetics* 19 (5): 1179–1187.

Choset, H., Lynch, K., Hutchinson, S. et al. (2005). *Principles of Robot Motion: Theory, Algorithms, and Implementation*. Cambridge, MA: Bradford Books/The MIT Press.

Davis, L. (1987). *Genetic Algorithms and Simulated Annealing*. Los Altos, CA: Morgan Kaufman Publishers, Inc.

Doucet, A., Godsill, S., and Andrieu, C. (2000). On sequential Monte Carlo sampling methods for Bayesian filtering. *Statistics and Computing* 10 (3): 197–208.

Hacohen, S., Shoval, S., and Shvalb, N. (2017). Applying probability navigation function in dynamic uncertain. *Robotics and Autonomous Systems* 237–246.

Hacohen, S., Shoval, S., and Shvalb, N. (2019). Probability Navigation Function for Stochastic Static Environments. *International Journal of Control, Automation and Systems*, 1–17. https://doi.org/10.1007/s12555-018-0563-2.

Kaelbling, L.P., Littman, M.L., and Moore, A.W. (1996). Reinforcement learning: a survey. *Journal of Artificial Intelligence Research* 4: 237–285.

Kaelbling, L.P., Littmann, M.L., and Cassandra, A.R. (1998). Planning and acting in partially observable stochastic domains. *Artificial Intelligence* 101 (2): 99–134.

Kagan, E. and Ben-Gal, I. (2013). *Probabilistic Search for Tracking Targets*. Chichester: Wiley.

Kagan, E. and Ben-Gal, I. (2015). *Search and Foraging. Individual Motion and Swarm Dynamics*. Boca Raton, FL: Chapman Hall/CRC/Taylor & Francis.

Khatib, O. (1986). Real-time obstacle avoidance for manipulators and mobile robots. *The International Journal of Robotics Research* 5 (1): 90–98.

Koditschek, D.E. and Rimon, E. (1990). Robot navigation functions on manifolds with boundary. *Advances in Applied Mathematics* 11 (4): 412–442.

Rimon, R. and Koditschek, D.E. (1992). Exact robot navigation using artificial potential functions. *IEEE Transactions of Robotics and Automation* 8 (5): 501–518.

Septier, F., Pang, S., Carmi, A., and Godsill, S. (2009). On MCMC-based particle methods for Bayesian filtering: application to multiagent tracking. In: *The 3rd IEEE Int. Workshop on Computational Advances in Multi-Sensor Adaptive Processing (CAMSAP)*, 360–363. Institute of Electrical and Electronics Engineers (IEEE).

Siciliano, B. and Khatib, O. (eds.) (2016). *Springer Handbook of Robotics*. Springer.

Siegwart, R., Nourbakhsh, I.R., and Scaramuzza, D. (2011). *Introduction to Autonomous Mobile Robots*, 2e. Cambridge, MA: The MIT Press.

Stone, L.D. (1975). *Theory of Optimal Search*. New York: Academic Press.

Sutton, R.S. and Barto, A.G. (1998). *Reinforcement Learning: An Introduction*. Cambridge, MA: The MIT Press.

5

GNSS and Robot Localization

Roi Yozevitch and Boaz Ben-Moshe

5.1 Introduction to Satellite Navigation

A mandatory demand for any mobile robot is to know its exact location in space. The Global Navigation Satellite System (GNSS) is the most famous and spread-out system for mobile localization outdoor. This chapter is devoted to satellite navigation and its application for mobile robots.

GNSSs enable an Earth-located receiver to determine its *absolute* position in space. This is done by measuring the receiver's distance to the navigation satellites and extracts its location from those distances. The most famous (and oldest) GNSS is the US global positioning system (GPS), which became operational at 1980. Nowadays, almost every receiver uses a GNSS, including the Russian system – GLONASS. Other GNSSs are the European GALILEO and the Chinese BeiDou2. Although not all GNSSs work exactly the same way, the principle is the same. But how does GNSS actually work?

5.1.1 Trilateration

Both GPS and GLONASS are navigation systems based on the trilateration method. As opposed to triangulation, trilateration is based on measuring the distance (not angle) between the receiver and three beacons at known location. The accuracy using this method is determined by the receiver's ability to measure the distance involved. The distances are calculated by means of the time of arrival (ToA) method. Each navigation satellite is equipped with four atomic clocks. When a signal is received, the arrival time is recorded and compared to the transmission time. The time difference between the transmission and arrival of the signal multiplied by the signal's velocity provides the receiver's distance to the specific satellite. Like any electromagnetic wave, the signal travels at the speed of light c ($\approx 3 \times 10^8$ m/s). This distance, *Dist*, is called **pseudorange,** where *Dist* is the distance from the transmitter to the receiver. This principle can be understood by using a simple, one-dimensional (1D) model. Let's say we want to determine a receiver's position along a log road. If both the transmitter's position and the sending time are

Autonomous Mobile Robots and Multi-Robot Systems: Motion-Planning, Communication, and Swarming,
First Edition. Edited by Eugene Kagan, Nir Shvalb and Irad Ben-Gal.

Companion website: www.wiley.com/go/kagan/robotsystems

exactly known, the receiver can extrapolate its distance by multiply Δt by c as depicted in Figure 5.1.

Figure 5.1 raises two important points. The first one is that the entire calculation is based on an accurate Δt computation. A 1 μ error traveling at c will cause a ≈ 300 m error. The second point is that the position is ambiguous, since the receiver can be either on the transmitter's right side or left side. One can ignore the left side (positions outside Earth) and get a definite position. However, overcoming the clock inaccuracies is more difficult. This is why the computed distance from the satellite to the receiver is called pseudorange; it is not the real distance due to the receiver's clock bias.

Any GNSS satellite is equipped with three Cesium-based very accurate atomic clocks. Alas, commercial receivers don't have this time-precision level. This is why an additional satellite is needed to obtain the receiver's position by the differential TOA (DToA). While the receiver clock has an unknown Δt bias, it can be eliminated by looking at the DToA from two satellite, as can be seen in Figure 5.2.

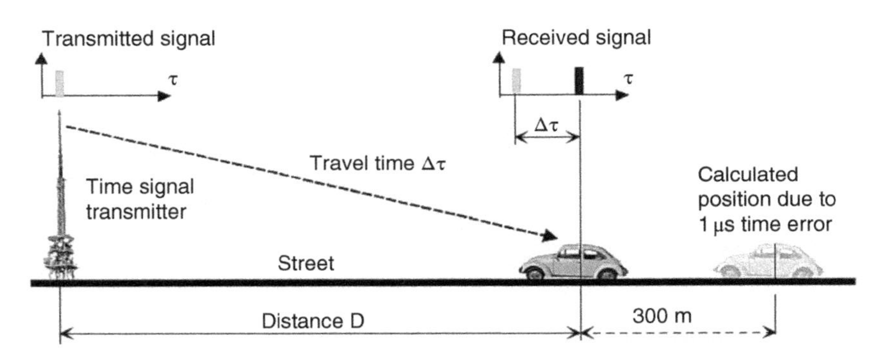

Figure 5.1 A 1D ToA trilateration example.

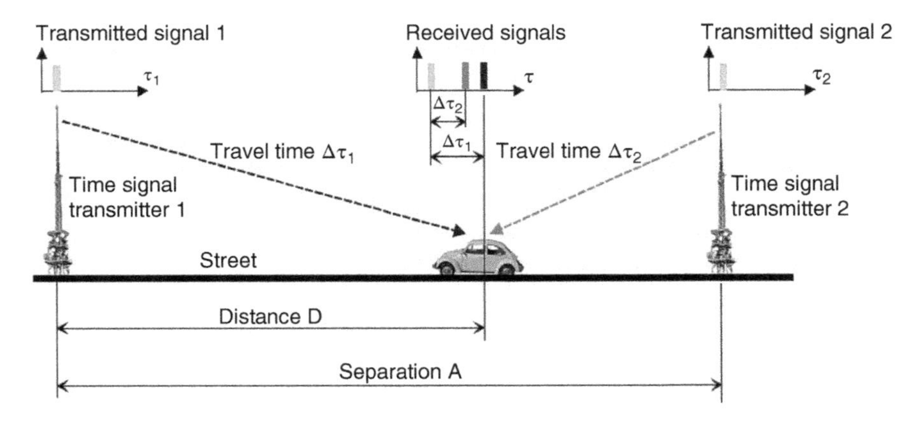

Figure 5.2 A DToA example. The additional satellite eliminates the need for accurate receiver clock.

5.2 Position Calculation

In theory, three satellites are sufficient in order to compute the receiver's position in 3D. Since each satellite creates a sphere of the distance to the receiver, the intersection of three spheres is sufficient for determining an injective position on Earth. However, an additional satellite is necessary to overcome the receiver's inherent clock bias (as demonstrated in Figure 5.2). In fact, the original GPS satellite constellation was designed so that from each point on Earth, there will always be at least four visible satellites (Zogg 2009). This is why four quadratic equations are needed for solving the receiver's 3D position (Hegarty 2006):

$$R_i = \sqrt{(X_{sati} - X_{user})^2 + (Y_{sati} - Y_{user})^2 + (Z_{sati} - Z_{user})^2} + b_u \tag{5.1}$$

where b_u is the user clock bias error expressed in meters. Note that b_u is the same for each satellite, since it's the receiver's inherent error. Those sets of equations can be solved by linearization and iteration. The satellites' positions are known, since each satellite's trajectory is known from the ephemeris – the mathematical description of its orbit. Four equations are needed in order to solve for X_{user}, Y_{user}, Z_{user}, and b_u. Thus, in a GNSS (mainly GPS and GLONASS) receiver, a minimum of four satellites is required in order to solve the receiver's position.

5.2.1 Multipath Signals

The term *visible* is crucial for solving the equations. This method of operation assumes line of sight (LOS) to the satellites. In reality, some satellites have non-line of sight (NLOS). Figure 5.3 demonstrates both LOS and NLOS signals.

One ramification of NLOS satellite is a bigger pseudorange due to the multipath phenomena (Ben-Moshe 2012). A NLOS satellite signal is usually detected via its reflection. Those reflections cause a bigger path for the signal to path (hence, the name *multipath*). A multipath signal can cause a position error of several dozen meters. When more than four visible satellites are available, modern GNSS receivers use all of them to gain a better position estimation (Groves 2013). However, in NLOS experiments (e.g., urban

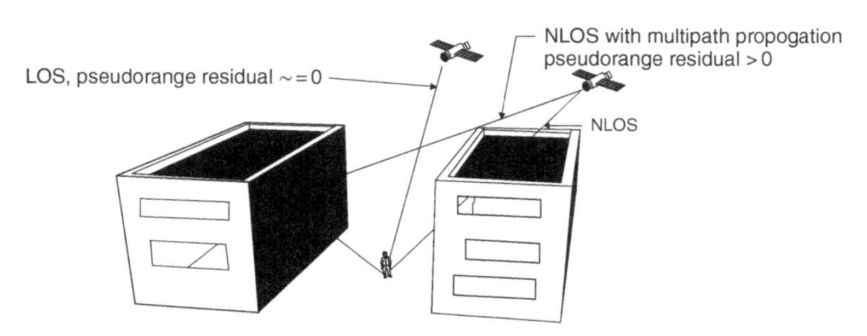

Figure 5.3 Illustration of optical visibility. A *LOS path is marked as a direct segment between the navigation satellite and the user receiver.* NLOS paths are shown as bouncing from the building walls or simply going through the building.

canyons), GNSS receivers produce poor position results. Section 5.4 is devoted to those scenarios.

5.2.2 GNSS Accuracy Analysis

A physical measurement without error bound is worthless. A mobile robot designer must know how accurate the receiver is. Alas, the error estimation GNSS receiver reports is far from being reliable. The maximum accuracy level that commercial GNSS receivers can reach is in the range of 2–5 m horizontal (latitude/longtime) and about 5–15 m vertical (altitude). This accuracy level can only be achieved in open-sky environments. Urban environments produce much less accurate results.

There are various causes of measurement errors. An important source of error is *ionospheric delay*. The ionosphere, a layer of charged particles surrounding the Earth, slows the GNSS signals as they pass through it. While contemporary receivers incorporate an ionospheric model to bypass this error, it still cannot be neglected. So how can we evaluate the quality of its receiver's reported position?

Two rules of thumbs can be offered for the receiver's accuracy. The first one is to increase the number of navigation satellites. Contemporary receivers can receive multiple GNSS constellations, including GPS, GLONASS, BeiDou, and Galileo. Combined, the number of tracked satellites may exceed 20 and even 30 navigation satellites. The more satellites participate in solving the position's equations, the better the expected accuracy. The second rule is to decrease the receiver's dilution of precision (DoP).

5.2.3 DoP

The DoP value can be thought of as the reciprocal value of the volume of a tetrahedron made up of the satellites' positions and the user, as demonstrated in Figure 5.4. The smaller the DoP, the more accurate the reported position. There are several DoPs, including

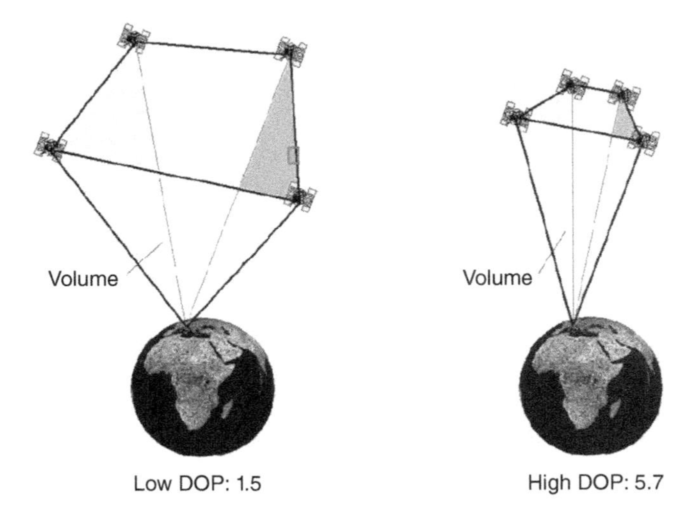

Figure 5.4 The larger the enclosed volume, the smaller the DoP.

vertical `dilution of precision` (VDoP) and horizontal dilution of precision (HDoP). HDoP represents the quality of the x-y solution and VDoP represents the quality in the altitude solution.

Those two rules of thumb are adopted by many real-time systems. In fact, many drones will not take off until their receiver can detect at least eight navigation satellites, and the HDoP figure is below a certain value (approximately 1–1.5).

5.3 Coordinate Systems

A mandatory demand for any coordinate system is to be consistent – the same position will always produce the same coordinates. Since the Copernican revolution, we know that the Earth's position is not fixed in space. The most common solution is to use a coordinates system where the Earth is fixed in place – an Earth-centered, Earth-fixed (*ECEF*) system. In ECEF, point (0,0,0) is center of Earth and every point in Earth will always have the same coordinates. Eq. (5.1) produces the receiver's absolute position in X–Y–Z ECEF coordinates. While GNSS utilize ECEF in the position computation, it is not intuitive to work with (e.g., z value does not represent height). The most used geographic coordinate system in GNSS receivers is based on latitude, longitude, and altitude (lat, long, alt). Another popular system for mobile robots is the grid-based Universal Transverse Mercator (UTM) projection.

5.3.1 Latitude, Longitude, and Altitude

Although the Earth's shape is not a perfect sphere (the radius at the equator is about 0.3% larger than the radius measured through the poles), it can be roughly modeled as a sphere with an average radius of approximately 6371 km. A more accurate model of the Earth's shape is somewhat improved using geodetic data such *WGS84* (World Geodetic System).

The *latitude* of a point on the Earth's surface is the angle between the equatorial plane and the straight line that passes through that point and through the center of the Earth. Thus, any place on the equator will have a latitude value of 0. The north and south poles have latitude coordinates of +90° and −90° , respectively.

The *longitude* of a point on the Earth's surface is the angle east or west from a reference meridian to another meridian that passes through that point. The longitude's values are centered (0) at the British Royal Observatory in Greenwich (in southeast London, England), and from there up to 180°east and up to −180° west.

Finally, the *altitude* is simply the height of a point (latitude and longitude) above (or below) sea level, as approximated by data (commonly WGS84).

While all GNSS receivers report their position in lat/long/alt, using such a non-Cartesian coordinate system is often confusing and might create technical issues in actual implementation. An alternative Cartesian global coordinate system is the UTM projection.

5.3.2 UTM Projection

The (UTM) conformal projection uses a two-dimensional Cartesian coordinate system to give locations on the surface of the Earth. The UTM projection bypasses the Earth's curvature by projecting it to a 2D plane. The UTM system divides the Earth into 60 zones,

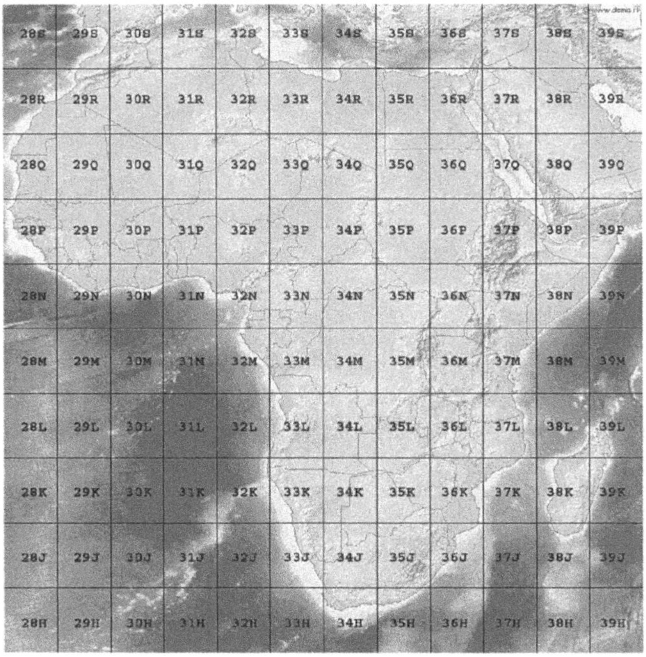

Figure 5.5 A UTM projection of Africa.

each being a six-degree band of longitude, and uses a secant transverse Mercator projection in each zone (Figure 5.5). Due to the projection, the UTM does not map the poles (both the south and the north). The advantage of the UTM system is threefold: simple, global, and widely in used in navigation systems.

5.3.3 Local Cartesian Coordinates

Another widely used coordination system is the local one. Assume we have a robot with a moving range of few kilometers. For most practical applications, the angular shape of Earth can be ignored; as a rule of thumb, in distances smaller than 100 km, the Earth's surface can be addressed as flat. This assumption allows us to use the simplified Cartesian coordinates system. The most popular one uses the X-axis to represent *east* coordination, Y-axis to represent the *north* coordination, and Z-axis to represent *up*. In a more formal manner, it is defined as follows:

i) Fix an origin point (*org*) and marks its global coordination (lat, long, alt).
ii) For any point p (which is close enough to *org*), computes its Δlatitude, Δlongitude, and Δaltitude (between p and *org* global coordinates).
iii) The local coordinates of p with respect to *org* can be compute as follows:

$$\Delta x = E_{rad} \sin(\Delta \text{longitude}) \cos(\text{org.latitude})$$
$$\Delta y = E_{rad} \sin(\Delta \text{latitude})$$
$$\Delta z = \Delta \text{altitude}$$

where E_{rad} is the Earth's radius. Notice that the reverse transformation can be computed easily from the same formula.

5.4 Velocity Calculation

The receiver's velocity vector, $v(t)$, can be acquired as the derivation of the position vector $X(t)$. While this approach is the most simple to implement, it suffers from meter per second level of accuracy due to the dependence on pseudorange-based position measurements. Therefore, GNSS receivers utilize a different (and orthogonal) approach to estimate $v(t)$. The velocity estimation is based on Doppler analysis. While the navigation satellites transmit their signal in a constant frequency, the observed frequency in the receiver is changes due to the relative velocity between the receiver and the satellite. Doppler frequency analysis of the received signal produced by satellite relative motion enables velocity accuracy of a few centimeters per second. While $v(t)$ is computed in 3D, the 2D magnitude is reported as speed over ground (SoG) and the 2D orientation as course over ground (CoG). It is important to understand that Doppler analysis is viable even for stationary receivers. Since the receiver is positioned on the Earth's surface, there is a considerable distance difference for horizon satellites (elevation 0) and satellite in the zenith (elevation 90). The basic principles of velocity computation in GNSS devices are as follows:

- Assume that the receiver has a general notion of the position and the time (1 km, 1 second accuracy is sufficient).
- The receiver computes a relative velocity to each navigation satellite (using the Doppler shift and the satellite known direction and orbit as obtained from the Ephemeris).

5.4.1 Calculation Outlines

- Each relative velocity (to satellite S_0) can be approximated as a plane that is the geometric place with a fixed distance to S_0.
- The receiver velocity can now be computed as the intersection of three (or more) planes.

Figure 5.6 presents the basic principle of the velocity vector computation of a GNSS receiver: The receiver computes the relative velocity to each of the satellites (sat_1, sat_2,

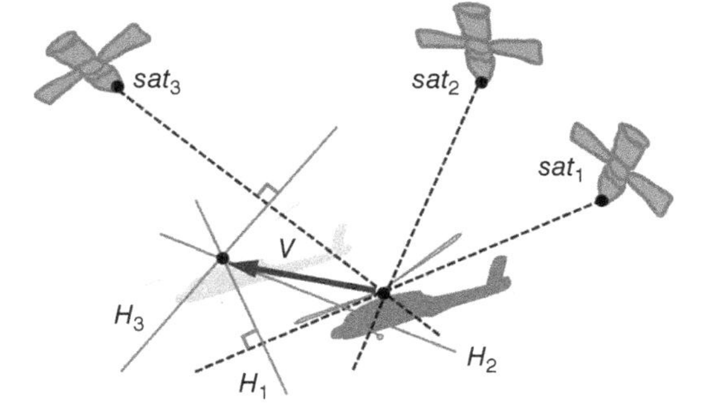

Figure 5.6 The velocity estimation process: V represents the distance vector that the GNSS receiver has moved during a short period of time. H_1, H_2, H_3 are planes representing the geometric constrain for the receiver position.

sat$_3$). Each of the planes H_1, H_2, and H_3 represents the possible position of the GNSS receiver after a short period of time (Δt). The intersection point of the three planes indicates the distance that the GNSS receiver had made during Δt, which is the receiver's velocity. Observe that an accurate velocity vector can be computed even if the receiver's position is only partly known (inaccurate).

5.4.2 Implantation Remarks

The orthogonal velocity computation approach and the fact that Doppler is more immune to multipath effects imply that the velocity vector can be utilizes even in areas where the position accuracy is insufficient. Although the position vector is unreliable in those areas, the COG/SOG is rather accurate. Figure 5.7 demonstrates this phenomena. On the left, one can see a pedestrian double square route, as was recorded by a GNSS receiver (the red tacks). On the right, the computed double *A*-*B*-*C*-*D* route is constructed *solely* utilizing the reported velocity values (COG/SOG). A two-square-like shape is definitely seen in the figure.

5.5 Urban Navigation

GNSS devices tend to perform poorly in urban canyon environments. As explained in the previous chapter, using four (or more) pseudoranges and their associated satellite locations, the GNSS receiver's *3D* location can be computed. This method of operation assumes LOS to the satellite. A GNSS device would simply use four (or more) strong enough signals to compute its estimated location. The computation assumes a LOS satellite, hence a correct pseudorange. Other GNSS devices (e.g., carrier phase GPS or

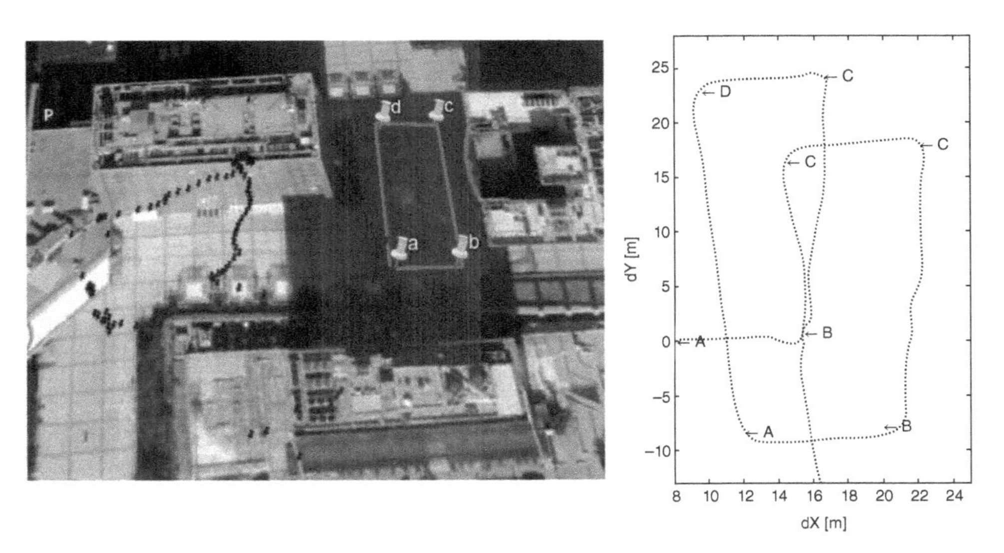

Figure 5.7 Velocity estimation. While the GNSS receiver cannot report the correct position (left figure), its velocity estimation is rather accurate (right figure).

geodetic GNSS receivers) can be accurate within centimeters. Those high-end devices, however, are expensive and require a special antenna; hence, we won't discuss them here. There is a lower, physical bound of 1 m accuracy in commercial devices (Zogg 2009). Errors caused by clock bias, atmosphere, and wrong ephemeris data cannot be eliminated completely. Those levels of accuracy, however, can only be achieved in an open sky scenario, which directly leads to the urban canyon problem.

Urban environments comprise both man-made and natural obstructions that hamper satellite availability. Hence, in dense urban canyon environments, there is often lack of four visible LOS satellites. As already noted, without four (or more) LOS satellites, the accuracy of the position deteriorates. Several methods have been introduced to tackle this problem, i.e. to estimate a degraded position using NLOS satellites (Bobrovsky 2012). Those methods, however, suffer from position inaccuracy. One major error is caused due to the multipath effect. As demonstrated in Figure 5.3, a satellite can be either LOS, blocked, or multipath (the latter two are considered NLOS).

5.5.1 Urban Canyon Navigation

Multipath satellites signals produce longer pseudoranges, which, in turn, lead to an error in the 3D position. When a GNSS receiver uses two or more NLOS-signals, the error can be increased to dozens of meters. Moreover, unlike the open-sky scenario, where satellites tend to slowly "shine/decline" (LOS/NLOS), urban environments are characterized by rapid changes between LOS/NLOS. The shine/decline analogy helps to understand the operation of GNSS satellites. Geostationary satellites appear motionless to ground observers. GNSS satellites, by contrast, complete an orbit in approximately 12 hours. The angular velocity as seen by ground observers is 1° every two minutes. This is very similar to the angular velocity of the sun, as can be seen from Earth – hence the analogy "shine/decline." During any route (vehicle or pedestrian) in an urban canyon, GNSS satellites seem to suddenly appear/disappear (LOS/NLOS). The ramification is that the position computation is done with different satellites; hence, there are jumps in the estimated position.This point should be clarified: In theory, there should be no difference in the result when solving the pseudorange equations for one set of satellites or for a completely different set. In practice, however, since some pseudoranges are incorrect (due to the multipath phenomena), there are differences. This change implies discontinuity in the position estimation. This discontinuity can be easily eliminated using the Kalman filter if the receiver's velocity is roughly known. Yet, the other side of the discontinuity elimination is that the positioning estimation tends to drift – leading to significant position errors.

5.5.2 Map Matching

As already noted, the classic approach to deal with the accuracy problem is to perform information fusion of the GNSS data with a roadmap database (map matching) (Liu 2008; Gao 2008) and to estimate the location via the velocity (dead reckoning). Map matching is basically a map-aided positioning. Since motor vehicles are restricted to roads (most of the time), knowing the map of the roads can assist in fine-tuning the receiver's estimated position. A vehicle is highly unlikely to be in the middle of a building while driving, so utilizing those road restrictions is the heart of the map-matching algorithm. In order to use road network data for map-aided positioning, both the map and the vehicle domains must be modeled. For example, the roads have a finite width, and each road is modeled by its center line (Scot 2004). Although this works very well most of the time, it fails to provide accurate vehicle navigation in dense urban environments. As every Waze user knows, getting a position fix in a dense city is nearly impossible at the very beginning of the journey. Moreover, in navigation scenarios that are not map restricted (mainly pedestrian navigation), map-matching algorithms fail to provide good results.

5.5.3 Dead Reckoning – Inertial Sensors

Dead reckoning measures the changes in the velocity/position and integrates them. This is added to the previous position in order to compute the current position. For vehicle navigation, 2D measurements are sufficient for most applications. Most commercial GNSS receivers now have built-in Kalman filters for accuracy improvement. The Kalman filter combines the previous position with the velocity to fine-tune the current position estimation. Dead reckoning is different in the fact that external hardware is used for that purpose. For example, in pedestrian navigation, an accelerometer can be used for step counting and for determining the step length. A magnetic compass can be used for velocity heading. Other dead-reckoning techniques may be image processing (e.g., Omni Skylines), laser landmark tracking, and Doppler sonar. In short, the dead-reckoning position solution is the sum of a series of relative position measurements (Groves 2013).

5.6 Incorporating GNSS Data with INS

One can either use a GNSS receiver in a stateless mode – for each time stamp the receiver calculates the position from scratch – or in a state mode. In this mode, the velocity vector (since its computation is orthogonal to the position computation) can aid the position computation. For 1 Hz receiver, $X(t+1) = X(t) + V(t) \times \Delta t$. This data fusion is implemented in most receivers via extended Kalman filter (EKF). Although this approach produces smooth results, it has an inherent ramification: When the position computation is done with different satellites, there are "jumps" in the estimated position.

In dense urban canyons, the use of EKF-like filters may help eliminate discontinuity in the positions, yet the positioning accuracy tends to drift, leading to significant position errors. Figure 5.7 shows a typical pedestrian measurement in urban environments. The average errors between the actual position and the GNSS computed position are typically 30–50 m with peaks up to 100 m and more. Figure 5.8 show a stationary measurement in the same region over a course of seven minutes. One can clearly see how multipath signals affect the position equations.

This problem can be solved by utilizing additional filters and fusing them with external sensors data. The most common approach for mobile robots is the particle filter.

5.6.1 Modified Particle Filter

A particle filter is nonparametric multimodal implementation of Bayesian filter. Similar to histogram filters, PF estimate the posterior by a finite number of parameters. The

Figure 5.8 A Google Earth recording of a stationary device over a period of 20 minutes. The line between p_1 and p_2 represents a 200 m "jump" during one second. The recording was conducted in Ramat Gan commercial center, Israel.

samples of the posterior distribution are called particles and are represented as $\chi_t{:=}x_t^{[1]},x_t^{[2]},\ldots.x_t^{[M]}$ where M is the number of particles. Each particle is represented by a belief function $bel(x_t)$. This belief function serves as a weight (importance) of each particle. Thus, $\left\{\left(w_t^{(L)},x_t^{(L)}\right):L\in\{1,M\}\right\}$.The importance weight is proportional to the likelihood of a specific particle (Thrun 2005):

$$X_t^{[L]} \sim p(x_t \mid z_{1:t},u_{1:t}) \times w_t^{(L)}$$

where $z_{1:t}$ and $u_{1:t}$ are the sense and action functions, respectively. The Bayesian property implies that the estimation in t_1 is derived from the estimation in t_0. The multimodal feature implies that unlike Kalman filters, there can be more than a single plausible solution. Thus, PF is well suited for scenarios where one need to estimate its location within a specific region of interest (ROI). Each particle is a set of position and velocity vectors $\longrightarrow X_t^{[L]} = \{position, velocity\}$.

The PF implementation for determining the position is called Monte Carlo localization (MCL) (Thrun 2005). In MCL, both the action function $u(t)$ and sense function $z(t)$ are derived from the robot's sensors. The action function is derived from the odometry (wheel encoding, pedometers).

5.6.2 Estimating Velocity by Combining GNSS and INS

In this subsection we elaborate how to actually contract a navigation system that combines both GNSS and inertial navigation system (INS). The general approach is using GNSS for global positioning and INS for velocity estimation while a particle filter framework is used in order to combine the sensors' reading into a single global positioning output. In order to estimate the velocity of a robot, there are currently five technologies:

1) *Inertial measurement unit (IMU) odometry*. IMU odometry is based on wheel encoding and global (or local) orientation.
2) *Barometric pressure*. Air pressure can indicate the receiver's relative height. While not accurate for absolute heights, barometers are very sensitive to height differentiations. Fused with GNSS receiver and accelerometers, they can report relatively accurate absolute height.
3) *Step counting*. For walking robots, a step counter, or pedometer, is often used instead of wheel-based odometry.
4) *Optic flow*. This method uses a camera, a range-finder sensor, and an orientation sensor in order to estimate the robot's movement in space. Those methods are commonly used on "flying-robots" such as drones. The visual sensor is often based on a high-speed camera that computes the changes in terms of pixels (like a computer mouse). The measurement from the range sensor is used in order to transform the movement from "pixels" to ground-based speed, while the orientation sensor is used to transform the movement from local to global coordinates.
5) *Visual odometry*. A camera (or several cameras) can be used to estimate the velocity. Here, both speed and local orientation are computed using the camera sensor. Such a system is often based on stereo cameras, which also compute what a 3D point could.

Following is a description of a positioning particle filter algorithm. The algorithm's input is the GNSS and other sensors data. Its output is an approximated position. Particle filters tend to converge so the solution in t_10 will be usually better (more accurate) than the solution in t_1.

5.7 GNSS Protocols

Algorithm 5.1 Simple Particle Filter Algorithm

Input: GNSS-raw measurements, IMU sensory data
Result: Improved GNSS position
Definitions:

ROI: Region of interest: the geographic area in which the current position is expected to be.
GNSS readings: Raw measurements – Doppler shift and pseudorange from each navigation satellite.
IMU: MEMS sensors that can approximate the current velocity.

1) **Init**: Distribute a set P of M particles in the ROI.
2) Estimate the velocity v_t from the GNSS and IMU readings.

3) **for each** particle p ∈ P **do**
   3.1 Approximate the **action function** $u_t(p)$ from v_t.
   3.2 Update p-position according $u_t(p)$.
   3.3 Evaluate the belief (weight) function: $w(p)$ based on the known landmarks and map restrictions (map matching).
4) Resample P according the likelihood (weight) of each particle.
5) Compute weighted center of $P \in ROI$ as $pos(P)$.
6) Report $pos(P)$.
7) Approximate the error estimation $err(P)$.
8) if($err(P) < ROI$) Goto 2 else Goto 1.

In order to implement this algorithm, we must determine both $u(t)$ and $z(t)$. Those function are mission dependent, and a pedestrian low-dynamic scenario is very different from a highly dynamic unmanned aerial vehicle (UAV) one.

In order to use any of the above technologies, we might also need to implement certain sensors, which have elements of both hardware and software. Modern robot designers utilize such sensors in flight control systems. Those system can perform very accurate sensor fusion for a wide variety of robots (rovers, drones etc.). The *PixHawk* Sensor is an open-source example of such system (see https://pixhawk.org).

Once the GNSS receiver has computed both its position and velocity vector, it must communicate it. This is done by GNSS protocols. The most famous one is the National Marine Electronics Association (NMEA). NMEA is a serial textual (UART) protocol. Each line in the protocol represents a specific message.

For the majority of applications, NMEA protocol is sufficient. There are dedicated messages for the receiver's position, velocity, and satellite information (azimuth, elevation, and signal to noise ratio (SNR)) of each navigation satellite. Moreover, NEMA-protocol works both with GPS and GLONASS. For example, message $GPGSA contains both the HDoP and VDoP values and message $GPGGA contains the receiver's position in lat/long/alt coordinates (Figure 5.9). However, raw data cannot be obtained utilizing NMEA protocol. The satellite's pseudorange and Doppler values are inaccessible using this protocol.

5.8 Other Types of GPS

Any GNSS receiver requires the satellite information in order to function properly. This section discusses three types of GPS systems: A-GPS, DGPS, and RTK.

5.8.1 A-GPS

When a device is inactive for more than few hours, satellite orbital data must be first downloaded in order to fix a position. The time period is called time to first fix (TTFF), and every GNSS manufacturer reports this data. The average TTFF is ≈18–36 seconds in

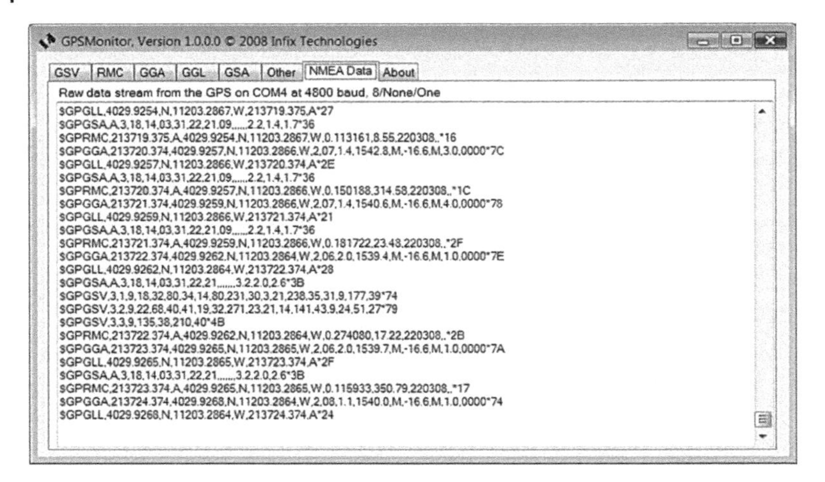

Figure 5.9 An example of NMEA output. Each message start with a new line and a $ sign.

open-sky scenarios. The TTFF can be shortened utilizing an aided-`global positioning system` (AGPS). The AGPS reduces the searching space by providing a coarse position and time, mostly via cellular network data.

5.8.2 DGPS Systems

As explained in Section 5.2, commercial code phased GNSS receivers have an accuracy limit of ≈2–5 m horizontal. A plausible solution is to use a differential `global positioning system` (DGPS). DGPS involves the cooperation of additional stationary receiver in a known location. Since the station position is known with very high precision, the received pseudoranges can be corrected. Those corrections are sent to nearby receivers, which update their position solution to gain more accuracy. One can assume that the pseudorange errors are the same in a small region of interest (ROI). Standalone robotics rarely use DGPS systems due to their complexity and need for a near transmitting base station.

5.8.3 RTK Navigation

One can think of real-time kinematic (RTK) navigation as the more sophisticated (and sexier) sister of the DGPS. While the former *only* utilizes the pseudoranges to form a correct position, RTK system rely on the carrier frequency and measured number of wavelengths. The carrier frequency of a commercial L1 GPS is 1.5472Ghz and the respective wavelength is ≈19 cm. In a nutshell, an RTK system is comprised of two GNSS receivers (similar to DGPS). The system detects the difference between the number of wave-phases in each receiver. However, unlike DGPS, both receivers can move. For many mobile robot applications, this is a crucial demand. A robot's ability to maintain a *fixed* distance from another mobile robot is very important.

5.9 GNSS Threats

In the last chapter, we covered the benefits of utilizing GNSS receivers in mobile robots. In the last section we focused on the drawbacks and danger of relying on those systems. As is widely known, GNSS signals are susceptible to many types of threats. If a certain receiver can be efficiently blocked or manipulated by a malicious attacker, this could lead to a real-life threat. In recent years, GNSS attackers have raised alerts and concerns. One very publicly known case is the radio frequency war between North and South Korea (Waterman 2012). Another well-communicated incident is the capture of a US drone in Iran (Mackenzie 2001). Additional dangerous attacks have even been made unintentionally by innocent people. In one case, a truck driver who wanted to deny his employer the ability to track his location used a jammer he bought on eBay for a few dollars. While he was driving near the Newark Liberty International Airport in August 2013, he disrupted the whole area's GNSS reception and almost caused an unfortunate accident (Matyszczyk 2013). In a nutshell, GNSS attacks can be divided into two groups: jamming and spoofing.

5.9.1 GNSS Jamming

GNSS jammers are devices that generate noise at a desired frequency in order to block or interfere with signal reception by decreasing its SNR (GNSS jammers transmit high power noise, saturate the receiver, and render it impossible to lock on any signal). As the GNSS signal received on Earth is already very weak, such a device is very easy to implement. What can be done against this kind of attack? One first needs to detect it. This is easy, since jamming attacks (intentional or accidental) cause the receiver to lose its "fix" – being unable to report its position. This is why incorporating GNSS with INS is crucial. One must never fully count on the GNSS receiver in critical missions.

5.9.2 GNSS Spoofing

A more sophisticated and malicious attack is called spoofing. Spoofers are transmitters that emit signals identical to the satellites signals in order to deceive the receiver and make it report an incorrect location fix. This type of attack is much harder to detect and its consequences are much more dangerous. Dealing with spoofing is beyond the scope of this book. The curious reader is referred to Humphreys (2008) and Warner (2003).

References

Ben-Moshe, B.C. (2012). Efficient model for indoor radio paths computation. *Simulation Modelling Practice and Theory* 29: 163–172.

Bobrovsky, E.T. (2012). A novel approach for modeling land vehicle kinematics to improve gps performance under urban environment conditions. *IEEE Transactions on Intelligent Transportation Systems* 99: 1–10.

Gao, Y.Z. (2008). A fuzzy logic map matching algorithm. In: *Fifth International Conference on Fuzzy Systems and Knowledge Discovery, 2008. FSKD'08*, vol. 3, 132–136. IEEE.

Groves, P. (2013). *Principles of GNSS, Inertial, and Multisensor Integrated Navigation Systems*. Artech house.

Hegarty, E.K. (2006). *Understanding GPS: Principles and Applications*. Artech House Publishers.

Humphreys, T. E. (2008). Assessing the spoofing threat: Development of a portable GPS civilian spoofer. *Radionavigation Laboratory Conference Proceeding*.

Liu, S. and Shi, Z. (2008). An urban map matching algorithm using rough sensor data. In: *Workshop on Power Electronics and Intelligent Transportation System, 2008. PEITS'08*, 266–271. IEEE.

Mackenzie, C. A. (2001). 'We hacked US drone': Iran claims it electronically hijacked spy aircraft's GPS and tricked aircraft into landing on its soil. *The Daily Mail*.

Matyszczyk, C. (2013). Truck driver has GPS jammer, accidentally jams Newark airport. *CNET News 11*.

Scot, C. A. (2004). Improved GPS Positioning for Motor Vehicles Through MapMatching. *Salt Palace Convention Center*, Salt Lake City, Utah*31, 6.,*

Thrun, S. (2005). Exploration. *Probabilistic Robotics*.

Warner, J.S. (2003). GPS spoofing countermeasures. *Homeland Security Journal* 25 (2): 19–27.

Waterman, S. (2012). North Korean jamming of GPS shows system's weakness. *Washington Times*.

Zogg, J. (2009). *GPS–essentials of satellite navigation*. Compendium.

6

Motion in Local Coordinates

Shraga Shoval

The chapter discusses the problem of online motion planning and navigation where the robot's position is defined either with respect to locally accessible landmarks or with respect to other robots in two and three dimensions. The discourse reviews the methods of motion planning for ground and aerial vehicles and, in particular, considers the methods of positioning by the use of local maps and localization with respect to the common target.

6.1 Global Motion Planning and Navigation

The motion planning problem is defined as the construction of a continuous path for a robot from a start configuration S to a target configuration T while satisfying a set of constraints according to the task's requirements (e.g. avoid contact with objects, go via a set of waypoints, perform specific tasks along the route, etc.). In general, the configuration of a robot (static or mobile) describes its position and shape in the working space, and the configuration complexity is determined by the robot and by the environment. For example, the configuration of a small mobile robot that moves on a flat surface consists of two dimensions (e.g., x and y position), while a larger unmanned aerial vehicle (UAV) may have six-dimension configuration (three lateral position and three orientation coordinates).

Motion can take place in a 2D or 3D Euclidean workspace and in various types of environments such as ground, underground, sea, undersea, and aerial. When the structure of the robot and the workspace are known a priori, and there are no uncertainties regarding the workspace features (e.g., the location of obstacles), the entire motion can be planned before the robot starts its motion. In this case, the complexity of motion planning is related to the computational complexity of finding a viable route from S to T at a finite time.

In case such a route does not exist, the motion planner should realize that such a route cannot be found at a finite time. Often, it is not enough just to find a viable route, but the motion planner is required to find an *optimal* path that accomplishes an objective function.

Autonomous Mobile Robots and Multi-Robot Systems: Motion-Planning, Communication, and Swarming, First Edition. Edited by Eugene Kagan, Nir Shvalb and Irad Ben-Gal.

Companion website: www.wiley.com/go/kagan/robotsystems

One of the most popular objective functions in motion planning is known as the "shortest path planning," in which the motion planner is searching for the path that minimizes the travel cost. The cost can represent the distance, time, risk, or any other parameters that are affected by the selection of the route. Many algorithms have been developed for determining the shortest path between initial and target configurations.

The Dijkstra algorithm (Dijkstra 1959) is one of the first motion planners, and it is widely used in many problems. Using the Dijkstra algorithm, the entire space is transformed into a network in which the nodes of the network represent points in the working space and the edges (the arcs) represent the motion from one node to the next. Commonly, each edge is associated with a non-negative value that represents the cost of moving from one node to another. A route is a collection of movements from the initial to the target node, and the route cost is the sum of all costs along the route. The Bellman-Ford algorithm (Bellman 1958) is more versatile than the Dijkstra algorithm, as it can handle networks with negative edges. The A^* algorithm (Hart et al. 1968) is one of the most common search methods used in motion planning for robots. In many ways, the A^* is an extension of the Dijkstra algorithm, and it has proved to be more efficient in terms of the computation time due to some heuristics that are implemented.

Another common objective in motion planning is the *optimal coverage path*. The objective in this case is to cover an area with minimal redundant traversals. Many practical applications can benefit from the optimal coverage path planning, such as cleaning, searching, and farming. Galceran and Carreras (2013) review some of the most popular methods for coverage path planning (CPP) and field applications of these methods. Among the common methods in the CPP are exact cellular decomposition (ECD), where the covered free space is decomposed into simple, nonoverlapping cells (Choset and Pignon 1998); Morse-based cellular decomposition (Acar et al. 2002), which is suitable for more complex environments that include nonpolygonal shaped obstacles; landmark-based topological coverage (Wong 2006), based on the detection of natural landmarks in the environment; and grid-based methods (Lee et al. 2011), in which the environment is divided into uniform grid cells (usually square or triangular shape), among others.

Often in mobile robotics it is not enough just to find a route from the initial to the target position, but it is required to also determine the speed along this route. A route that also determines the speed is called *trajectory*. A full robotic motion-planning system determines the geometric features of the route, as well as motion commands (speeds) along that route.

The construction of a trajectory that considers the structure of the robot and the workspace is traditionally done in the configuration space (also known as the C-space), in which the robot structure in the Cartesian space is transformed into a simpler structure (preferably a point), and the workspace is updated accordingly (Lozano-Perez 1983). The configuration space is different from the Cartesian space both in terms of its dimensions and topology. The dimension is determined by the number of parameters required to uniquely define the configuration of the robot, and the topology of the workspace is determined by the transformation between the two spaces. Although additional effort is required for the transformation from the Cartesian workspace to the C-space, the benefits for motion planning go beyond the required efforts. The C-space is usually divided into the C_{free} – the free zones in the configuration space, in which the robot is safe to move – and the C_{obs}, which represents the obstacles in the C-space.

Many methods have been proposed for constructing a route from initial to target configuration, such as the visibility graph (de Berg et al. 2000), Voronoi diagrams (Takahashi and Schilling 1989), and cell decomposition (Zhu and Latombe 1991), just to name a few. A more detailed collection of fundamental algorithms and methods for motion planning can be found in Latombe (1991), Sharir (1995), and Lavalle (2011). A flowchart describing the entire motion planning process is shown in Figure 6.1.

First, a safe path τ is generated based on the environment model, using one of the above-mentioned algorithms. Notice that the path defines only the geometrical properties of route. The next stages attach kinematic and dynamic properties to the path, turning it into a trajectory. The safe path is updated according to kinematic constraints that may limit possible maneuvers of the robot. For example, a trajectory that includes sharp curves may need to be adjusted by smoothing the curves according to the robot's nonholonomic constraints (e.g., motion planning for a Dubin car [Balluchi et al. 1996]). Next, the dynamic constraints are considered. These constraints include limitations on the robot's speeds and accelerations, as well as on the dynamic interaction between the robot and the environment (Shiller and Gwo 1991). For example, a trajectory over an inclined surface may become unstable when the robot travels at certain speeds, resulting in possible slippage, lost contact with the ground, or even tip over. Based on the updated trajectory, a set of control commands are produced to drive the robot along the desired trajectory.

Many motion planners can provide optimal trajectories as long as the robot and the workspace models are accurate and known a-priori. In addition, the process shown in Figure 6.1 assumes that the robot can perform the required motions without significant errors, and that the workspace remains stationary. Unfortunately, most robotic applications involve some uncertainties due to inaccuracies and errors in the dynamics of the robots and its sensor measurements. This problem becomes critical when the workspace models are dynamic (e.g., new objects appear in the space, existing obstacles change their location or the robot's model changes before completing the entire trajectory). These uncertainties can result in suboptimal performance, or even in unsafe behaviors, such as unstable motions or collisions with objects. The motion planning is such cases must

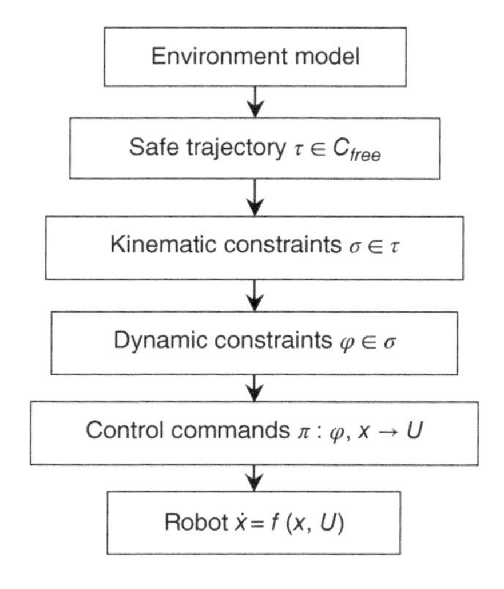

Figure 6.1 A flowchart describing the motion planning process.

operate in an online fashion, by reacting to the changing operating condition. The following section discusses such motion planners.

6.2 Motion Planning with Uncertainties

As discussed above, uncertainties require additional consideration when planning a robot's trajectory. Furthermore, the increasing uses of mobile robotics technology, and in particular the integration of autonomous vehicles in diverse applications that share the workspace with humans and other robots, introduce additional uncertainties to the motion-planning process.

Unmanned cars operating in standard highways (Fisher 2013), travel aids for the disables (Shoval et al. 2003), or UAVs flying in close proximity to manned aircrafts (Hemmerdinger 2013) must react to the unpredicted changes in the workspace and update their trajectory in order to avoid possible risks to themselves, and to the objects in the environment. Due to the uncertainties in the workspace, the robot must be equipped with sufficient sensing capabilities as well as an appropriate motion planner that can accommodate these uncertainties. Let us describe briefly the sources of uncertainties for unmanned ground, as well as UAVs.

6.2.1 Uncertainties in Vehicle Performance

Uncertainties in the vehicle's performance can be divided into two main categories: internal dynamic uncertainties and external dynamic uncertainties.

6.2.1.1 Internal Dynamic Uncertainties

Internal dynamic uncertainties refer to unanticipated reaction of the robotic system to the incoming control commands. In an ideal world, a control action u applied to a robot at configuration s_i results in a new configuration s_{i+1} according to a deterministic control function f such that

$$s_{i+1} = f(s_{i+1}, u). \tag{6.1}$$

However, the unanticipated reaction can bring the robot to a different configuration due to limited accuracy of the actuators and other robot's components. For example, applying identical voltage on two identical driving motors of an asynchronous ground mobile robot would ideally result in the robot moving at a straight line (as the two driving wheels rotate theoretically at identical speed). However, backlashes in the driving gears, variations in the wheels' diameters, wheels' misalignment, or small differences in the electronic circuits result in a curved motion that can cause erroneous localization. A more realistic approach would be to use a probabilistic function (Thrun et al. 2005) that provides a probabilistic value for the robot to arrive from its current configuration to the new configuration, given a control command such that

$$p = P(s_{i+1} \mid s_i, u), \tag{6.2}$$

where p represents the probability of the robot to arrive to configuration s_{i+1} from configuration s_i giving input command u. Configuration s_i consists of the robot configuration, as well as of the state of the relevant components that affect its dynamic reaction.

6.2.1.2 External Dynamic Uncertainties

External uncertainties are usually caused by environmental disturbances that diverge the robot form its nominal trajectory. Unmanned ground vehicles traveling on rough, slippery, or irregular surface are subject to nonsystematic navigation errors (Borenstein et al. 1996). These errors are hard to predict and are caused by uneven surfaces, bumps and holes, unexpected external forces, and slippage due to smooth surfaces or skidding.

External uncertainties in UAVs are usually due to atmospheric instability that can cause disturbances, such as winds or irregular air pressure. The effect of atmospheric instability on the UAV dynamics is particularly critical in small UAVs, as their slower speed and limited propulsion expose them to even the smallest disturbances (Solovyev et al. 2015). While the response of a robot to internal uncertainties can be estimated and can be used in a probabilistic model, external dynamic uncertainties are hard to predict, and require additional sensing capabilities.

6.2.2 Sensors Uncertainties

In an ideal world, all the sensors used by the robot to determine its location, as well as the locations of the objects in the environment, are accurate and consistent. In this ideal world, the locations of the robot and the objects can be uniquely determined using a deterministic model such that

$$x_i = g(d_i), \tag{6.3}$$

where x_i is the updated location (of the robot and/or objects), and d_i is the set of most updated sensor data.

However, in the real world, all sensors have limited accuracy, and are usually subjected to external disturbances. For example, inertial navigation systems (INSs) that are commonly used by mobile robots for lateral and angular measurements are subject to integration drift that results in accumulated unbounded errors. Another common localization device – the global positioning system (GPS) – is sensitive to atmospheric effects such as ionospheric and tropospheric delays, as well to jamming devices (intentional and/or unintentional). In addition, the use of GPS is limited to open spaces, and it has limited accuracy in urban environment.

The location of the robot and other objects can also be determined using a probabilistic model such as

$$p = P(x_i \mid x_{i-1}, d_i) \tag{6.4}$$

that determines the probability of an object to be at a specific location x_i given its previous location and the available sensor data.

Combining the probabilistic models of the robot's dynamics and the sensory localization provides approximation of the robot and objects location distributions, using probabilistic Bayesian filters such as Bayesian belief networks (BBNs) (Neapolitan 2004), Kalman filters (Gibbs 2011), or hidden Markov models (Fraser 2008). The outcome of the Bayesian filters are the set of locations of the robot and objects, commonly represented by a set of particles, which is the collection of possible distributed locations.

6.2.3 Motion-Planning Adaptation to Uncertainties

Nature provides ample examples for systems that operate efficiently without complete knowledge about the environment. The behavior of an individual working ant in a colony proves how limited knowledge about the environment and a relatively simple set of control rules are sufficient for accomplishing the complex task of foraging and transporting food to the net. With limited vision (sometimes no vision at all, e.g., termites), hearing, and tactile sensing capabilities, and more advanced smell sensors, ants manage to perform efficient global motion planning in challenging environments. Similarly to the ants, many mobile robotic systems have limited local information about the environment.

However, the robot controller can use its pervious sensors and actuation information to construct the *history information state* (*I*-state). Applying a set of control rules based on the *I*-state, an updated operation command can be generated. Given the uncertainty of the system (robot and environment), the motion planner can perform a greedy search that maximizes the immediate payoff based on a small subset of the I-state, in what is often known as *reactive* or *sensory control.*

In this type of control, the robot reacts only to the most recent sensory data, ignoring previous information. On the other hand, a non-greedy search can provide a better long-term payoff by considering the sequence of possible control commands given all available current and previous information. Although the non-greedy algorithms provide better long-term outcomes, their computational complexity is far more demanding than the greedy algorithms, and they may become unmanageable in intensive scenarios that require fast adaptation.

To illustrate the use of these concepts, consider the search task shown in Figure 6.2.

The figure shows a two dimensional area that consists of 11 × 11 tile matrix. Each tile in the matrix is either empty (white) or busy (black). A target is located randomly on one of the tiles, and the robot is also placed randomly on another tile with random orientation.

In the system shown in Figure 6.2a, the robot's sensor can detect objects only in the close proximity in front of it (shown by the gray triangle). The robot can perform only

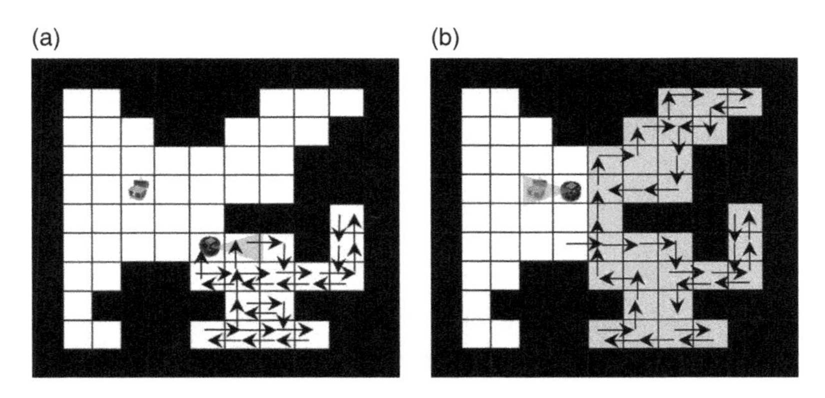

Figure 6.2 Setup for the search task with no memory (a) and with full *I*-state (b).

four types of basic movements: forward, backward, and 90° left and right turns. Since the robot has no information about its location and the locations of the target and the obstacles, applying a reactive algorithm that does not consider the *I*-state (no memory) will result in an inefficient search. Furthermore, this search does not guarantee that the robot will find the target even after a very long time.

In the path shown in Figure 6.2a, the motion planner guides the robot along an infinite route that will never reach the target. Figure 6.2b, on the other hand, shows a motion planner with a full *I*-state. The motion planner constructs a histogram map of the traveled route as well as the obstacles, and detects the target with 38 steps.

In both cases the motion planner uses an identical simple set of behaviors based on the wall-following algorithm (Katsev et al. 2011). However, in the second case, the robot does not enter an area that has already been scanned in previous tours. The two trajectories shown in Figure 6.2 were generated using the NetLogo multi-agent programmable modeling environment (Tisue and Wilensky 2004). In the next section, we present some of the common online motion planners and their implementation in ground and aerial autonomous vehicles.

6.3 Online Motion Planning

When a robot is required to travel in an environment with partial or no information regarding the location of the obstacles, it must have sufficient sensors to accommodate the lack of such information. If the initial and target configurations are known before the robot starts its motion, a global trajectory can be generated using all available information, and then, based on updated sensory data, the trajectory is modified to avoid possible risks to the robot and its surroundings.

For example, Lumelsky et al. (1990) propose an approach where the motion planner initially assumes there are no obstacles along the trajectory and therefore the robot can move along a straight line from the initial to the target position. When a new obstacle is detected, the robot circumvents the object by detecting the object's edges, and moving around it until a new direct trajectory is constructed toward the goal. If no new valid trajectory is found, a new global trajectory is generated. While such an approach is complete, in the sense that it eventually generates a trajectory to the goal (if such a trajectory exists), it is suboptimal, as there might be better trajectory in terms of the cost.

Stents (1994) suggests an algorithm for generating an optimal path using a dynamic navigation map that is constantly being updated by the robot's sensors. The algorithm, known as the D^* is, in many ways, similar to the A^* algorithm, but it can cope with dynamic changes of the navigation map. According to the D^* algorithm, the space is formulated as a set of discrete states that are connected by directional arcs, each associated with a positive cost. Each state has a pointer to the next state, and if two states have a connecting arc with a defined positive cost, they are considered to be neighboring states. The OPEN list contains the information about the costs of the arcs from each state to its neighbors, and is constantly being updated according to new data obtained by the sensors. The algorithm consists of two main procedures: the PROCESS-STATE procedure that computes an optimal path from the initial to the goal position, and the MODIFY-COST procedure that updates the cost function of the arcs and their effect on the OPEN

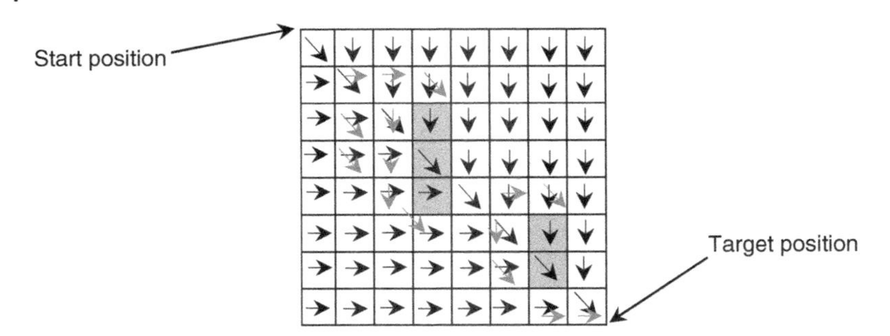

Figure 6.3 Illustration of the operation of the D^* algorithm with two objects detected on-line.

list. Experimental results show how the algorithm handles changes in the cost function during the robot's motion to produce optimal and efficient trajectories. Figure 6.3 demonstrates the D^* algorithm where the environment consists of 64 states.

Initially, the algorithm determines an optimal path from the target location to the start position of the robot assuming no obstacles exists. This path is determined based on the initial OPEN list that determines the costs of the arcs between neighboring states (the black arrows). As the robot moves along this path, it detects the first object, and the OPEN list is updated such that all the arcs from all states that are affected by the new data are adjusted (the gray arrows at the upper left side). A new path, from the goal to the current robot location, is constructed and the robot continues along that path. The same process repeats when the second object is detected (the gray arrows at the bottom right side).

Bornstein and Koren (1991) present the vector field histogram (VFH), which was originally developed for online motion planning using ultrasonic sensors. These sensors are subject to poor directionality, frequent misreading due to noise from other electronic devices in the close vicinity, and specular reflections characteristics of the objects. They use a certainty grid formulation in which each cell in the space is assigned with a certainty value (CV) that represents the relative certainty that an object is within that cell. They use a conical probabilistic distribution (based on the geometrical field of view of an ultrasonic sensor) to continuously update the certainty grid. They then apply a potential field method (Khatib 1986), in which objects apply virtual repulsive forces that "push" the robot away from that object, while the target applies an attractive force that "pulls" the robot toward the target. The repulsive and attractive forces are then summed, and the resultant force guides the robot toward the target while avoiding contact with obstacles.

6.3.1 Motion Planning with Differential Constraints

While the global motion planning discussed so far has considered the geometric features of the robot and its environment, there is another set of local constraints that need to be considered – namely, the differential constraints. These constraints reflect the limited maneuverability of the robot due to the allowable speeds and accelerations.

Differential constraints are expressed by the robot kinematic and dynamic models, and should be considered during the motion-planning process rather than a supplementary task. Consider a trajectory τ_{free} that links initial to target positions while avoiding obstacles, and a robot velocity vector

$$\dot{q} = f(q, u), \tag{6.5}$$

where the current robot state is q and u is a command set, also known as the input set. Assuming no differential constraints, a trajectory can be constructed according to the equation

$$\tau_{\text{free}} = \int_0^t f(q, u)dt \tag{6.6}$$

known as the configuration transition equation (Lavalle 2011).

However, adding differential constraints to the configuration transition equation may interrupt with the construction of the trajectory, necessitating replanning. Figure 6.4 illustrates how kinematic differential constraints affect trajectory planning of a tricycle-like robot that moves on a flat surface.

The command set for this robot is $u = \{u_l, u_\emptyset\}$, where u_l is the longitudinal command (linear speed) and $u_\emptyset$ is the steering command of the front wheel. The differential model of this robot is therefore given by the equation

$$\dot{q} = \begin{Bmatrix} \dot{x} \\ \dot{y} \\ \dot{\theta} \end{Bmatrix} = \begin{Bmatrix} u_l \cos\theta \\ u_l \cos\theta \\ \frac{u_l}{d} \tan u_\emptyset \end{Bmatrix}. \tag{6.7}$$

Since u_l and $u_\emptyset$ have lower and upper bounds, τ_{free} is limited by these constraints. For example, it is clear that the set of states $\begin{Bmatrix} 0 \\ \dot{y} \\ * \end{Bmatrix}$ is invalid for any value of $\dot{y}$, as the robot cannot perform a pure lateral motion. Similarly, the set $\begin{Bmatrix} 0 \\ 0 \\ \dot{\theta} \end{Bmatrix}$ is invalid for any value of $\dot{\theta}$ if $|\emptyset| < \frac{\pi}{2}$ (no pure rotation around the robot center is possible). As a result of these constraints, the trajectory shown in Figure 6.4b may need to be modified according to the kinematic differential constraints, and may even result in an empty set, where the given robot cannot reach the target.

The example shown in Figure 6.4 considers only the kinematic differential constraints. However, the dynamic model of the robot may introduce additional constraints that result in higher-order differential equations in the form of

$$\ddot{q} = g(q, \dot{q}, u). \tag{6.8}$$

Although higher-order differential equations are more challenging, a simple expansion of the robot state can reduce the order of the differential equation. For example, defining $x_1 = q$ and $x_2 = \dot{q}$ results in a set of lower-order differential equations in the form of $\dot{x} = h(x, u)$ that, although they involve more variables, are easier to solve.

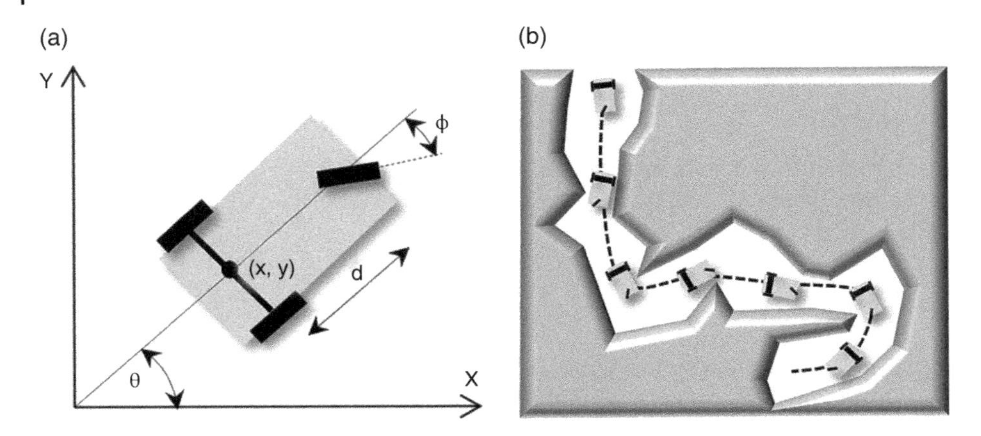

Figure 6.4 Differential kinematic constraints of tricycle-like robot (a) moving in a cluttered environment (b).

A common method for solving the set of differential equations is to discretize the time and space in which the time is divided into small time intervals Δt, and space is divided into a predetermined homogeneous cell grid (see Figures 6.2 and 6.3) or according to specific features of the environment. The differential equations are then transformed into discrete equations in the form of $x_{k+1} = h(x_k, u_k)$, and the entire space is represented by a directional graph in which the nodes are specific configurations of the robot, and the edges are the transitions from one configuration to another. Possible trajectories are determined using graph search methods, and an optimal trajectory can be found using optimization algorithms (e.g., dynamic planning).

6.3.2 Reactive Motion Planning

In reactive motion planning (RMP), the next action of the robot is determined quickly with little or no preprocessing efforts, based on its current state. It is particularly beneficial when the environment consists of fast dynamic changes.

The RMP process decomposes the global motion planning into more simple subproblems, while exploring the close neighborhood space. Typical implementations of such a motion planner are reflected in the potential function $\emptyset$ that takes a minimal value at the target configuration, maximal values at prohibited configurations (e.g., obstacles) and determines the discrete trajectory such that $\emptyset(x_{k+1}) < \emptyset(x_k)$. Figure 6.5 shows such a

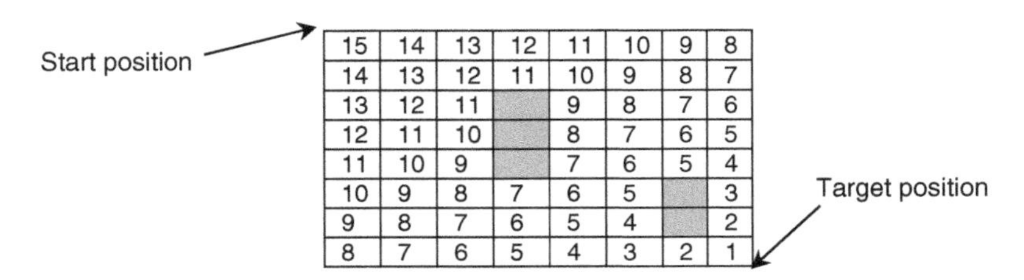

15	14	13	12	11	10	9	8
14	13	12	11	10	9	8	7
13	12	11		9	8	7	6
12	11	10		8	7	6	5
11	10	9		7	6	5	4
10	9	8	7	6	5		3
9	8	7	6	5	4		2
8	7	6	5	4	3	2	1

Figure 6.5 Potential function representing the distance to the target position.

function (also known as navigation function, NF) (Rimon and Koditschek 1992), where the potential value is proportional to the distance of each cell to the target position.

In this simple case, the motion planner advances the robot to the neighboring cell that reduces the potential value. An extension of the NF for dynamic environments with uncertainties (Hacohen et al. 2017) considers the distribution probabilities of the locations of the robot and the obstacles by formulating a probabilistic density function (PDF). Using the PDF, a safe trajectory is constructed using a potential function that limits to the probability for collision between the robot and the objects in the environment.

6.4 Global Positioning with Local Maps

Mobile robots commonly have the means to gather information from their close proximity, and based on that information, to construct a local map. By incrementally extending the boundaries of the map, the robots can determine their location within the extended map in what is known as simultaneous localization and mapping (SLAM).

Early theoretical work by Smith and Cheesman (1986) and by Durrant-Whyte (1988) combined with work on visual navigation (Ayache and Faugeras 1988) and sonar-based navigation (Crowley 1989; Chatila and Laumond 1985) laid the foundation for this conceptual breakthrough in autonomous navigation in unknown environments. Estimation of the trajectory the robot is traversing, and the relative measurements to significant landmarks along that trajectory, enables the robot to determine its location without any prior knowledge.

To illustrate this concept, consider the robot and the set of three landmarks shown in Figure 6.6, where the sharp images of the robot and the landmarks represent their actual locations, while the blared images represent the estimated locations. The dashed line in Figure 6.6 represents the estimated robot's trajectory.

The robot moves along a trajectory shown by the solid line, where X_k is the actual robot's location at time k. During its motion, the robot measures the relative locations of the landmarks, given by $Z_{k,\,j}$ where k is the time and j is the landmark number, U_{k+1} is the control command to the robot to move from location X_k to X_{k+1} and L_i is the real location of landmark i.

Then, the set $X_{0:k} = \{x_0, x_1, \ldots x_k\}$ represents all previous locations of the robot, and the set $U_{0:k} = \{u_0, u_1, \ldots u_k\}$ consists of all previous control commands. The set $Z_{0:k} = \{z_0, z, \ldots z_k\}$ is the history of all previous measurements to the landmarks, and $M_{0:n} = \{L_0, L_1, \ldots L_n\}$ is the set of all the landmarks that construct the map.

Given the above formulations, the SLAM procedure is determined (in a probabilistic manner) by the following conditional equation:

$$P(x_{k+1}, m \mid Z_{0:k+1}, U_{0:k}, X_{0:k}, u_{k+1}).$$

It provides a probabilistic distribution of both the robot and the landmarks' locations at time $k + 1$, given the previous robot's locations and control commands, current measurements to landmarks, as well as the current control command.

The observation model of the SLAM procedure is given by $P(z_{k+1} \mid x_{k+1}, m)$, which determines the probability distribution of the relative measurement from the robot to the landmarks, given the estimated robot's location and landmarks' locations.

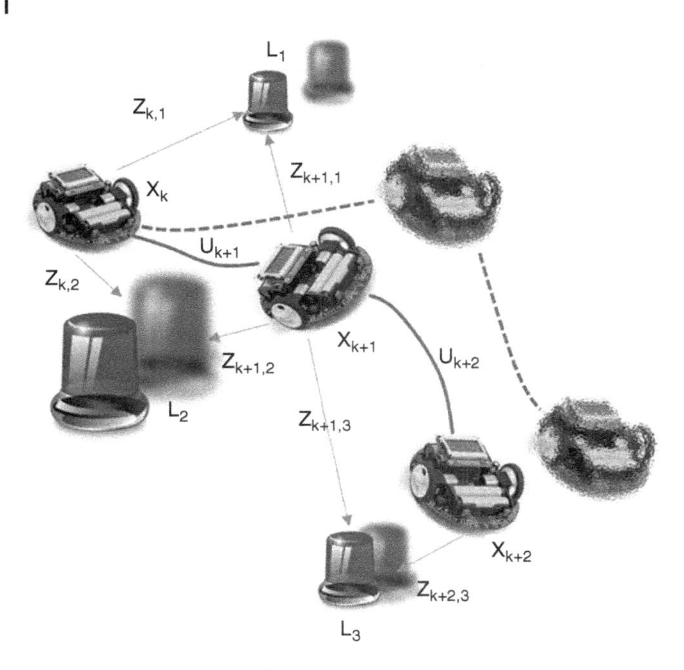

Figure 6.6 Description of the SLAM principles.

The motion model is given simply by $P(x_{k+1} \mid x_k, u_{k+1})$ (the next robot's location depends on its current location and control command).

The SLAM procedure is implemented in two simultaneous recursive steps:

1) Time prediction given by

$$(x_{k+1}, m \mid Z_{0:k}, U_{0:k}, X_{0:k}, u_{k+1}) = \int P(x_{k+1} \mid x_k, u_{k+1}) \times (x_k, m \mid Z_{0:k}, U_{0:k}, X_{0:k}) dx_k,$$

2) Recursive measurement update given by

$$P(x_{k+1}, m \mid Z_{0:k+1}, U_{0:k}, X_{0:k}) = \frac{P(z_{k+1} \mid x_{k+1}, m) P(x_{k+1}, m \mid Z_{0:k}, U_{0:k}, X_{0:k})}{P(z_{k+1} \mid z_{0:k}, U_{0:k}, u_{k+1})}.$$

An important observation of the SLAM procedure, assuming the landmarks are static, is that the relative position of landmarks is also static, and therefore the correlation between all landmarks' estimates increases with the number of measurements performed by the robot (Dissanayake and Gamini 2001).

Notice that the distance between the sharp (actual) and blare (estimate) position of the landmarks in Figure 6.6 decreases as the robot progresses along the trajectory. Obviously, when the robot detects a new landmark, its correlation with other landmarks is relatively low, but given the high correlation between the previous landmarks, the correlation of the new landmarks with the other landmarks increases with the additional measurements. Similarly, the correlation between the robot's estimated location and the locations of the landmarks increases (the variance of the probabilistic distribution decreases).

Several algorithms have been suggested for practical solutions of the SLAM procedures. Among the popular ones are the extended Kalman filter algorithm – EKF-SLAM and the particle filter algorithm – FastSLAM (Bailey and Durrant-Whyte 2006). The next section discusses the implementation of SLAM is UAVs.

6.5 UAV Motion Planning in 3D Space

The general motion-planning methods discussed in the previous sections are applicable for ground and aerial autonomous vehicles. However, UAVs introduce additional challenges for motion planning, particularly in time-variant environments that include uncertainties. Most UAVs have limited payload and computation capacities, which may limit their on-board sensing capabilities and autonomous control performance. In addition, UAVs are more profound due to differential constraints and to environmental disturbances than autonomous ground vehicles. Finally, UAVs operate in a 3D space, often in close proximity to other fast-moving UAVs and obstacles. As a result, the selection of a motion planner for a UAV must consider the specific platform, tasks, and environments.

A typical UAV has either two or four degrees of freedom with significant differential constraints that result in a problem dimension in the range of 5–12. The solution of this type of problems is proved to be nonpolynomial (NP) even with simplifications, approximations, and generalizations (Canny 1988). Furthermore, in the search for an optimal solution, additional factors such as complexity of computation, smoothness of the trajectory, motion duration, and energy consumption must be considered.

Finding an optimal solution is substantially more complex even in the two-dimensional space, and in general requires approximations and heuristics. In sampling-based trajectory planning (SBTP), the entire workspace is discretizes to a 3D matrix (Reif and Hongyan 2000), and a time-optimal trajectory is approximated by searching through this matrix for a safety corridor. In another SBTP method, known as state-space navigation function (SSNF), a navigation function is constructed. The gradient of this function approximates a time-optimal trajectory (Lavalle and Konkimalla 2001).

The rapidly exploring random tree (RRT) algorithm uses a stochastic search method over the configuration space to produce feasible trajectories while considering the dynamic constraints (Redding et al. 2007). In decoupled trajectory planning (DTP), a trajectory is first constructed using one of the conventional motion-planning algorithms (e.g., A^*, potential field, Voronoi map, etc.). This trajectory is then modified to accommodate the dynamic constraints of the UAV, as well other specific requirements.

The decomposition of motion planning into two stages simplified the computation complexity, although there is no guarantee of optimality. The hierarchical decoupled planning and control (HDPC) algorithms integrate waypoints into the UAV control system. The waypoints are selected to avoid collision with obstacles and to generate a smooth a trajectory according to the UAV dynamic constraints. Due to its simplicity and ease of implementation, this algorithm is one of the most common motion planner for UAVs. Its modular hierarchical structure defines the functional decomposition of the system, so it can be implemented by various types of platforms and operating systems.

Many commercial UAVs come with built-in waypoint control systems, allowing efficient integration with sensing and obstacle avoidance methods. Scherer et al. (2008) developed a 3D motion planner for a helicopter operating at low elevations in cluttered environments by combining an online sensing system with obstacle avoidance. Running at relatively high speed (10 m s^{-1}) at elevations in the range of 5–10 m above ground, their UAV managed to safely avoid small objects (6 mm wires). In a different approach, the 3D workspace is decomposed into several 2D planes, each containing the initial and goal positions. A 2D motion planner (e.g., Voronoi roadmap planner) constructs a 2D trajectory on each plane. The planes are ranked according to the cost function and then adjusted to the dynamic constraints of the UAV (Whalley et al. 2005).

The finite state model (FSM) approach reduces the computational complexity of the motion-planning problem by relaxing the constraints on control and time and creating motion primitives. A full trajectory consists of a collection of several motion primitives, similar to the way human pilots control aircrafts by combining trim trajectories and 3D maneuvers, in what is known as maneuver automation (MA). Schouwenaars et al. (2004) implement the concept of MA in an agile UAV. The motion planner constructs the trajectory from pre-programmed agile maneuvers using a maneuver scheduler. The dynamic constraints are simplified to a set of linear time-invariant modes, and a mixed-integer linear programming (MILP) algorithm determines the optimal trajectory. This way, the motion-planning problem is converted to a numerical optimization problem, and given reasonable initial conditions, the algorithm converges in a polynomial time.

The receding horizon control (RHC) is another popular motion planning method for UAVs that reduces the computational complexity by reducing the time horizon of the problem. A trajectory is constructed for each time horizon, and an optimal numerical solver (e.g., MILP) constructs the global optimal trajectory. Only the relevant environmental data is incorporated with each time horizon calculation, thus reducing the required computational resources. Shim and Sastry (2006) implement this method into the Berkeley UAV flight control system in a fixed wing, as well as rotorcraft UAVs in various missions.

RMP algorithms are used in UAVs when global data is either unavailable or the uncertainty level is too large. In such cases, the local data is considered for the construction of last-minute obstacle avoidance maneuvers. RMP algorithms are particularly crucial when operating in dynamic environments, where the location of objects (obstacles and/or other UAVs) changes in time. RMP algorithm does not aim at constructing a global trajectory, and therefore must operate in conjunction with other global motion planners.

The motion description language (MDL) (Brockett 1990) is an efficient tool for constructing a reactive motion planner by assigning vehicle maneuvers to specific sensor inputs. As extension of the MDL, known as MDLe (Manikonda et al. 1999) integrates the UAV's differential constraints in the reactive behaviors. These behaviors are formalized in terms of kinetic state machine, where transitions between states are governed by real time sensory data.

Variations of the MDL include the use of fuzzy logic (Zavlangas et al. 2000) and natural human performance (Hamner et al. 2006). The fuzzy motion planner repels to the updated sensory data and the global target location and construct an updated obstacle avoidance maneuvers. Using a neural network training algorithm, it can be tuned to actual task's requirements. Observing natural human operator reactive performance while avoiding obstacles along a predetermined trajectory leads to the formulation of reactive control behaviors that can then be incorporated in UAVs.

Finally, many RMP algorithms are inspired by biosystems such as ant colony optimizations (ACO) (Dorigo et al. 1999), firefly algorithms (FA) (Yang 2010a), bat algorithms (BA) (Yang 2010b), artificial bee colony algorithms (ABC) (Karaboga 2010), bacterial foraging optimization algorithms (BFOA) (Passino 2002), pigeon-inspired optimization algorithms (PIOA) (Duan and Qiao 2014), glowworm swarm optimization algorithms (GSOA) (Krishnanand and Ghose 2005), and more. Bio-inspired motion planners imitate features from the behaviors creatures in their natural habitat. In particular, bio-inspired algorithms are beneficial when a swarm of UAVs operates toward a collective goal with decentralized motion planning and minimal communication between members of the swarm. Each UAV follows a set of simplistic motion behaviors while observing its immediate environment (obstacles as well as other UAVs). Bio-inspired motion planners require relatively low computation resources and are usually robust to local failures of some of the swarm's members.

6.6 Summary

In the chapter, we considered the methods of online motion planning and navigation where the robot's position is defined either with respect to locally accessible landmarks or with respect to other robots in two and three dimensions. The chapter included a brief overview of the methods of motion planning for ground and aerial vehicles and provides a review of the related literature in the field of robots' positioning and localization.

In particular, the chapter dealt with the following issues:

1) The methods of creating the robot's trajectory in 2D and 3D physical spaces and the methods of configuration space, which allows a natural and effective consideration of the motion in arbitrary environment and obstacle avoidance. Special attention was paid to motion planning with uncertainties that are the internal and external dynamical uncertainties and the sensor uncertainties and adaptation to the uncertain environment.
2) Following general classification of the motion-planning techniques, the chapter considered online motion planning (especially, the popular D^* algorithm), motion planning with constraints, and RMP based on the potential field method. In addition, it noted that the last method can be extended to motion planning with uncertainties.
3) The chapter presented a general idea of SLAM methods that allow navigation of the robots in initially unknown environment and creating the environmental maps during the motion.
4) It overviews the methods and algorithms of navigation in 3D space, and in particular it mentions the rapid-expanding tree algorithm, the decoupled trajectory planning and control algorithms, as well as the methods for reducing computational complexity of the motion-planning problem like FSM and RHC.
5) The chapter mentioned the bio-inspired techniques of motion planning that mimic the behavior of animals and are widely used for navigation of the groups of mobile robots.

References

Acar, E., Choset, H., Rizzi, A. et al. (2002). Morse decompositions for coverage tasks. *International Journal of Robotics Research* 21 (4): 331–344.

Ayache, N. and Faugeras, O.D. (1988). Building, registrating, and fusing noisy visual maps. *International Journal of Robotics Research* 7 (6): 45–65.

Bailey, T. and Durrant-Whyte, H. (2006). Simultaneous localization and mapping (SLAM): part II. *IEEE Robotics and Automation Magazine* 13 (3): 108–117.

Balluchi, A., Bicchi, A., Balestrino, A., and Casalino, G. (1996). Path tracking control for Dubin's cars. In: *Proc. IEEE Int. Conf. Robotics and Automation*, vol. 4, 3123–3128. IEEE.

Bellman, R. (1958). On a routing problem. *Quarterly of Applied Mathematics* 16: 87–90.

de Berg, M., van Kreveld, M., Overmars, M., and Schwarzkopf, O. (2000). Visibility graphs. In: *Computational Geometry*, 307–317. Berlin: Springer-Verlag.

Borenstein, J. and Koren, Y. (1991). The vector field histogram-fast obstacle avoidance for mobile robots. *IEEE Transactions on Robotics and Automation* 7 (3): 278–288.

Borenstein, J., Everett, B., and Feng, L. (1996). *Navigating Mobile Robots: Systems and Techniques*. Wellesley: A. K. Peters, Ltd.

Brockett, R.W. (1990). Formal languages for motion description and map making. *Robotics* 41: 181–191.

Canny, J. (1988). *The Complexity of Robot Motion Planning*. Cambridge MIT Press.

Chatila, R. and Laumond, J.-P. (1985). Position referencing and consistent world modeling for mobile robots. In: *Proc. IEEE Int. Conf. Robotics and Automation*, vol. 2, 138–145. IEEE.

Choset, H. and Pignon, P. (1998). Coverage path planning: The boustrophedon cellular decomposition. In: *Field and Service Robotics* (ed. A. Zelinsky), 203–209. London: Springer.

Crowley, J.L. (1989). World modeling and position estimation for a mobile robot using ultrasonic ranging. In: *Proc. Int. IEEE Conf. Robotics and Automation*, 674–580. IEEE.

Dijkstra, E.W. (1959). A note on two problems in connexion with graphs. *Numerische Mathematik* 1: 269–271.

Dissanayake, M.W. and Gamini, M. (2001). A solution to the simultaneous localization and map building (SLAM) problem. *IEEE Transactions on Robotics and Automation* 17 (3): 229–241.

Dorigo, M., Di Caro, G., and Gambardella, L.M. (1999). Ant algorithms for discrete optimization. *Artificial Life* 5 (2): 137–172.

Duan, H. and Qiao, P. (2014). Pigeon-inspired optimization: a new swarm intelligence optimizer for air robot path planning. *International Journal of Intelligent Computing and Cybernetics* 7 (1): 24–37.

Durrant-Whyte, H.F. (1988). Uncertain geometry in robotics. *IEEE Journal of Robotics and Automation* 4 (1): 23–31.

Fisher, A. (2013). Inside Google's quest to popularize self-driving cars. *Popular Science*.

Fraser, A.M. (2008). *Hidden Markov Models and Dynamical Systems*. Philadelphia, PA: SIAM.

Galceran, E. and Carreras, M. (2013). A survey on coverage path planning for robotics. *Robotics and Autonomous Systems* 61 (12): 1258–1276.

Gibbs, B.P. (2011). *Advanced Kalman Filtering, Least-Squares and Modelling: A Practical Handbook*. Hoboken, NJ: Wiley.

Hacohen, S., Shoval, S., and Shvalb, N. (2017). Applying probability navigation function in dynamic uncertain environments. *Robotics and Autonomous Systems* 87: 237–246.

Hamner, B., Singh, S., and Scherer, S. (2006). Learning obstacle avoidance parameters from operator behavior. *Journal of Field Robotics* 23 (11–12): 1037–1058.

Hart, P.E., Nilsson, N.J., and Raphael, B. (1968). Formal basis for the heuristic determination of minimum cost paths. *IEEE Transactions of Systems Science and Cybernetics* 4 (2): 100–107.

Hemmerdinger, J. (2013). FAA approves bid for UAV flights in civil airspace. *Flight International* 184 (5412): 10.

Karaboga, D. (2010). Artificial bee colony algorithm. *Scholarpedia* 5 (3): 6915.

Katsev, M., Yershova, A., Tovar, B. et al. (2011). Mapping and pursuit-evasion strategies for a simple wall-following robot. *IEEE Transactions on Robotics* 27 (1): 113–121.

Khatib, O. (1986). Real-time obstacle avoidance for manipulators and mobile robots. In: *Autonomous Robot Vehicles*, 396–404. New York: Springer.

Krishnanand, K.N. and Ghose, D. (2005). Detection of multiple source locations using a glowworm metaphor with applications to collective robotics. In: *Proc. IEEE Symp. Swarm Intelligence. SIS 2005*, 84–91. IEEE.

Latombe, J. (1991). *Robot Motion Planning*. Boston: Kluwer Academic.

Lavalle, S.M. (2011). Motion planning. *IEEE Robotics and Automation Magazine* 18 (2): 108–118.

Lavalle, S.M. and Konkimalla, P. (2001). Algorithms for computing numerical optimal feedback motion strategies. *International Journal of Robotics Research* 20 (9): 729–752.

Lee, T.-K., Baek, S.-H., Choi, Y.-H., and Oh, S.-Y. (2011). Smooth coverage path planning and control of mobile robots based on high-resolution grid map representation. *Robotics and Autonomous Systems* 59 (10): 801–812.

Lozano-Perez, T. (1983). Spatial planning: a configuration space approach. *IEEE Transactions on Computers* 32 (2): 108–120.

Lumelsky, V., Mukhopadhyay, S., and Sun, K. (1990). Dynamic path planning in sensor-based terrain acquisition. *IEEE Transactions on Robotics and Automation* 6 (4).

Manikonda, V., Krishnaprasad, P.S., and Hendler, J. (1999). Languages, behaviors, hybrid architectures, and motion control. In: *Mathematical Control Theory*, 199–226. New York: Springer.

Neapolitan, R.E. (2004). *Learning Bayesian Networks*. Harlow: Prentice Hall.

Passino, K.M. (2002). Biomimicry of bacterial foraging for distributed optimization and control. *IEEE Transactions of Control Systems* 22 (3): 52–67.

Redding, J., Jayesh, N.A., Boskovic, J.D. et al. (2007). A real-time obstacle detection and reactive path planning system for autonomous small-scale helicopters. In: *Proc. AIAA Navigation, Guidance and Control Conf.*, 6413. Hilton Head, SC. Curran Associates Inc.

Reif, J.H. and Hongyan, W. (2000). Nonuniform discretization for kinodynamic motion planning and its applications. *SIAM Journal on Computing* 30 (1): 161–190.

Rimon, R. and Koditschek, D.E. (1992). Exact robot navigation using artificial potential functions. *IEEE Transactions of Robotics and Automation* 8 (5): 501–518.

Scherer, S., Singh, S., Chamberlain, L., and Elgersman, M. (2008). Flying fast and low among obstacles: methodology and experiments. *International Journal of Robotics Research* 27 (5): 549–574.

Schouwenaars, T., Mettler, B., Feron, E., and How, J.P. (2004). Hybrid model for trajectory planning of agile autonomous vehicles. *JACIC* 1 (12): 629–651.

Sharir, M. (1995). Robot motion planning. *Communications on Pure and Applied Mathematics* 48 (9): 1173–1186.

Shiller, Z. and Gwo, Y. (1991). Dynamic motion planning of autonomous vehicles. *IEEE Transactions on Robotics and Automation* 7 (2): 241–249.

Shim, D.H. and Sastry, S. (2006). A situation-aware flight control system design using real-time model predictive control for unmanned autonomous helicopters. In: *Proc. AIAA Guidance, Navigation, and Control Conference*, vol. 16, 38–43. Curran Associates Inc.

Shoval, S., Ulrich, I., and Borenstein, J. (2003). NavBelt and the GuideCane. *IEEE Robotics and Automation Magazine* 10 (1): 9–20.

Smith, R.C. and Cheeseman, P. (1986). On the representation and estimation of spatial uncertainty. *International Journal of Robotics Research* 5 (4): 56–68.

Solovyev, V.V., Finaev Valery, A., Zargaryan et al. (2015). Simulation of wind effect on a quadrotor flight. *ARPN Journal of Engineering and Applied Sciences* 10 (4): 1535–1538.

Stentz, A. (1994). Optimal and efficient path planning for partially-known environments. In: *Proc. IEEE Int. Conf. Robotics and Automation*, vol. 4, 3310–3317. Institute of Electrical and Electronics Engineers (IEEE).

Takahashi, O. and Schilling, R. (1989). Motion planning in a plane using generalized Voronoi diagrams. *IEEE Transactions on Robotics and Automation* 5 (2): 143–150.

Thrun, S., Burgard, W., and Fox, D. (2005). *Probabilistic Robotics*. The MIT Press.

Tisue, S. and Wilensky, U. (2004). Netlogo: A simple environment for modeling complexity. In: *International conference on complex systems*, vol. 21, 16–21. Springer Verlag.

Whalley, M., Freed, M., Harris, R. et al. (2005). Design, integration, and flight test results for an autonomous surveillance helicopter. In: *Proc. AHS Int. Specialists' Meeting on Unmanned Rotorcraft*. Vertical Flight Society (VFS).

Wong, S. (2006). *Qualitative Topological Coverage of Unknown Environments by Mobile Robots. Ph.D. Thesis*. The University of Auckland.

Yang, X.S. (2010b). A new metaheuristic bat-inspired algorithm. In: *Nature Inspired Cooperative Strategies for Optimization*, 65–74. Berlin, Heidelberg: Springer.

Yang, X.S. (2010a). Firefly algorithm, stochastic test functions and design optimisation. *International Journal of Bio-Inspired Computation* 2 (2): 78–84.

Zavlangas, P.G., Tzafestas, S.G., and Althoefer, K. (2000). Fuzzy obstacle avoidance and navigation for omnidirectional mobile robots. In: *Proc. European Symp. Intelligent Techniques*, 375–382. Aachen, Germany. ERUDIT Service Center.

Zhu, D. and Latombe, J. (1991). New heuristic algorithms for efficient hierarchical path planning. *IEEE Transactions on Robotics and Automation* 7 (1): 9–26.

7

Motion in an Unknown Environment

Eugene Kagan

This chapter considers the methods of probabilistic motion planning in the unknown environment widely known as probabilistic robotics. It introduces the concept of belief space, considers basic estimation and prediction methods and additional methods of the environment mapping. In addition, the chapter presents the simplest learning methods and their implementations for the robots' control.

7.1 Probabilistic Map-Based Localization

In contrast to the fixed-base arm manipulators, which act in artificial well-structured environments and because of practically unlimited power supply are equipped with robust heavy-duty gear, mobile robots usually act in real unstructured worlds and have relatively simple economical and consequently inaccurate equipment that leads to errors and uncertainties in the mobile robots behavior. In the widely accepted systematization, the sources of such uncertainties are listed as follows (Thrun, Burgard, and Fox 2005):

1) *Environments*. The real-world environments, in which the mobile robots act, are highly unpredictable and usually change in time.
2) *Sensors*. The sensors are limited in their perception abilities, and the obtained measurements are perturbed both by the environmental noise and by the internal errors.
3) *Actuators*. The actuators usually driven by motors are not accurate and are perturbed by external noise. Additional errors are introduced by the control noise.
4) *Models*. Any model provides noncomplete description of the real-world processes and events that introduces errors in the available image of the robot and its environment.
5) *Computation*. The on-board computers and controllers have limited abilities and usually the implemented algorithms are approximate that results in inaccurate computation results.

The treating of the inaccuracy of the robot's motion led by the indicated uncertainties can be conducted in several ways. If the uncertainties are defined in the terms of possible

Autonomous Mobile Robots and Multi-Robot Systems: Motion-Planning, Communication, and Swarming, First Edition. Edited by Eugene Kagan, Nir Shvalb and Irad Ben-Gal.

Companion website: www.wiley.com/go/kagan/robotsystems

errors and their probabilities with the known or assumed distributions, and the motion is planned by the standard stochastic control methods (see, e.g., Aoki (1967) and Astrom (1970)) in such a manner that it minimizes the resulting error at the end of the mission. The other way to describe the uncertainties in this case is application of some kind multi-valued or fuzzy logic that allows making uncertain decisions, which counterbalance the uncertainties in the available data (Cuesta and Ollero 2005). The other way of treating the uncertainties implements the probabilistic methods of decision-making, e.g., Markov decision processes (MDV) (White 1993), or more general methods, which also utilize the history of the agent's activity (Bertsecas and Shreve 1978; Bertsekas 1995).

Probabilistic robotics follows the probabilistic decision-making approaches that are applied for navigation of the robot in stochastic environment with taking into account the other indicated uncertainties, which can differ with respect to the robot's types and abilities. In general, this decision-making process is defined as follows.

Let $\mathfrak{S} = \{\mathfrak{s}_1, \mathfrak{s}_2, \mathfrak{s}_3, \ldots\}$ be a finite or countable set of possible states of the agent and $\mathfrak{A} = \{\mathfrak{a}_1, \mathfrak{a}_2, \mathfrak{a}_3, \ldots\}$ be a set of its possible actions. In addition, let $\mathfrak{M} = \{\mathfrak{m}_1, \mathfrak{m}_2, \mathfrak{m}_3, \ldots\}$ be a set of possible results of the measurements of the environment of the agent; for example, the measurements can represent the results of checks of the observed areas or the characteristics of the neighboring agents (see search algorithms and flocking procedures in Chapter 10). Then, general activity of the agent in discrete time $t = 0, 1, 2, \ldots$ is represented by the following operations (Kagan and Ben-Gal 2013):

1) Set $t = 0$ and start with initial state $\mathfrak{s}(0) \in \mathfrak{S}$.
2) Being in the state $\mathfrak{s}(t) \in \mathfrak{S}$, do
   2.1 Observe the environment and obtain the measurement result $\mathfrak{m}(t) \in \mathfrak{M}$.
   2.2 Choose action $\mathfrak{a}(t) \in \mathfrak{A}$ with respect to the state $\mathfrak{s}(t)$ and the measurement result $\mathfrak{m}(t)$.
3) Apply action $\mathfrak{a}(t)$ and obtain new state $\mathfrak{s}(t+1) \in \mathfrak{S}$.
4) Set $t = t + 1$.
5) Continue with line 2.

Certainly, if at time t the chosen action $\mathfrak{a}(t)$ is *terminate*, then the state $\mathfrak{s}(t+1)$ is *stop* and the agent finishes its mission. Similarly, if the time is limited, then the agent finishes its mission when the time reaches this limit. The activity of the agent is illustrated in Figure 7.1.

As a result, the activity of the agent up to finite of infinite time T is described by three sequences ($t = 0, 1, 2, \ldots, T$) (Thrun, Burgard, and Fox 2005; Siegwart, Nourbakhsh, and Scaramuzza 2011):

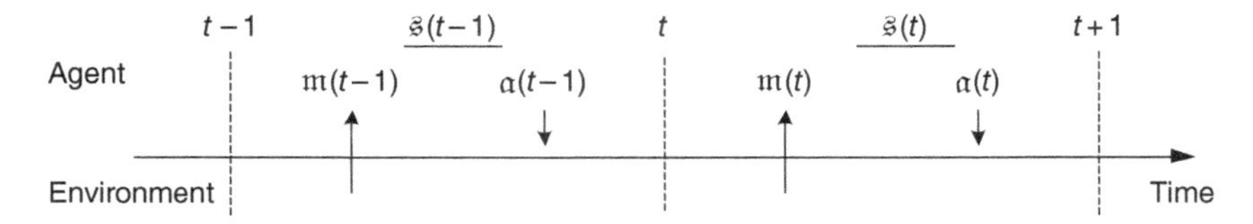

Figure 7.1 General activity of the agent in discrete time.

- the sequence $\mathfrak{s}_{0:T} = \langle \mathfrak{s}(0), \mathfrak{s}(1), \mathfrak{s}(2), \ldots, \mathfrak{s}(T) \rangle$ of states $\mathfrak{s}(t) \in \mathfrak{S}$,
- the sequence $\mathfrak{m}_{0:T} = \langle \mathfrak{m}(0), \mathfrak{m}(1), \mathfrak{m}(2), \ldots, \mathfrak{m}(T) \rangle$ of measurements $\mathfrak{m}(t) \in \mathfrak{M}$, and
- the sequence $\mathfrak{a}_{0:T} = \langle \mathfrak{a}(0), \mathfrak{a}(1), \mathfrak{a}(2), \ldots, \mathfrak{a}(T) \rangle$ of actions $\mathfrak{a}(t) \in \mathfrak{A}$,

such that the current state $\mathfrak{s}(t)$ of the agent is defined by its previous states, measurements and actions that is

$$\mathfrak{s}(t) \longleftarrow (\mathfrak{s}_{0:t-1}, \mathfrak{m}_{0:t-1}, \mathfrak{a}_{0:t-1}). \tag{7.1}$$

Similarly, the result $\mathfrak{m}(t)$ of the current measurement is defined by the agent's states up to the current state and previous measurements and actions:

$$\mathfrak{m}(t) \longleftarrow (\mathfrak{s}_{0:t}, \mathfrak{m}_{0:t-1}, \mathfrak{a}_{0:t-1}), \tag{7.2}$$

and the current action $\mathfrak{a}(t)$ id defined by the agent's states and measurements up to the current ones and previous actions that is

$$\mathfrak{a}(t) \longleftarrow (\mathfrak{s}_{0:t}, \mathfrak{m}_{0:t}, \mathfrak{a}_{0:t-1}). \tag{7.3}$$

Notice that in the presented formulation it is assumed that the conducted action $\mathfrak{a}(t)$ moves the agent into the next state $\mathfrak{s}(t+1)$. In the other but certainly equivalent formulations (see, e.g., Thrun, Burgard, and Fox (2005)) it is assumed that been in the state $\mathfrak{s}(t)$ the agent at first executes the chosen action $\mathfrak{a}(t)$, then obtains the measurement result $\mathfrak{m}(t)$ and moves to the next state $\mathfrak{s}(t+1)$.

In the case of deterministic motion planning, especially for the fixed-base arm manipulators, the rules defined by Eqs. (7.1) and (7.2) are specified by definite transition functions and the decision-making regarding the actions defined by Eq. (7.3) is provided by usual optimal control methods. For the mobile robots, in contrast, the environmental uncertainties and possibly erroneous measurements as well as the limited models and restricted abilities of computation and actuating lead to the uncertainties in specification of the states and choosing the actions. As a result, transitions between the states and measurements are governed by certain stochastic processes, and instead of Eqs. (7.1) and (7.2) the conditional probabilities (Thrun, Burgard, and Fox 2005; Siegwart, Nourbakhsh, and Scaramuzza 2011)

$$p(\mathfrak{s}(t) \mid \mathfrak{s}_{0:t-1}, \mathfrak{m}_{0:t-1}, \mathfrak{a}_{0:t-1}), \tag{7.4}$$

$$p(\mathfrak{m}(t) \mid \mathfrak{s}_{0:t}, \mathfrak{m}_{0:t-1}, \mathfrak{a}_{0:t-1}), \tag{7.5}$$

and the actions $\mathfrak{a}(t)$ are chosen using appropriate methods of statistical decisions (Wald 1950; DeGroot 1970). The next sections consider such probabilistic motion planning and localization of mobile robots.

7.1.1 Beliefs Distribution and Markov Localization

Consider the motion of the mobile robot in a certain coordinate domain X, which can include obstacles and over which certain potentials or probability distributions are defined. The points of the domain are denoted by $\vec{x} \in X$. If $X \subset \mathbb{R}$ is a one-dimensional line, then the point represents the coordinate on one axis, that is $\vec{x} = x$; if $X \subset \mathbb{R}^2$ is two-dimensional, then each point of X is a pair of coordinates $\vec{x} = (x, y)$, and so far. While considering the discrete domain $X = \left\{\vec{x}_1, \vec{x}_2, \ldots, \vec{x}_n\right\}$ of the size n with the points $\vec{x} = (x, y)$ it is assumed that it is a square grid such that each point $\vec{x}_i$, $i = 1, 2, \ldots, n$, is defined by a pair (i_x, i_y) of indices $i_x = 1, 2, \ldots, n_x$ and $i_y = 1, 2, \ldots, n_y$, where $n = n_x \times n_y$ and $i = (i_x - 1)n_x + i_y$.

The state $\mathfrak{s}(t)$ of the robot at time t can be defined in different manners with respect to the tasks and motion planning methodology. In the most cases, it is assumed that the state of the robot represents the coordinate $\mathfrak{s}(t) = \vec{x}(t)$, while if, for example, the motion of the robot is defined using the phase space, the state includes both the robot's coordinate and velocity and is defined as a pair $\mathfrak{s}(t) = (\vec{x}(t), \vec{v}(t))$. The measurement's result $\mathfrak{m}(t)$ is usually considered as a result $z(t)$ of observation of the local environment $a(t) \subset X$ of the robot, values of potential functions or probabilities. Finally, the action $\mathfrak{a}(t)$ of the robot is defined as its motion or turn, e.g. (see Section 10.3.1), in the case of discrete time and gridded space the actions are specified as movements or steps $\delta(t) \in \mathcal{D} = \{\delta_1, \delta_2, \delta_3, \delta_4, \delta_5\}$, where δ_1 = *move forward*, δ_2 = *move backward*, δ_3 = *move right*, δ_4 = *move left* and δ_5 = *stay in the current point*; the mentioned above action *terminate* is, certainly, included into the possible set of actions, but since it is not used for navigation of the robots, usually it is considered separately.

As indicated above, because of the listed uncertainties the robot is not able to know its real location $\vec{x}(t)$ at time t. However, after observation of the available environment $a(t) \subset X$ with obtaining the observation result $z(t-1)$ and then conducting the chosen motion step $\delta(t-1)$, the robot can estimate its location $\vec{x}(t)$. The best estimation of the robot's location is often called *belief*, which following the probabilistic approach and is defined as a conditional probability (Thrun, Burgard, and Fox 2005; Siegwart, Nourbakhsh, and Scaramuzza 2011):

$$\overline{bel}(\vec{x}(t)) = p(\vec{x}(t) \mid z_{0:t-1}, \delta_{0:t-1}), \tag{7.6}$$

where, similar to above, $z_{0:t-1}$ and $\delta_{0:t-1}$ stand for the history of observations and of movements, respectively.

The belief $\overline{bel}(\vec{x}(t))$ does not incorporate the observation result $z(t)$, which is obtained at the location $\vec{x}(t)$. If, in contrast, the observation result at time t is taken in to account, then the belief is denoted by $bel(\vec{x}(t))$ and is defined by the conditional probability

$$bel(\vec{x}(t)) = p(\vec{x}(t) \mid z_{0:t}, \delta_{0:t-1}). \tag{7.7}$$

The distributions of the conditional probabilities $\overline{bel}(\vec{x})$ and $bel(\vec{x})$, $\vec{x} \in X$, are called belief distributions, and the set of all possible belief distributions over the domain X with respect to the abilities of the mobile robot is considered as a beliefs space. However, notice that often the domain with certain belief values in the points is also considered as a beliefs space.

Let $\mathcal{M}$ be a real or estimated map of the environment that represents locations of the obstacles, topography of the domain or distribution of potentials or probabilities over a domain. For example, the map $\mathcal{M}$ can be considered as a set of points $\vec{x} \in X$ together with the information, whether the point is occupied by some obstacle, or with the probability of the target's location in this point, and similar. Denote by $p(z(t) \mid \vec{x}(t), \mathcal{M})$ the probability of obtaining the observation result $z(t)$ given that the robot is in the location $\vec{x}(t)$ and observes the area $a(t)$ of the domain X with the map $\mathcal{M}$. Then the relation between the beliefs $bel(\vec{x}(t))$ and $\overline{bel}(\vec{x}(t))$ is straightforward:

$$bel(\vec{x}(t)) = \eta(\vec{x}(t))p(z(t) \mid \vec{x}(t), \mathcal{M})\overline{bel}(\vec{x}(t)), \tag{7.8}$$

where $\eta(\vec{x}(t)) > 0$ is a normalization coefficient such that the overall belief on the domain X is equal to unit.

Notice that if the robot starts with the void map $\mathcal{M}$ and builds it during the motion (or if the domain does not include obstacles or any other information and provides a clean arena for the robot's motion so that the map $\mathcal{M}$ is always void), then the beliefs $\overline{bel}$ and bel involve the Markov property and instead of Eqs. (7.6) and (7.8) are used the following definitions:

$$\overline{bel}(\vec{x}(t)) = p(\vec{x}(t) \mid z(t-1), \delta(t-1)), \tag{7.9}$$

$$bel(\vec{x}(t)) = p(\vec{x}(t) \mid z(t), \delta(t-1)), \tag{7.10}$$

in which it is assumed that the beliefs $\overline{bel}$ and bel regarding the robot's location at time t do not depend on the complete observations' and movements' history and take into account only the last observation and motion of the robot.

Finally, given the beliefs $\overline{bel}(\vec{x}(t-1))$ and $bel(\vec{x}(t-1))$ at time $t-1$, the beliefs at time t are defined using general probabilistic approach and are specified as follows (Thrun, Burgard, and Fox 2005; Siegwart, Nourbakhsh, and Scaramuzza 2011):

- continuous domain:

$$\overline{bel}(\vec{x}(t)) = \int_X p(\vec{x}(t) \mid z(t-1), \delta(t-1))bel(\vec{x}(t-1))d\vec{x}, \tag{7.11}$$

- discrete domain $X = \{\vec{x}_1, \vec{x}_2, \ldots, \vec{x}_n\}$:

$$\overline{bel}(\vec{x}_i(t)) = \sum_{j=1}^{n} p(\vec{x}_j(t) \mid z(t-1), \delta(t-1))bel(\vec{x}_j(t-1)). \tag{7.12}$$

The presented definitions allow direct formulation of the Markov localization algorithm. This procedure is based on the Bayesian filter, which updates of the beliefs regarding the robot's location with respect to such beliefs $bel(\vec{x}(t-1))$, $\vec{x}(t-1) \in X$, at the previous time, the previous observation result $z(t-1)$, and the movement $\delta(t-1)$. The Bayesian filter is defined by the following procedure (Thrun, Burgard, and Fox 2005; Choset et al. 2005; (Siegwart, Nourbakhsh, and Scaramuzza 2011):

bayesian_filter $(z(t), z(t-1), \delta(t-1), bel(\vec{x}(t-1)))$: $bel(\vec{x}(t))$

1) For all points $\vec{x}(t) \in X$ do:
2) Set belief $\overline{bel}(\vec{x}(t))$:
   continuous X: $\overline{bel}(\vec{x}(t)) = \int_X p(\vec{x}(t) \mid z(t-1), \delta(t-1))bel(\vec{x}(t-1))d\vec{x}$,
   discrete X: $\overline{bel}(\vec{x}_i(t)) = \sum_{j=1}^{n} p(\vec{x}_j(t) \mid z(t-1), \delta(t-1))bel(\vec{x}_j(t-1))$.
3) End for.
4) For all points $\vec{x}(t) \in X$ do:
5) Calculate normalization coefficient $\eta(\vec{x}(t))$ by the Bayes rule.

6) End for.
7) For all points $\vec{x}(t) \in X$ do:
8) Set belief $bel(\vec{x}(t)) = \eta(\vec{x}(t))p(z(t) \mid \vec{x}(t), \mathcal{M})\overline{bel}(\vec{x}(t))$.
9) End for.
10) Return $bel(\vec{x}(t))$, $\vec{x}(t) \in X$.

It is clear that in line 2 this procedure applies the update rules defined by the Eqs. (7.11) or (7.12), and in line 8 – the definition given by Eq. (7.8). Using the Bayesian filter the Markov localization algorithm is outlined as follows.

Algorithm 7.1 (Markov localization). Given the domain X with the map $\mathcal{M}$ and termination time T, do:

1) Initialize location $\vec{x}(0) \in X$ and observation result $z(0)$.
2) Initialize beliefs $bel(\vec{x}(0))$.
3) Choose motion $\delta(0)$.
4) Set $t = 1$.
5) While $t \leq T$ do:
6) Move to the new location $\vec{x}(t)$ with respect to the chosen motion $\delta(t-1)$.
7) Obtain observation result $z(t)$.
8) Set $bel(\vec{x}(t)) = bayesian\_filter(z(t), z(t-1), \delta(t-1), bel(\vec{x}(t-1)))$, $\vec{x}(t-1) \in X$.
9) Choose motion $\delta(t)$.
10) Set $t = t + 1$.
11) End while.

The Markov localization algorithm is the most general procedure, which implements only the basic assumptions regarding probabilities. The next example illustrates the actions of this algorithm.

Example 7.1 (based on the example by Siegwart, Nourbakhsh, and Scaramuzza (2011)) Assume that the robot moves linearly in two-dimensional domain $X \subset \mathbb{R}^2$, which includes certain equivalent landmarks. The landmarks are represented on the map $\mathcal{M}$ of the environment and this map is available to the robot. The robot is equipped with the sensor such that it can observe areas $a \subset X$ and perceive an existence of a landmark in the local environment of the robot. However, since the landmarks are equivalent, the robot is not able to recognize near which landmark it is. For simplicity, assume that the landmarks are located linearly and the robot moves in parallel to the landmarks' locations. The movement of the robot is shown in Figure 7.2, where it is assumed that the robot starts with the uniform belief $bel(\vec{x}(0))$. Then, at the location $\vec{x}(1)$ next to the first landmark it calculates the belief $bel(\vec{x}(1))$ and moves to the location $\vec{x}(2)$ next to the second landmark. At this point it calculates the belief $bel(\vec{x}(2))$ and finally moves to the location $\vec{x}(3)$ next the third landmark, in which the belief is $bel(\vec{x}(3))$.

Starting from the uniform $bel(\vec{x}(0))$, the robot moves up to sensing one of the landmarks. At time $t = 1$ the robot senses one of three equivalent landmarks; hence its belief

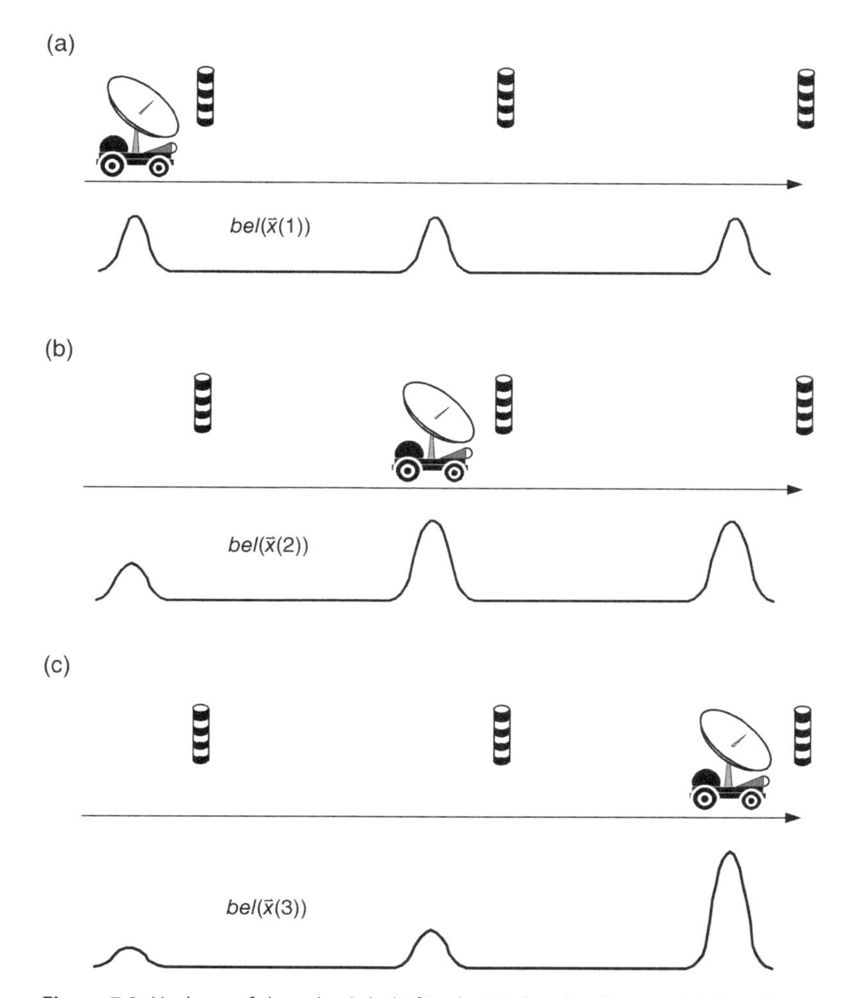

Figure 7.2 Updates of the robot's belief in the Markov localization: (a) the robot senses the landmark, but cannot recognize, next to which one of three possible landmarks it is located; (b) after moving one step right, the robot sense the landmark, which, because of the robot's step, is certainly not the previous one, but cannot recognize whether it is the second or the third one; (c) after the next step to the right the robot again senses the landmark and it is certainly not one the previously observed landmarks; thus, at this step the location of the robot is next to the third landmark.

$bel(\vec{x}(1))$ includes three equivalent peaks with respect to the landmarks on the map $\mathcal{M}$. At time $t = 2$, after moving one step right, the robot is in the location $\vec{x}(2)$, which differs from the previous location $\vec{x}(1)$. At this location, the robot senses the landmark, which is obviously differs from previously observed landmark. Thus, it is one of two remaining landmarks on the map $\mathcal{M}$; hence the robot's belief $bel(\vec{x}(2))$ includes two equivalent peaks corresponding to these remaining landmarks, while the probability that the robot observes the previous landmark decreases. Finally, the robot moves next one step right and at time $t = 3$ again observes the landmark, which, according to the map $\mathcal{M}$, is the last one. Thus, after updating, the belief $bel(\vec{x}\,(3))$ includes only one high peak corresponding to the last landmark, next to which the robot is located. Small but nonzero probabilities of observing the other landmarks represent the uncertainties in sensing and motion of the robot. ■

The presented Markov localization algorithm is the basic and the simplest one from a wide family of localization and mapping methods. It directly implements the Bayesian rule for updating belief space with respect to the given map of the environment and observation results obtained during the motion. In addition, notice that this algorithm does not considers decision-making process regarding the next motion step of the robot, which is assumed to be predefined separately before starting the algorithm.

7.1.2 Motion Prediction and Kalman Localization

The Markov localization algorithm utilizes information obtained by the robot during its motion and updates its beliefs following general probabilistic methods. Additional information, which can be used in the localization tasks, is provided by the robot's motion and, consequently, by possible prediction of the robot's locations and of the result of observations. Under the assumptions that the robot's motion is defined by some linear dynamical system and that the uncertainties are led by the Gaussian noise, such prediction is conducted using the methods of linear prediction known as Kalman filtering (Kalman 1960). A methodology of application of the Kalman filter to the robot's localization problem is presented in the paper by Roumeliotis and Bekey (2000) and in the book by Thrun, Burgard, and Fox (2005) (see also the books by Choset et al. (2005) and by Siegwart, Nourbakhsh, and Scaramuzza (2011)); a simple explanation of the principles of the Kalman filtering is presented by Faragher (2012).

In general terms of mobile robot localization problem, the Kalman filter is defined as follows. Assume that the robot's motion and the observations are governed by two linear equations:

$$\vec{x}\,(t) = \mathrm{L}(t)\,\vec{x}\,(t-1) + \mathrm{M}(t)\delta(t-1) + \mathcal{G}_t^x, \tag{7.13}$$

$$z(t) = \mathrm{O}(t)\,\vec{x}\,(t) + \mathcal{G}_t^z, \tag{7.14}$$

where $\mathrm{L}(t)$ is a state transition matrix, $\mathrm{M}(t)$ is a motion control matrix, $\mathrm{O}(t)$ is an observation control matrix, and $\mathcal{G}_t^x$ and $\mathcal{G}_t^z$ are the Gaussian noise terms. Matrix $\mathrm{L}(t)$ maps the previous state to the current state taking into account the parameters of the state that in the considered case are coordinates of the robot. Matrix $\mathrm{M}(t)$ maps the chosen action into the coordinates of the robot and matrix $\mathrm{O}(t)$ maps the robot's location into the observation result. In other words, matrix $\mathrm{L}(t)$ represents the behavior of the robot without sensors, matrix $\mathrm{M}(t)$ defines the influence of the control parameters including the perceived information, and matrix $\mathrm{O}(t)$ specifies the observation results given the robot's location. The noise terms $\mathcal{G}_t^x$ and $\mathcal{G}_t^z$ represent the uncertainties in the robot's locations and sensing; regarding these terms, it is assumed that they are defined by the processes with zero means and covariance matrices Cov_t^x and Cov_t^z, respectively.

Given linear equations (7.13) and (7.14), which govern the robot's behavior, the predicted location of the robot is defined as follows. Denote by $\tilde{x}(t \mid z_{0:t-1})$ the estimated location of the robot at time t given the history $z_{0:t-1}$ of observations up to the time $t-1$ and by $\sigma(t \mid z_{0:t-1})$ the error covariance associated with the estimated location $\tilde{x}(t \mid z_{0:t-1})$. Then, with respect to Eqs. (7.13) and (7.14), the Kalman filter equations are the following (Choset et al. 2005; Siegwart, Nourbakhsh, and Scaramuzza 2011):

1) Prediction of the next robot's location and corresponding error covariance given current location estimate and observations $z_{0:t}$:

$$\tilde{x}(t+1 \mid z_{0:t}) = \mathrm{L}(t)\tilde{x}(t \mid z_{0:t}) + \mathrm{M}(t)\delta(t), \tag{7.15}$$

$$\sigma(t+1 \mid z_{0:t}) = \mathrm{L}(t)\sigma(t \mid z_{0:t})\mathrm{L}(t)^T + \mathrm{Cov}_t^x, \tag{7.16}$$

where A^T stands for transposition of the matrix A.

It is clear that the first equation is a direct application of the Eq. (7.13) of the robot's motion to the estimated robot's coordinates, and the second equation defines straightforward calculation of the error covariance using the is a state transition matrix L(t) and taking into account the uncertainty in specification of the robot's location.

After the movement, the robot is in the location that is associated with the predicted location. At this time $t = t+1$ the robot observes the environment and obtains the observation result $z(t)$, which allows an update of the predicted location.

2) Update of the robot's location and corresponding error covariance after the observation $z(t)$:

$$\tilde{x}(t \mid z_{0:t}) = \tilde{x}(t \mid z_{0:t-1}) + \mathrm{R}(t \mid z_{0:t-1})\mathrm{V}(t \mid z_{0:t}), \tag{7.17}$$

$$\sigma(t \mid z_{0:t}) = \sigma(t \mid z_{0:t-1}) - \mathrm{R}(t \mid z_{0:t-1})\mathrm{O}(t)\sigma(t \mid z_{0:t-1}), \tag{7.18}$$

where

$$\mathrm{R}(t \mid z_{0:t-1}) = \sigma(t \mid z_{0:t-1})\mathrm{O}(t)^T\left[\mathrm{O}(t)\sigma(t \mid z_{0:t-1})\mathrm{O}(t)^T + \mathrm{Cov}_t^z\right]^{-1}$$

$$\mathrm{V}(t \mid z_{0:t}) = z(t) - \mathrm{O}(t)\tilde{x}(t \mid z_{0:t-1})$$

In these equations, the updating of the robot's location and error covariance is conducted by correcting the predicted values with the current observation results and by additional uncertainties introduced by observations. In the equations, the value of R can be considered as a weighting factor, which represents the difference between the predicted locations and the observation noise. The greater values of R indicate that the observations are more accurate that the predictions, and the smaller values of R show that the observations are less accurate. Consequently, the Kalman filter applies the greater values of R and ignores the smaller values that results in optimal estimation of the robot's coordinates such that the expected difference between the actual location $\vec{x}\,(t)$ of the robot and the estimated one $\tilde{x}(t \mid z_{0:t})$ is minimized at every time t.

Following the Kalman filter prediction and update equations, the Kalman localization algorithm is outlined as follows.

Algorithm 7.2 (Kalman localization). Given the domain X with the map $\mathcal{M}$ and termination time T, do:

1) Initialize estimated location $\tilde{x}(0 \mid z_{0:0}) \in X$ and observation result $z(0)$.
2) Initialize beliefs $bel(\tilde{x}(0 \mid z_{0:0}))$.
3) Choose motion $\delta(0)$.
4) While $t \leq T$ do:
5) Predict the next location by Eq. (7.15): $\tilde{x}(t+1 \mid z_{0:t}) = \mathrm{L}(t)\tilde{x}(t \mid z_{0:t}) + \mathrm{M}(t)\delta(t)$.
6) Set $t = t + 1$.
7) Move to the new location $\vec{x}\,(t)$ with respect to the chosen motion $\delta(t-1)$.
8) Obtain observation result $z(t)$.
9) Update the predicted location by Eq. (7.17):
$$\tilde{x}(t \mid z_{0:t}) = \tilde{x}(t \mid z_{0:t-1}) + \mathrm{R}(t \mid z_{0:t-1})\mathrm{V}(t \mid z_{0:t}).$$
10) Set $bel(\tilde{x}(t \mid z_{0:t})) = bayesian\_filter(z(t), z(t-1), \delta(t-1), bel(\tilde{x}(t \mid z_{0:t-1})))$.
11) Choose motion $\delta(t)$.
12) End while.

In the Kalman localization, the robot at first predicts its next location and then, after the motion, corrects this prediction with respect to the obtained observation result. Consequently, the robot's belief regarding its location is calculated using the estimated location.

Notice that the Kalman localization can also apply the predicted observation result $\tilde{z}(t+1)$ given predicted $\tilde{x}(t+1 \mid z_{0:t})$ location of the robot and the environmental map $\mathcal{M}$ (Siegwart, Nourbakhsh, and Scaramuzza 2011). Denote by h the function, which executes such prediction. Then the value $\tilde{z}(t)$ is defined as:

$$\tilde{z}(t+1) = h(\tilde{x}(t+1 \mid z_{0:t}), \mathcal{M}). \quad (7.19)$$

Formally, the function h is a combination of the Eqs. (7.13) and (7.14) and uncertainties in its resulting value are governed by the probabilities $p(\tilde{z}(t) \mid \tilde{x}(t+1 \mid z_{0:t}), \mathcal{M})$ applied to the predicted locations of the robot.

The actions of the Kalman localization algorithm are illustrated by the following example.

Example 7.2 (based on the example by Siegwart, Nourbakhsh, and Scaramuzza (2011)). Similar to Example 7.1, assume that the robot moves linearly in two-dimensional domain $X \subset \mathbb{R}^2$ with three equivalent landmarks, which appear on the available map $\mathcal{M}$, and that the robot is able to recognize a landmark but is not able to recognize which one of three landmarks it is. The movement of the robot is shown in Figure 7.3, where the stating estimated belief $bel(\tilde{x}(0 \mid z_{0:0}))$ is assumed to be uniform (cf. Example 7.1). Then, since the landmarks are equivalent and not distinguishable, after arriving to the first landmark the estimated robot's belief $bel(\tilde{x}(1 \mid z_{0:1}))$ represents these landmarks by equivalent peaks. The robot's location and its estimated belief are shown in Figure 7.3a. Now, staying in its current location, the robot applies the Kalman filter and predicts its next location given that it chooses to move right. The estimated belief $bel(\tilde{x}(2 \mid z_{0:1}))$ given the predicted location $\tilde{x}(2)$ is shown in Figure 7.3b. Notice that

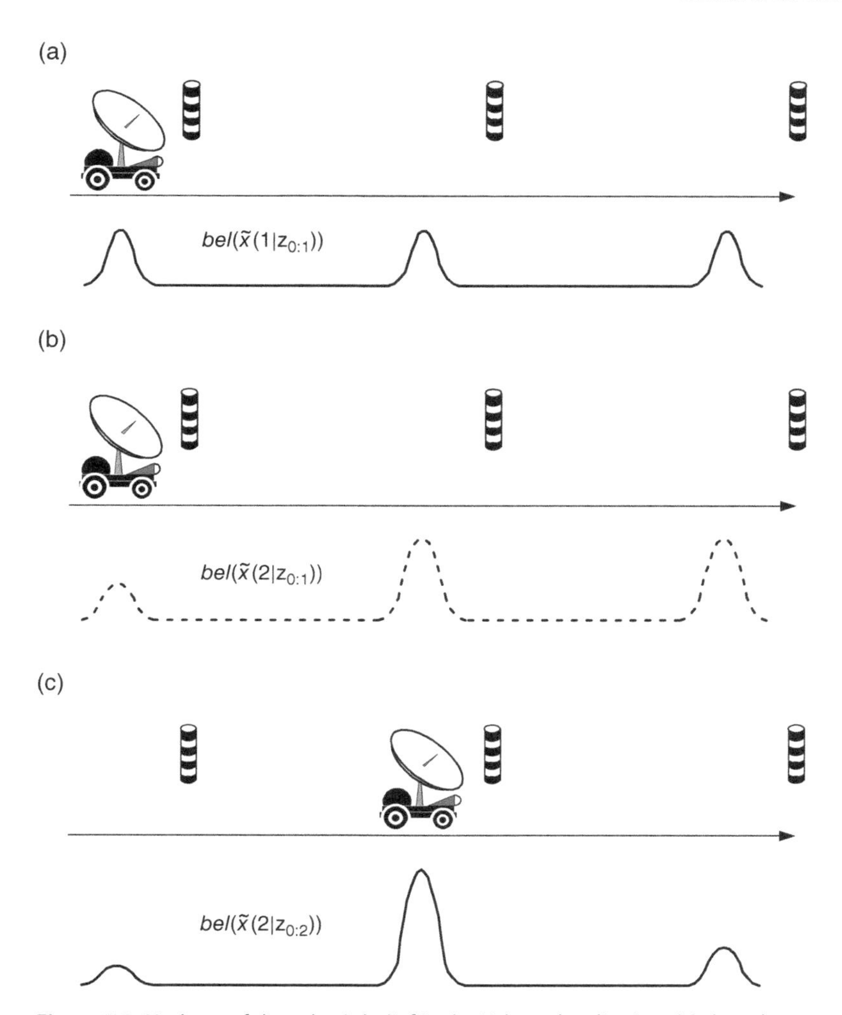

Figure 7.3 Updates of the robot's belief in the Kalman localization: (a) the robot senses the landmark, but cannot recognize, next to which one of three possible landmarks it is located; (b) after prediction of the next location with respect to the chosen movement (one step right), the robot estimates the belief, which does not includes the current landmark, but includes the second and the third landmarks equivalently; (c) after the actual motion with respect to the chosen movement, the robot senses the landmark and updates the estimated belief with respect to the observation result.

with respect to noise the estimated belief $bel(\tilde{x}(1\,|\,z_{0:1}))$ and the predicted belief $bel(\tilde{x}(2\,|\,z_{0:1}))$ are, correspondingly, equivalent to the beliefs $bel(\vec{x}\,(1))$ and $bel(\vec{x}\,(2))$ calculated by the Markov localization algorithms. Finally, the robot moves forward according to the chosen motion, updates its measurement and calculates belief $bel(\tilde{x}(2\,|\,z_{0:2}))$. This belief is shown in Figure 7.3c. The obtained belief includes two lower peaks, which represent the first and the last landmarks, and one higher peak, which corresponds to the landmark, next to which the robot is located.

It is seen that in contrast to the Markov localization considered in Section 7.1.1 (see Algorithm 7.1), the Kalman localization Algorithm 7.2 because of the prediction stage provides more quick positioning of the robot with respect to the environment. However, notice that such prediction and further update of the robot's location and belief can be conducted under the assumption that the uncertainties in the robot's motion and sensing are governed by Gaussian noise. Additional assumptions address the rules, which govern the robot's motion and observations, in the considered case these are defined by the linear equations (7.13) and (7.14), and the existence and availability of the environmental map $\mathcal{M}$.

It is clear that in spite of the limitations resulted by the assumptions regarding the noise and the linearity of the robot's dynamics in many situations they are reasonable enough and do not change the statement of the problem. In contrast, the assumption regarding an existence of the environmental map is crucial and follows from the definition of the robot's mission. The next section considers the basic methods of mapping and of localization with simultaneous map building. The methods of Kalman localization with nonlinear equations of the robot's motion can be found in the books by Choset et al. (2005) and by Siegwart, Nourbakhsh, and Scaramuzza (2011). Detailed consideration of the Kalman localization and probabilistic methods in robotics is presented in the book by Thrun, Burgard, and Fox (2005). ■

7.2 Mapping the Unknown Environment and Decision-Making

The considered algorithms of the robot's localization imply that the robot is able to recognize or estimate its position with respect to the landmarks that are represented in the environmental map, while the map defines the relation of the landmarks with some global coordinates system. In such a setup, the robot updates its belief either according to the actual observation (Algorithm 7.1) or by comparison of the predicted and observed locations of the landmarks (Algorithm 7.2). Certainly, it is assumed that the map has been built beforehand and is available to the robot during the mission. An essentially different problem arises when the map is not available and the robot is required to build the environmental map during its motion in the environment.

Formally, there are several types of the maps, each of which stress different aspect of the environment with respect to the robot's abilities. The main maps are the following (Dudek and Jenkin 2010):

- *Metric maps* define the absolute coordinates of the objects (robots and obstacles) in global coordinated space or estimates of such coordinates with respect to global coordinates system.
- *Topological maps* represent the connectivity between the objects in the terms of possible paths without specification of actual or estimated distances; often such map are also called relational maps.

Constricting maps is based on the data perceived by the robot during its motion with further recognition of the meaningful features and geometric information. Usually such a process is conducted in five stages (Dudek and Jenkin 2010); however, often these states do not distinguished explicitly.

1) Perception of *sensorial raw data* that includes certain images of the environment and their transformation with respect to the robot's motion.

2) Extraction of *geometric information* from the perceived data and constructing or recognizing two- or three-dimensional objects.
3) Specification of *local relations* between the constructed or recognized objects, especially their relative positions with respect to each other.
4) Implication of global *topological relations* between the objects and their groups all over the considered environment.
5) Designation of the objects by *semantic labels* understandable by the robot and, it need, by human operator.

Certainly, the last three stages require additional information and assumptions regarding the possible environment, while the first two stages can be treated using general approaches and depend only on the abilities of the robot and available sensors (Everett 1995; Borenstein, Everett, and Feng 1996; Castellanos and Tardos 1999). The next section considers basic mapping methods, which act either independently of the robot's localization or together with its localization. The last methods are known as simultaneous localization and mapping (SLAM).

7.2.1 Mapping and Localization

The indicated stages of constructing the environmental map stress that mapping is a process of checking and learning of the occupancy of the space in which the robot moves, with further representation of the learned occupancy in the form suitable for the considered task (Dudek and Jenkin 2010). In particular, the resulting map may be used for external purposes, for example, in the tasks of exploration search, in which the objective is to define positions of the target objects, or for internal needs of the robot's motion, such as in the problems of localization of the robot in the unknown or partially unknown environment. In the first case, the mapping is considered as an independent process and the problem is known as mapping without localization (Elfes 1987, 1990; Konolige 1997), and in the second case, it is treated together with the localization and is called SLAM (Leonard and Durrant-Whyte 1991; Thrun and Leonard 2008); see also the SLAM tutorial (Durrant-Whyte and Bailey 2006; Bailey and Durrant-Whyte 2006). For a detailed overview of the mapping methods including historical remarks, see the paper by Thurn (2002).

Let us start with the methods of mapping that do not imply simultaneous localization and positioning of the robot and provide the map of the space occupancy. The key ideas of such mapping were suggested by Elfes (1987, 1990) and are known as occupancy grid representation (Dudek and Jenkin 2010). In this representation the robot specifies the probabilities of occupancy of the environment with respect to its own position and the abilities of applied sensor.

Assume that the robot is equipped with the sonar (ultrasonic) distance sensor, which measures the distance between the robot and the obstacle; for simplicity, we do not distinguish between the sensor's and the robot's location. Denote by r_{sense} a result of distance measurement obtained by the sensor; it is assumed that the value r_{sense} is bounded by certain minimal $r_{\text{sence}}^{\min}$ and maximal $r_{\text{sence}}^{\max}$ distances, that is $r_{\text{sense}} \in \left[r_{\text{sence}}^{\min}, r_{\text{sence}}^{\max}\right]$. An inaccuracy of the measurement is specified by the measurement error and its maximal value is denoted by ε, $0 \leq \varepsilon \ll 1$. Finally, denote by ω the solid angle subtending the sensitivity lobe of the sonar.

Let $X \subset \mathbb{R}^2$ be a two-dimensional grid $X = \left\{\vec{x}_1, \vec{x}_2, \ldots, \vec{x}_n\right\}$ with the points $\vec{x}_i = (i_x, i_y)$, $i_x = 1, 2, \ldots, n_x$, $i_y = 1, 2, \ldots, n_y$, $i = (i_x - 1)n_x + i_y$ and $n = n_x \times n_y$. Assume that the robot is in the point $\vec{x} \in X$, and let $\vec{y} \in X$ be some other point in the grid. The distance between $\vec{x}$ and $\vec{y}$ is denoted by $d(\vec{x}, \vec{y}) = \| \vec{x} - \vec{y} \|$ and the angle between the main axis of the beam of the sonar located in $\vec{x}$ and the point $\vec{y}$ is denoted by $\theta(\vec{x}, \vec{y})$.

Following Elfes (1987), the probability $p_{\text{empty}}(\vec{x}, \vec{y})$ that the area near the point $\vec{y}$ observed by the sonar sensor of the robot located in the point $\vec{x}$ is empty is defined as follows:

$$p_{\text{empty}}(\vec{x}, \vec{y}) = Pr\left\{\text{point } \vec{y} \text{ observed from the point } \vec{x} \text{ is empty}\right\}$$

$$= \left(1 - \left(\frac{2\theta(\vec{x}, \vec{y})}{\omega}\right)^2\right)\begin{cases} 1 - \left(\dfrac{d(\vec{x}, \vec{y}) - \mathrm{r}_{\text{sence}}^{\min}}{\mathrm{r}_{\text{sense}} - \varepsilon - \mathrm{r}_{\text{sence}}^{\min}}\right)^2 & \text{if } \mathrm{r}_{\text{sence}}^{\min} \le d(\vec{x}, \vec{y}) \le \mathrm{r}_{\text{sense}} - \varepsilon, \\ 0 & \text{otherwise,} \end{cases} \tag{7.20}$$

where the angle $\theta(\vec{x}, \vec{y})$ is bounded as $-\omega/2 \le \theta(\vec{x}, \vec{y}) \le \omega/2$.

Similarly, the probability $p_{\text{occupied}}(\vec{x}, \vec{y})$ that the area near point $\vec{y}$ is occupied by some obstacle is

$$p_{\text{occupied}}(\vec{x}, \vec{y}) = Pr\left\{\text{point } \vec{y} \text{ observed from the point } \vec{x} \text{ is occupied}\right\}$$

$$= \left(1 - \left(\frac{2\theta(\vec{x}, \vec{y})}{\omega}\right)^2\right)\begin{cases} 1 - \left(\dfrac{d(\vec{x}, \vec{y}) - \delta_{\text{sense}}}{\varepsilon}\right)^2 & \text{if } \mathrm{r}_{\text{sense}} - \varepsilon \le d(\vec{x}, \vec{y}) \le \mathrm{r}_{\text{sense}} + \varepsilon, \\ 0 & \text{otherwise,} \end{cases} \tag{7.21}$$

where, as above, $-\omega/2 \le \theta(\vec{x}, \vec{y}) \le \omega/2$.

During the mapping process these probabilities are conjoined with the points $\vec{x} \in X$ and provide initial information for building of the occupancy map of the grid X. In the Elfes implementation (1987), the empty and occupied points are signed by the values from the ranges $[-1, 0)$ and $(0, 1]$, respectively; the points over which the information was not obtained are signed with 0. Formally, the mapping process is outlined as follows:

Algorithm 7.3 (mapping without localization). Given the grid X, appropriate sonar sensors and termination time T, do:

1) Sign all points $\vec{x} \in X$ as the points with unknown occupancy and with zero probabilities $p_{\text{empty}}(\vec{x})$ and $p_{\text{occupied}}(\vec{x})$.
2) Set $t = 0$.
3) Start with initial position $\vec{x}(t)$.
4) While $t \le T$, do:
5) For all points $\vec{y} \in X$ do

6) Observe the environment and obtain the probabilities $p_{\text{empty}}(\vec{x}(t), \vec{y})$ and $p_{\text{occupied}}(\vec{x}(t), \vec{y})$ according the formulas (7.20) and (7.21).
7) Set $p_{\text{empty}}(\vec{y}) = p_{\text{empty}}(\vec{y}) + p_{\text{empty}}(\vec{x}(t), \vec{y}) - p_{\text{empty}}(\vec{y}) \times p_{\text{empty}}(\vec{x}(t), \vec{y})$.
8) Set
$p_{\text{occupied}}(\vec{y}) = p_{\text{occupied}}(\vec{y}) + p_{\text{occupied}}(\vec{x}(t), \vec{y}) - p_{\text{occupied}}(\vec{y}) \times p_{\text{occupied}}(\vec{x}(t), \vec{y})$.
9) Set occupation value of the point $\vec{y}$ with respect to the probabilities $p_{\text{empty}}(\vec{y})$ and $p_{\text{occupied}}(\vec{y})$.
10) End for.
11) Set $t = t + 1$.
12) Choose the next position $\vec{x}(t)$ and move to it.
13) End while.

Notice again that in the original Elfes implementation, the occupation value (see line 9 of the algorithm) for empty points is in the range $[-1, 0)$, for occupied points it is in the range $(0, 1]$, and for the points with unknown occupancy the occupation value is 0. General scheme of mapping based on the occupancy grid and consequent building of geometric and symbolic maps (cf. the listed above five stages of mapping) is shown in Figure 7.4 (based on the architecture of sonar mapping and navigation presented by Elfes (1987).

In the figure, the sensorial row data is obtained by the *sensor control* module that is an interface between the sensor and the robot's controller. *Scanner* conducts initial preprocessing and filtering of the obtained sensorial data and defined the correspondence between the sensor data and the robot's current *position and orientation*, which is required by the mapping Algorithm (7.3), which is executed by the *mapper*. As a result, the *sensor map* (occupancy grid), which represents the information about occupancy of the grid points, is created. The sensor map is sent to the high level analysis, which starts from the *object's extraction* module, which recognizes the objects and their geometric relations and creates the *geometric map*. This module can apply different recognition algorithms that depend on the assumptions regarding the robot's environment, the executed task, and on the available information obtained from the other sensors. The *graph-building* module labels the recognized objects and generates the symbolic description of the environment that is represented by the *symbolic map*.

The created *sensor map*, *geometric map*, and *symbolic map* are used by the *path-planner*, which creates the robot's *path* using high-level symbolic information for specification of the goal positions and general restrictions, intermediate-level geometric relations for definition of more accurate directions of further steps, and, finally, the low-level sensorial data for creating the detailed paths using certain usually A^*-based path-planning algorithms over graphs. For basic methods and techniques of pattern recognition see, e.g., Tou and Gonzales (1974), and for information regarding A^*-based path-planning algorithms see Pearl (1984) and Kagan and Ben-Gal (2013) and references therein. The created path is passed to the *navigator* that translates the path into actual locomotion commands and is then passed to the *conductor* – that is, an interface between the algorithmic issues and actual motion control. In addition, the conductor receives the

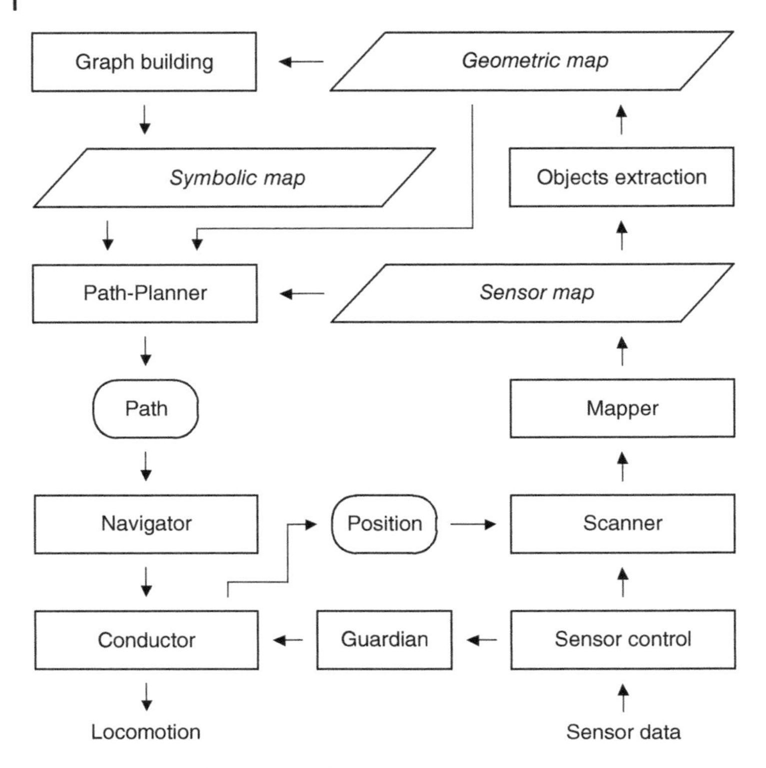

Figure 7.4 The Elfes scheme of mapping and navigation using occupancy grid. In the scheme, the sensorial data is directly used for creating the sensor map (see Algorithm 7.3). This map follows the modules of objects recognition and builds the geometric map. The further labeling of the objects and their topological relations results in the symbolic map. Finally, the obtained three maps are used for path planning and navigation of the robot.

information from the *guardian*, which checks the sensor readings and guaranties collision avoidance for the obstacles, which were not detected at previous stages.

The presented Algorithm 7.3 and the Elfes scheme shown in Figure 7.4 provide a general framework of the robot's navigation with the goal of exploring the unknown environment and providing the environmental maps. At the high-level analysis and during the path-planning and navigation, the robot can implement different algorithms of objects' recognition and navigation based on learning methods that result in smarter behavior and decreases the neediness of supervisor. However, notice that in spite of the generality of the navigation scheme, the mapping algorithm is essentially based on the global positioning of the robot in the environment and the knowledge of the robot's location and orientation in the grid. If such information is unavailable, the robot should localize itself using the maps, which it creates during the motion.

The methods that solve this problem are known as algorithms of SLAM or as dynamic map building (Leonard, Durrant-Whyte, and Cox 1990); next, we consider the basic principles and algorithms of such localization and mapping. The detailed description of SLAM methods have also been investigated (Durrant-Whyte and Bailey 2006; Bailey and Durrant-Whyte 2006; Thrun 2002; Thrun and Leonard 2008; Choset et al. 2005; Thrun, Burgard, and Fox 2005; Dudek and Jenkin 2010; Siegwart, Nourbakhsh, and Scaramuzza 2011). An extension of these methods for the teams of mobile robots is presented in the paper by Thrun (2001) and in the book chapter by Fox et al. (2002) and for the mapping of the dynamic environment – in the paper by Hähnel et al. (2003).

As above, assume that the robot moves over a discrete two-dimensional gridded domain $X = \left\{\vec{x}_1, \vec{x}_2, \ldots, \vec{x}_n\right\}$ with the points $\vec{x} = (x, y)$. The map of the domain X that is a set of the landmarks' coordinates $\vec{m}$ is denoted by $\mathcal{M}$. The robot's position at time $t = 0, 1, 2, \ldots$ is denoted by $\vec{x}(t)$, the observation result by $z(t)$, and the chosen motion or step by $\delta(t)$. Accordingly, the robot's path up to the time t is defined by the sequence $\underset{0:t}{\vec{x}} = \left\langle \vec{x}(0), \vec{x}(1), \ldots, \vec{x}(t) \right\rangle$, and histories of the observation results and of the movements are represented by the sequences $z_{0:t} = \langle z(0), z(1), \ldots, z(t)\rangle$ and $\delta_{0:t} = \langle \delta(0), \delta(1), \ldots, \delta(t)\rangle$, respectively.

In general, the problem that is addressed by the SLAM algorithms is a construction of the environmental map $\mathcal{M}$ and definition of the robot's path given the observation results and the movements of the robot. Certainly, in the probabilistic setup that is led by the uncertainties in the robot's motion and observations, the construction of the exact map is impossible and the goal of SLAM methods is to estimate the map $\mathcal{M}$ with minimal errors. In such a setup, usually there are distinguished two types of the SLAM problem (Thrun, Burgard, and Fox 2005; Siegwart, Nourbakhsh, and Scaramuzza 2011):

1) *Online SLAM problem* deals with estimating the next robot's step $\vec{x}(t)$, together with recovering the map $\mathcal{M}$ given the history of observation results $z_{0:t}$ and of the robot's movements $\delta_{0:t}$ up to the time t; that results in the posterior joint probability

$$p(\vec{x}(t), \mathcal{M} \mid z_{0:t}, \delta_{0:t})$$

that the robot will occupy the point $\vec{x}(t)$ and the estimated map is $\mathcal{M}$.

2) The *full SLAM problem* deals with definition of the robot's path $\underset{0:T}{\vec{x}}$ up to certain termination time T together with recovering the map $\mathcal{M}$, given the history of observation results $z_{0:T}$ and of the robot's movements $\delta_{0:T}$ up to the time T; that results in the posterior joint probability

$$p(\underset{0:T}{\vec{x}}, \mathcal{M} \mid z_{0:T}, \delta_{0:T})$$

that the robot's path is $\underset{0:T}{\vec{x}}$ and the recovered map is $\mathcal{M}$.

Most of the algorithms for solving both online and full SLAM problems follow the same approach as the localization methods already discussed and implement certain mobility and measurement models and prediction techniques. Below, we briefly consider the method of SLAM that is based on the Kalman filter techniques and directly extends localization techniques considered in Section 7.1.2; usually, such method is called the extended Kalman filter (EKF) SLAM (Choset et al. 2005; Thrun, Burgard, and Fox 2005; Siegwart, Nourbakhsh, and Scaramuzza 2011).

Similar to the Kalman filter localization, the EKF-SLAM implements two models, which represent the robot's abilities (Durrant-Whyte and Bailey 2006; Bailey and Durrant-Whyte 2006).

- The motion model

$$\vec{x}(t) = f(\vec{x}(t-1), \delta(t)) + \mathcal{G}_t^x, \tag{7.22}$$

where $\vec{x}(t)$ is the location of the robot at time t, which is specified with respect to the movement $\delta(t)$ at this time and previous location $\vec{x}(t-1)$ of the robot. Function f represents the robot's motion abilities and Gaussian noise $\mathcal{G}_t^x$ represents the errors of the motion.
- The observation model (cf. Eq. (7.19))

$$z(t) = h(\vec{x}(t), \mathcal{M}) + \mathcal{G}_t^z, \tag{7.23}$$

where $z(t)$ is the observation result at time t given that the robot's location at this time is $\vec{x}(t)$ and the actual map of the environment is $\mathcal{M}$. Function h represents the sensing abilities of the robot and Gaussian noise $\mathcal{G}_t^z$ represents the observation errors.

Following these models, the estimated robot's location $\tilde{x}(t+1 \mid z_{0:t})$ given the observations history $z_{0:t}$ and expected observation result $\tilde{z}(t)$ are defined as follows:

1) *Prediction* of the next robot's location (cf. Eqs. (7.15) and (7.16)):

$$\tilde{x}(t+1 \mid z_{0:t}) = \mathfrak{f}(\tilde{x}(t \mid z_{0:t}), \delta(t)), \tag{7.24}$$

$$\sigma(t+1 \mid z_{0:t}) = \nabla\mathfrak{f}\sigma(t \mid z_{0:t})\nabla\mathfrak{f}^T + \mathrm{Cov}_t^x, \tag{7.25}$$

where $\nabla\mathfrak{f}$ is the Jacobian of $\mathfrak{f}$ at the estimation $\tilde{x}(t \mid z_{0:t})$ and Cov_t^x, as above, is the covariance matrix of the process $\mathcal{G}_t^x$.
2) *Update* of the robot's expected location $\tilde{x}(t \mid z_{0:t})$ given the observations history $z_{0:t}$ and of the obtained observation result $\tilde{z}(t)$ given the previous observation result $z(t-1)$ and the map $\mathcal{M}_{t-1}$, which was built to the time $t-1$ (cf. Eqs. (7.17) and (7.18)):

$$\begin{bmatrix} \tilde{x}(t \mid z_{0:t}) \\ \tilde{z}(t) \end{bmatrix} = \begin{bmatrix} \tilde{x}(t \mid z_{0:t-1}) \\ z(t-1) \end{bmatrix} + \mathrm{R}(t \mid z_{0:t-1})\mathrm{V}(t \mid z_{0:t}), \tag{7.26}$$

$$\sigma(t \mid z_{0:t}) = \sigma(t \mid z_{0:t-1}) - \mathrm{R}(t \mid z_{0:t-1})\nabla\mathfrak{h}\sigma(t \mid z_{0:t-1}), \tag{7.27}$$

where $\nabla\mathfrak{h}$ is the Jacobian of $\mathfrak{h}$ at the estimation $\tilde{x}(t \mid z_{0:t})$ and

$$\mathrm{R}(t \mid z_{0:t-1}) = \sigma(t \mid z_{0:t-1})\nabla\mathfrak{h}^T\left[\nabla\mathfrak{h}\sigma(t \mid z_{0:t-1})\nabla\mathfrak{h}^T + \mathrm{Cov}_t^z\right]^{-1},$$

$$\mathrm{V}(t \mid z_{0:t}) = z(t) - \mathfrak{h}(\tilde{x}(t \mid z_{0:t-1}), \mathcal{M}_{t-1}),$$

where Cov_t^z is the covariance matrix of the process $\mathcal{G}_t^z$.

It is clear that the main difference between these equations and Eqs. (7.15)–(7.18), which specify Kalman localization without mapping, is the dependence of the expected observations on the estimated map created up to the current time. In contrast to general transition and control matrices used in the Kalman localization without mapping, functions $\mathfrak{f}$ and $\mathfrak{h}$ appearing in the Eqs. (7.24)–(7.27) depend on the robot's motion and perception abilities and should be specified with respect to the type of the robot and the considered task. The examples of such functions depending on the robots kinematics are presented in the indicated books (Choset et al. 2005; Thrun, Burgard, and Fox 2005; Siegwart, Nourbakhsh, and Scaramuzza 2011). In addition, in these sources and references therein (see also the book Dudek and Jenkin (2010) and the papers by Durrant-Whyte and Bailey (2006), (Bailey and Durrant-Whyte 2006) other methods are presented, such as Rao-Blackwell and particle filters that are widely used for solving the SLAM problems.

The considered methods of Kalman localization and mapping provide a general framework for navigation of mobile robots in unknown environments such that the uncertainties are represented by certain Gaussian noise and the mobility of the robot is defined by the known usually linear equations. However, notice that both dynamical equations (7.15) and (7.24) as well as the outlined localization Algorithms 7.1 and 7.2 include the motion $\delta(t)$ chosen by the robot at corresponding time t. The next chapter considers the decision-making processes that govern such choices.

7.2.2 Decision-Making under Uncertainties

Let us return to general activity of the mobile robot moving in the domain X with the map $\mathcal{M}$. As indicated in Section 7.1.1, in the terms of the robot's locations $\vec{x}(t) \in X$, observation results $z(t) \in \mathcal{M}$ and movements $\delta(t) \in \mathcal{D}$, where $\mathcal{D}$ is a set of possible direct movements or steps of the robot, Eqs. (7.1)–(7.3) define the activity of the robot as follows:

- Location $\vec{x}(t)$ of the robot at time t is defined with respect to its path $x_{0:t-1}$ up to the previous time $t-1$ and to the histories of observations $z_{0:t-1}$ and movements $\delta_{0:t-1}$ up to the previous time (see Eq. (7.1)).
- Observation result $z(t)$ at time t depends on the path $x_{0:t}$ of the robot up to the current time t and on the histories of observations $z_{0:t-1}$ and movements $\delta_{0:t-1}$ up to the previous time $t-1$ (see Eq. (7.2)).
- The movement $\delta(t)$ of the robot at time t is chosen with respect to the robot's path $x_{0:t}$ and the history of observations $z_{0:t}$ up to the current time t and to the history of movements $\delta_{0:t-1}$ up to the previous time $t-1$ (see Eq. (7.3)).

If the robot acts under uncertainties, then locations, observation results, and movements are unknown and are accordingly defined with the probabilities

$$p\left(\vec{x}(t)x_{0:t-1},z_{0:t-1},\delta_{0:t-1}\right),\ p\left(z(t)x_{0:t},z_{0:t-1},\delta_{0:t-1}\right),\ p\left(\delta(t)x_{0:t},z_{0:t},\delta_{0:t-1}\right),$$

which depend on the robot's path, previous observations and movements. It is clear that the first two probabilities are the same as the probabilities defined by the Eqs. (7.4) and (7.5) and the third equation is a probabilistic variant of the Eq. (7.3). As already indicated, such probabilities can be considered using general theory of statistical decisions (Wald 1950; DeGroot 1970) and optimization techniques (Bertsekas 1995).

Finally, similar to the consideration of the robot's beliefs in Section 7.1.1, assume that the indicated probabilities meet the Markov property and do not depend on the complete path of the robot and the complete histories of observations and chosen movements. Then these probabilities are specified as follows:

- Location transition probability

$$\rho_{ij}^{x}(\delta_l) = Pr\left\{\vec{x}(t+1) = \vec{x}_j \mid \vec{x}(t) = \vec{x}_i, \delta(t) = \delta_l\right\}, \tag{7.28}$$

  which is the probability that at the next time $t+1$ the robot's location will be $\vec{x}(t+1) = \vec{x}_j$ given that its current location is $\vec{x}(t) = \vec{x}_i$ and the chosen movement is $\delta(t) = \delta_l$; in addition, here, in accordance with convenient terminology used in decision-making theory, the location probability is defined for the next time $t+1$ (cf. Eqs. (7.1) and (7.4));
- Observation probability

$$p_k^{z}(\vec{x}_i) = Pr\left\{z(t) = z_k \mid \vec{x}(t) = \vec{x}_i\right\}, \tag{7.29}$$

  which is the probability that being at time t in the location $\vec{x}(t) = \vec{x}_i$ the robot obtains the observation result $z(t) = z_k$;
- Movement or action probability

$$p_l^{\delta}(\vec{x}_i, z_k) = Pr\left\{\delta(t) = \delta_l \mid \vec{x}(t) = \vec{x}_i, z(t) = z_k\right\}, \tag{7.30}$$

  which is the probability that being at time t in the location $\vec{x}(t) = \vec{x}_i$ and obtaining the observation result $z(t) = z_k$ the robot chooses the movement $\delta(t) = \delta_l$.

Notice that since the probability of the movement $\delta(t)$ is defined with respect to the observation result $z(t)$ (see Eq. (7.30)), in Eq. (7.28) the term $z(t)$ is omitted and the location transition probabilities are defined only with respect to the chosen movement $\delta(t)$.

It is clear that the observation probabilities describe the uncertainties introduced by the sensors and the location transition probabilities characterize the internal uncertainties of the robot's controller and activators and the dependence of the robot's motion on the environmental conditions. Usually, these probabilities are specified by certain well-defined functions, which do not vary with the robot's motion; an example of such functions is provided by the Kalman filter, which defines the expected robot's locations and observations. In contrast, the movement or action probabilities characterize the "intelligence" of the robot, and the rule that governs these probabilities defines the process of making decisions regarding the movement $\delta(t)$ with respect to the current location $\vec{x}(t)$ and observation result $z(t)$.

Additional uncertainty appears if the robot is not aware of its current location and is required to choose the movement $\delta(t)$ based only on the observation result $z(t)$. Then instead of Eq. (7.30), the probability of the movement is defined as

$$p_l^{\delta}(z_k) = Pr\{\delta(t) = \delta_l \mid z(t) = z_k\}, \tag{7.31}$$

which is the probability that at time t based on the observation result $z(t) = z_k$, the robot chooses the movement $\delta(t) = \delta_l$.

The Markov process defined by Eq. (7.28), which governs the robot's motion in the environment, together with the decision-making process regarding the one-step movements defined by Eq. (7.30), forms a MDP (Derman 1970; White 1993; Ross 2003) also known as controlled Markov process (Dynkin and Yushkevich 1979). If the choice of movements is conducted basing on the uncertain observation results, then the decision-making process defined by Eqs. (7.28), (7.29) and (7.31) is a partially observable Markov decision process (POMDP) (Aoki 1967; Monahan 1982; White 1993). Then,

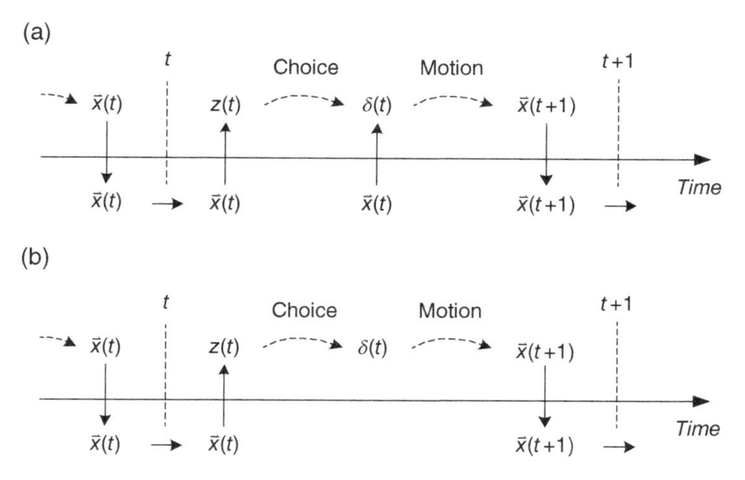

Figure 7.5 Decision-making processes with known (figure (a)) and unknown (figure (b)) locations: (a) decision regarding the next movement resolves inconsistency between observed and actual location; and (b) the movement is chosen based only on the observation result.

the Markov process defined by Eq. (7.28) is considered as a hidden Markov process and the model is called hidden Markov model (HMM) (Rabiner 1989; Ephraim and Merhav 2002). The decision-making processes with known and unknown locations are illustrated in Figure 7.5.

From the figure, it follows that the choices of the movement with known and unknown robot's location have a rather different nature. In the first case (Figure 7.5a), the decision-making process requires resolving possible inconsistency between the known actual location and the observed environment, while in the second case (Figure 7.5b), making decisions deals only with the observation results and the uncertainty in the choice of the movement is led by the uncertainty in environment sensing.

A brief introduction into MDP and additional information about optimization techniques is presented in Kagan and Ben-Gal (2013). In the context of artificial intelligent research these issues are considered in Kaelbling, Littman, and Moore (1996), Kaelbling, Littmann, and Cassandra (1998), and in the framework of mobile robots navigation (Kagan and Ben-Gal 2008). While considering Markov decision-making processes, we follow these sources and also Filar and Vrieze (1997).

Formally, the choices of the robots movements follow a general decision-making framework under uncertainty (Luce and Raiffa 1957; Raiffa 1968) and in the terms of MDP and POMDP the techniques of are defined as follows.

As above, let $X = \left\{\vec{x}_1, \vec{x}_2, \ldots, \vec{x}_n\right\}$ be a discrete domain associated with the finite set of states, $\mathcal{D} = \{\delta_1, \delta_2, \ldots, \delta_m\}$ be a finite set of possible one-step movements of the robot that is considered as a set of actions and $\mathcal{M} = \{z_1, z_2, \ldots, z_o\}$ be an estimated or actual map of the environment that is the finite set of possible observations of the landmarks in X. The probability $\rho_{ij}^{x}(\delta_l)$ of motion from location $\vec{x}_i \in X$ to location $\vec{x}_j \in X$ when the chosen movement is $\delta_l \in \mathcal{D}$ is defined by Eq. (7.28) and the probability $p_k^z(\vec{x}_i)$ of obtaining the observation result $z_k \in \mathcal{M}$ while the robot is in the location $\vec{x}_i \in X$ is defined by

Eq. (7.29). Notice that in general POMDP framework (Kaelbling, Littmann, and Cassandra 1998), the observation probability is often defined with respect to both the action and state that in the considered case gives

$$p_k^z(\delta_l, \vec{x}_i) = Pr\left\{z = z_k \mid \delta = \delta_l,\, \vec{x} = \vec{x}_i\right\}, z_k \in \mathcal{M}, \delta_l \in \mathcal{D}, \vec{x}_i \in X.$$

In addition, let $r : X \times \mathcal{D} \to [0, \infty)$ be a reward function that defines the immediate non-negative income $r(\vec{x}(t), \delta(t))$, which is obtained if at time t the robot is located in the point $\vec{x}(t) \in X$ and chooses the movement $\delta(t) \in \mathcal{D}$. Then, the problem is to find the sequence $\delta_{0:T} = \langle \delta(0), \delta(1), \delta(2), \dots, \delta(T) \rangle$ of the robot's movements such that the expected total reward $\nabla(\delta_{0:T})$ over the robot's path up to time T reaches its maximum. The sequence of movements that provides the maximal total expected reward is denoted by $\delta^*_{0:T}$.

For the finite termination time $T < \infty$ that defines the finite horizon optimization problem, the expected reward is specified as a sum of immediate rewards:

$$\boldsymbol{r}(\delta_{0:T}) = \mathcal{E}\left(\sum_{t=0}^{T} r(\vec{x}(t), \delta(t))\right). \tag{7.32}$$

For infinite horizon with $T = \infty$ we usually use either the discounted reward model, where the expected reward is defined using the discount factor $\gamma \in (0, 1)$

$$\boldsymbol{r}(\delta_{0:\infty}) = \mathcal{E}\left(\sum_{t=0}^{\infty} \gamma^t r(\vec{x}(t), \delta(t))\right), \tag{7.33}$$

or the average reward model (Bertsekas 1995; Kaelbling, Littman, and Moore 1996), where expected reward is obtained as an infinite limit in time:

$$\boldsymbol{r}(\delta_{0:\infty}) = \lim_{t \to \infty} \mathcal{E}\left(\sum_{\tau=0}^{t} \frac{1}{t} r(\vec{x}(\tau), \delta(\tau))\right). \tag{7.34}$$

Notice that in the recursive methods of finding optimal path $\delta_{0:T}$, the finite horizon model also includes the discount factor γ that governs the influence of future rewards.

Let us start with the deterministic choice of the movements that unambiguously prescribes the relation between the robot locations or observation results and the conducted movements. In the other words, such choice implies that the movement probabilities defined by Eqs. (7.30) or (7.31) are $p_l^\delta(\vec{x}_i, z_k) = 1$ or $p_l^\delta(z_k) = 1$ for some pairs of indices (l, i) or (l, k), respectively, and are zero for all other indices. For simplicity, let us concentrate on the choice of the movements $\delta(t)$ based on the robot's locations $\vec{x}(t)$. A function $\pi_t : X \to \mathcal{D}$ that prescribes the choice of the movement with respect to the robot's location is called policy. The policy, which does not depend on time, is called stationary policy; in its notation the index t is omitted and it is denoted by π. Otherwise, the policy is called nonstationary policy.

Given a policy π_t, denote by $V_{0:t}(\vec{x}_j \mid \pi_t)$, $j = 1, 2, \dots, n$, the expected sum of the rewards that are obtained while the robot is in the location $\vec{x}(t) = \vec{x}_j$ and apply the policy π_t up to the time t. At $t = 0$, this sum is equal to the immediate reward:

$$V_{0:0}(\vec{x}_j \mid \pi_0) = r(\vec{x}_j, \pi_0(\vec{x}_j)), \tag{7.35}$$

and for further times $t = 1, 2, \dots, T$ it is defined by induction as follows:

$$V_{0:t}(\vec{x}_j \mid \pi_t) = r(\vec{x}_j, \pi_t(\vec{x}_j)) + \gamma \sum_{i=1}^{n} \rho_{ij}^{x}(\pi_{t-1}(\vec{x}_i)) V_{0:t-1}(\vec{x}_i \mid \pi_{t-1}). \tag{7.36}$$

In the other words, at each time t the value $V_{0:t}(\vec{x}_j \mid \pi_t)$ includes an immediate reward obtained by choosing the movement $\pi_t(\vec{x}_j)$ added with the discounted expected value of the rewards obtained up to time $t - 1$. In the machine learning and artificial intelligence literature, the function $V_{0:t}$ is usually called the *value function,* and the goal of the robot is to choose such a policy π_t^* that the value $V_{0:t}(\vec{x}_j \mid \pi_t^*)$ is maximal over all available policies π_t.

In some cases, especially for the learning automata, the policy π_t^* can be defined directly by certain logical rules. For example, for the Tsetlin automaton (1973) (Kaelbling, Littman, and Moore 1996), the choice is defined by the logical *not – XOR* gate; in the framework of mobile robots, it prescribes to continue motion on the chosen direct if the immediate reward is 1 and to change the direction of the movement to opposite if the immediate reward is 0. In the extended form (Kagan et al. 2014), such automaton is also applicable to the values from the interval [0, 1] and controls the changes of the movement's direction in the range $[0, 2\pi]$ (see Section 11.3.1). However, in general the specification of the policy π_t^* requires certain algorithmic solutions and optimization techniques.

Usually, the choice of the policy π_t^* follows the reinforcement approach and is conducted by value iteration or policy iteration algorithms (Kaelbling, Littman, and Moore 1996; Sutton and Barto 1998). The value iteration algorithm deals with the value function and is outlined as follows.

Algorithm 7.4 (value iteration). Given the domain X, the set of movements $\mathcal{D}$, immediate reward function r, locations transition function ρ^x, discount factor $\gamma \le 1$ and strictness $\epsilon > 0$, do:

1) For each location $\vec{x}_j \in X$, $j = 1, 2, \dots, n$, do:
2) Set initial value $V_{0:0}(\vec{x}_j \mid \pi_0) = 0$.
3) End for.
4) Do:
5) Set $t = t + 1$.
6) For each location $\vec{x}_j \in X$, $j = 1, 2, \dots, n$, do:
7) For each movement $\delta \in \mathcal{D}$ do:
   Set $Q_t(\vec{x}_j \mid \delta) = r(\vec{x}_j, \delta) + \gamma \sum_{i=1}^{n} \rho_{ij}^{x}(\delta) V_{0:t-1}(\vec{x}_i \mid \pi_{t-1})$.
8) End for.
9) Set value $V_{0:t}(\vec{x}_j \mid \pi_t) = \max_{\delta \in \mathcal{D}} Q_t(\vec{x}_j \mid \delta)$.
10) End for.
11) While $\left| V_{0:t}(\vec{x}_j \mid \pi_t) - V_{0:t-1}(\vec{x}_j \mid \pi_{t-1}) \right| \ge \epsilon$ for all locations $\vec{x}_j \in X$, $j = 1, 2, \dots, n$.

In this algorithm, the policies $\pi_0, \dots, \pi_t$ for each time t are obtained indirectly by varying the values $V_{0:t}(\vec{x}_j \mid \pi_t^*)$ for the points $\vec{x}_j \in X, j = 1, 2, \dots, n$, and the termination rule of the algorithm defined in line 12 are also specified with respect to these values.

In contrast, the policy iteration algorithm directly deals with the policies $\pi_0, \dots, \pi_t$ and converges to the one that cannot be improved in the sense of its value $V_{0:t}(\vec{x}_j \mid \pi_t), j = 1, 2, \dots, n$. This algorithm is outlined as follows.

Algorithm 7.5 (policy iteration). Given the domain X, the set of movements $\mathcal{D}$, immediate reward function r, locations transition function ρ^x and discount factor $\gamma \le 1$, do:

1) Start with $t = 0$.
2) For each location $\vec{x}_j \in X, j = 1, 2, \dots, n$, do:
3) Set arbitrary initial policy $\pi_t^*(\vec{x}_j)$.
4) End for.
5) Do:
6) Set $t = t + 1$.
7) Set candidate policy $\pi_t = \pi_{t-1}^*$.
8) Compute the value for the candidate policy π_t by solving linear equations $V_{0:t}(\vec{x}_j \mid \pi_t) = r(\vec{x}_j, \pi_t(\vec{x}_j)) + \gamma \sum_{i=1}^{n} \rho_{ij}^x(\pi_t(\vec{x}_i)) V_{0:t}(\vec{x}_i \mid \pi_t)$.
9) For each location $\vec{x}_j \in X, j = 1, 2, \dots, n$, do:
10) Set $\pi_t^*(\vec{x}_j) = \arg\max_{\delta \in \mathcal{D}} (r(\vec{x}_j, \delta) + \gamma \sum_{i=1}^{n} \rho_{ij}^x(\delta) V_{0:t}(\vec{x}_i \mid \pi_t))$.
11) End for.
12) While $\pi_t(\vec{x}_j) \ne \pi_{t-1}^*(\vec{x}_j)$ for any location $\vec{x}_j \in X, j = 1, 2, \dots, n$.

The outlined algorithms are the basic instruments for finding optimal policies in a wide range of AI problems (Kaelbling, Littmann, and Cassandra 1998; Kaelbling, Littman, and Moore 1996) and in the optimization tasks and machine learning in general (Bertsecas and Shreve 1978; Sutton and Barto 1998). In contrast to the Kalman filter approach, these algorithms do not consider a model of the system's dynamics and deal only with the value function $V_{0:t}$ defined accordingly to the considered task.

Probably the most illustrative example of application of such algorithms in the framework of mobile robots is a planning of the robot's path in the problem of the probabilistic search for a moving target (Kagan and Ben-Gal 2013; Kagan and Ben-Gal 2015). In particular, the implementation of the Stewart procedure (Stewart 1979) for such problem resulted in the well-known Forward and Backward (FAB) algorithm suggested by Washburn (1980, 1983); a similar algorithm was independently developed by Brown (1980). The idea of this algorithm is as follows.

Let $X = \{\vec{x}_1, \vec{x}_2, \dots, \vec{x}_n\}$ be a finite discrete domain, and assume that in this domain these is a target moving according to the Markov process with the transition probability

$$\rho_{ij}(t) = Pr\{\vec{x}(t+1) = \vec{x}_j \mid \vec{x}(t) = \vec{x}_i\}, \tag{7.37}$$

and initial probabilities of the target's location in the points $\vec{x}_i$, $i = 1, 2, \dots, n$, at time $t = 0$

$$p_i(0) = Pr\left\{\vec{x}(0) = \vec{x}_i\right\}, \sum_{i=1}^{n} p_i(0) = 1. \tag{7.38}$$

The searcher moves over the domain X with the aim of finding the target during a given finite period $[0, T]$. It is assumed that the searcher is aware of the target's location probabilities $p_i(t)$ and transition probabilities $\rho_{ij}(t)$, $i, j = 1, 2, \dots, n$, $t = 0, 1, \dots, T$. Then the goal of the searcher is to maximize the probability $P(T)$ of finding the target up to time T or, that is the same, to minimize the probability $Q(T)$ of nondetecting the target up to this time.

Following the general approach used in the search theory (Koopman 1956–1957; Stone 1975), in the FAB algorithm the policies π_t, $t = 0, 1, \dots, T$ of the searcher are associated with the distributions of the search efforts over the domain X. In the case of search by a single mobile robot, such distributions specify the probabilities of detecting the target while the robot observes certain area $a \subset X$ of the domain. If at each time t the observed area $a(t)$ includes a single point $\vec{y}(t) \in X$ in which the robot is located, then the problem of search is formulated as in the problem of motion planning with the decision-making process based on observation results and known locations (see Figure 7.5a).

Denote by $q_i(t)$ the probability of nondetecting the target while at time t the searcher is located in the point $\vec{y}_i \in X$, $i = 1, 2, \dots, n$, $t = 0, 1, \dots, T$. In addition, let $\mathbb{I}(t)$ be a random variable independent on the target's locations $\vec{x}(t)$ and represent the search conditions. For example, if $\mathbb{I}(t) = 0$ for all $t < T$ and $\mathbb{I}(T) = 1$, then the expected value $\mathcal{E}\left(\sum_{t=0}^{T}\mathbb{I}(t)\prod_{\tau=0}^{t} q_{\vec{y}(t)}(\tau)\right)$ is the probability of nondetecting the target up to the termination time T. Denote by $c_i(t) = \mathcal{E}(\mathbb{I}(t) \,|\, \vec{x}(t) = \vec{x}_i)$ the expected value of the random variable $\mathbb{I}(t)$ given that at time t the target is located in the point $\vec{x}(t) = \vec{x}_i$. In contrast to the probability $q_i(t)$, the value $c_i(t)$ is the probability of nondetecting the target up to time t while at this time it is located in the point $\vec{x}_i$, $i = 1, 2, \dots, n$. Then, the dynamics of the system "searcher-target" is governed by the following forward and backward equations:

- forward equations given initial target's location probabilities $p_i(0)$ at time $t = 0$:

$$f_i(0) = p_i(0), i = 1, 2, \dots, n, \tag{7.39}$$

$$f_j(t+1) = \sum_{i=1}^{n} \rho_{ij}(t) q_i(t) f_i(t), t = 0, 1, \dots, T-1; \tag{7.40}$$

- Backward equations given the probabilities $c_i(t)$ of nondetecting the target up to time t:

$$g_i(T) = c_i(T), i = 1, 2, \dots, n, \tag{7.41}$$

$$g_i(t) = c_i(t) + \sum_{j=1}^{n} \rho_{ij}(t) q_j(t+1) g_j(t+1), t = 0, 1, \dots, T-1. \tag{7.42}$$

Forward equations (7.39) and (7.40) define the evolution of the target's location probabilities as they are perceived by the searcher starting from their initial distribution at time $t = 0$ up to the termination time T, and backward equations (7.41) and (7.42) define these probabilities starting from their expected distribution at the end of search at time T down to the beginning of the process.

Finally, notice that the probability $q_i(t)$ of nondetecting the target at time t given that the searcher is in the point $\vec{y}_i \in X$, $i = 1, 2, \ldots, n$, $t = 0, 1, \ldots, T$, is unambiguously defined by the sensing abilities of the searcher and its location. For example, it can be specified by the well-known Koopman function (1956–1957) $q(t) = \exp(-\kappa \vec{y}(t))$, where κ specifies the abilities of the searcher's sensors. Hence, the choice of the (near) optimal policies π_t^* of distribution of the search efforts, which in the considered case are the locations of the searcher, is equivalent to the choice of the probability distributions $q^*(t)$, which leads to minimum of the probability $Q(T)$ of nondetecting the target up to time T and so to the maximum of the probability $P(T) = 1 - Q(T)$ that the target will be detected.

The algorithm to search for a moving target that implements the policy iteration Algorithm 7.5 for finding the distributions $q^*(t)$ of the searcher's locations over the domain X is outlined as follows (Washburn 1980, 1983; see also Brown 1980).

Algorithm 7.6 (FAB algorithm of search for a moving target). Given the domain X, the target's transition probabilities $\rho_{ij}(t)$, initial target's location probabilities $p_i(0)$ and the probabilities $c_i(t)$ of non-detecting the target up to time t, $t = 0, 1, \ldots, T$, $i, j = 1, 2, \ldots, n$, do:

1) For each location $\vec{x}_i \in X$, $i = 1, 2, \ldots, n$, do:
2) Compute $g_i(t)$ using backward equations (7.41) and (7.42).
3) End for.
4) Set $t = 0$.
5) For each location $\vec{x}_i \in X$, $i = 1, 2, \ldots, n$, do:
6) Initialize $f_i(t)$ using forward equation (7.39).
7) End for.
8) While $t < T$ do:
9) For each location $\vec{x}_i \in X$, $i = 1, 2, \ldots, n$, do:
10) Set $q_i^*(t) = \arg\min_{q(t)} \left(\sum_{j=1}^{n} f_j(t) q_j(t) g_j(t)\right)$, where minimum is taken over all feasible distributions of the searcher's locations at time t.
11) End for.
12) Set $t = t + 1$.
13) For each location $\vec{x}_i \in X$, $i = 1, 2, \ldots, n$, do:
14) Compute $f_i(t)$ using forward equation (7.40).
15) End for.
16) End while.

The considered FAB algorithm of search for a target moving in the discrete domain according to the Markov process provides a simple illustrative example of the probabilistic motion planning for the mobile agent acting under uncertainties with finite horizon. Also, in 2003, Singh and Krishnamurthy (2003) reported the search algorithm, which implements the indicated above infinite horizon scheme. For a detailed overview of these algorithms and similar methods, see Kagan and Ben-Gal (2013, 2015). For general consideration of the methods of finding optimal policies for MDP and POMDP see the books by Ross (2003) and by White (1993) and for the reinforcement learning techniques see the book by (Sutton and Barto 1998).

Certainly, these basic decision-making techniques do not exhaust the wide range of the existing algorithms and methods applied for motion planning under uncertainties. In particular, since these algorithms implement recursive computations of the policies, they, in contrast to general Bayesian methods, cannot be applied for online motion planning. In the examples presented in the next section, we return to Bayesian motion planning and consider the algorithms that resolve the uncertainties using informational heuristics.

7.3 Examples of Probabilistic Motion Planning

The considered above algorithms implement different decision-making techniques: for the known model of the robot's motion, the motion planning is based on the predicted positions of the robot and observation results, and for unknown motion model the motion planning is based on general Bayesian updates of the robot's beliefs or of the occupancy grid. In the case of offline potion planning, an optimal or near optimal path of the robot is usually obtained following general optimization techniques, for example by value iteration or policy iteration algorithms, and for online motion planning some heuristic methods are applied. Next, we consider examples of online motion planning and mapping using informational heuristics.

7.3.1 Motion Planning in Belief Space

As an example of motion planning in the belief space, we consider an algorithm of search for a single target (Kagan, Goren, and Ben-Gal 2010). The search is conducted in two-dimensional domain $X \subset \mathbb{R}^2$ such that each point $\vec{x} \in X$ is a pair $\vec{x} = (x,y)$ of its Cartesian coordinates. Similar to the Markov localization (Section 7.1.1) and occupancy grid approach (Section 7.2.1), it is assumed that the domain X is a two-dimensional grid $X = \left\{\vec{x}_1, \vec{x}_2, \ldots, \vec{x}_n\right\}$ and each point $\vec{x}_i$, $i = 1, 2, \ldots, n$, is defined by a pair (i_x, i_y) of indices $i_x = 1, 2, \ldots, n_x$ and $i_y = 1, 2, \ldots, n_y$ such that $n = n_x \times n_y$ and $i = (i_x - 1)n_x + i_y$.

The robot starts at time $t = 0$ with initial beliefs $bel(\vec{x}\,(0))$ regarding the target's location that are defined by the target's location probabilities $p(\vec{x},0)$ and the during its motion observes certain areas $a(t) \subset X$, $t = 0, 1, 2, \ldots$, and obtains exact observation result $z(t) \in \{0, 1\}$. Result $z(t) = 1$ is interpreted as detection of the target in the area $a(t)$ and the search terminated; otherwise, the robot chooses the movement $\delta(t) \in \mathcal{D}$ from the set $\mathcal{D}$ of possible movements, decreases the target's location probabilities $p(\vec{x},t)$ for the points $\vec{x} \in a(t)$ down to zero, and updates the probabilities for the points in the area $X\,a(t)$ according to the Bayes rule. As in Section 7.1.1, it is assumed that $\mathcal{D} = \{\delta_1, \delta_2, \delta_3, \delta_4, \delta_5\}$, where δ_1 = *move forward*, δ_2 = *move backward*, δ_3 = *move right*, δ_4 = *move left* and δ_5 = *stay in the current point*. An algorithm that implements such search, is outlined as follows.

Algorithm 7.7 (probabilistic search; developed in collaboration with Gal Goren). (Kagan, Goren, and Ben-Gal 2010) Given a domain X and initial map of the target's location probabilities $p(\vec{x},0)$, $\vec{x} \in X$, do:

1) Start with $t = 0$ and initial target's location probabilities $p(\vec{x},t)$, $\vec{x} \in X$.
2) Choose observed area a such that $\sum_{\vec{x} \in a} p(\vec{x},t) > 0$ and such that the observation of a and decrease of the probabilities in the points $\vec{x} \in a$ down to zero will result in maximal difference between current probability map and the map, which will be obtained after the observation.
3) If the choice in line 2 is impossible because of zero probabilities in all points of all available observed areas, then apply diffusion process to the map of the target's location probabilities. The time of diffusion process is specified by the weighted distance between the agent and the center of distribution of the target's location probabilities. The weight is defined as a lack of information about the target's location. After finishing diffusion process, return to choosing the observed area in line 2.
4) Move to the chosen area a, observe it, and obtain observation result $z(t)$.
5) If $z(t) = 1$ (the target is found), then terminate and return area a. Otherwise, decrease the probabilities $p(\vec{x},t)$ in all points $\vec{x} \in a$ down to zero that is: for all $\vec{x} \in X$ set $p(\vec{x},t) = 0$, and then update the probabilities $p(\vec{x},t)$, $\vec{x} \in X$, such that $\sum_{\vec{x} \in X} p(\vec{x},t) = 1$.
6) Increase time t and continue starting from line 2.

The presented algorithm follows the destructive search scenario (Viswanathan et al. 2011), in which the agent during the search changes the environment states, in the considered case – the probability map, and then applies these changes for decision-making in the next steps. From the path-planning point of view, the algorithm implements the Elfes's mapping and navigation scheme (see Figure 7.4) excluding object's extraction and graph-building parts.

The presented algorithm follows general approach of maximum gradient search with informational heuristic for resolving uncertainties in the movements' choices. The next routine clarifies the implementation of the algorithm (developed in collaboration with Gal Goren).

Given the grid $X = \{\vec{x}_1, \vec{x}_2, \ldots, \vec{x}_n\}$, initial target location probabilities $p(\vec{x}_i,0)$, $i = 1, 2, \ldots, n$, transition probabilities matrix ρ and the set $\mathcal{D}$ of possible movements, this routine is outlined as follows.

probabilistic_search $\left(\left\{\vec{x}_1, \vec{x}_2, \ldots, \vec{x}_n\right\}, \mathcal{D}, \varphi, \rho, \left\{p(\vec{x}_1, 0), \ldots, p(\vec{x}_n, 0)\right\}\right)$

Start search

1) Set $t = 0$.
2) Set initial location of the agent with the corresponding observed area is $a(t)$.
3) Observe $a(t)$ and obtain $z(t)$.
   Continue search up to detecting the target.
4) While $z(t) = 0$ do:
   Set current observed probabilities.
5) Set $\left\{p(\vec{x}_1, t), \ldots, p(\vec{x}_n, t)\right\} =$
   current_observed_probabilities$\left(a(t), \left\{p(\vec{x}_1, t), \ldots, p(\vec{x}_n, t)\right\}\right)$.
   Calculate common map of next location probabilities.
6) Set $\left\{p(\vec{x}_1, t+1), \ldots, p(\vec{x}_n, t+1)\right\} =$
   next_location_probabilities$\left(\rho, \left\{p(\vec{x}_1, t), \ldots, p(\vec{x}_n, t)\right\}\right)$.
   Choose next observed area.
7) Set $a(t+1) =$ next_*observed_area*$\left(\vec{x}(t), \mathcal{D}, \rho, \left\{p(\vec{x}_1, t+1), \ldots, p(\vec{x}_n, t+1)\right\}\right)$.
   Increment time.
8) Set $t = t + 1$.
   Move to the next location and observe the chosen area.
9) Set location $\vec{x}(t)$ with respect to the area $a(t)$.
10) Observe $a(t)$ and obtain $z(t)$.
11) End while.
    Return the target's location.
12) Return the area $a(t)$, for which $z(t) = 1$.

The routine uses three functions. Function *current_observed_probabilities*(...) implements line 5 of the algorithm that is it decreases the probabilities in the observed area and updates the probabilities n the other points of the grid. Function *next_location_probabilities*(...) is a generic function, which takes into account the target's motion abilities; in the case of a static target, its transition probability matrix ρ is a unit matrix, so the function *next_location_probabilities*(...) does not changes the input probabilities, and in the case of moving target it defines the probabilities of the target's locations given the observation result. In addition, this function is used for implementing the diffusion process. Finally, function *next_observed_area*(...) implements the decision-making process. These functions are outlined as follows:

current_observed_probabilities $\left(a(t), \left\{p(\vec{x}_1, t), \ldots, p(\vec{x}_n, t)\right\}\right)$: $\left\{p(\vec{x}_1, t), \ldots, p(\vec{x}_n, t)\right\}$

1) For $i = 1 \ldots n$ do:
2) If $\vec{x}_i \in a(t)$ set $p(\vec{x}_i, t) = 0$.
3) End for.
4) Set $sum = 0$;

5) For $i = 1 \ldots n$ do:
6) Set $sum = sum + p(\vec{x}_i, t)$.
7) End for.
8) For $i = 1 \ldots n$ do:
9) Set $p(\vec{x}_i, t) = p(\vec{x}_i, t)/sum$.
10) End for.
11) Return $\left\{p(\vec{x}_1, t), \ldots, p(\vec{x}_n, t)\right\}$.

The next function is based on the indicated assumption that in the grid X each point is $\vec{x}_i = (i_x, i_y)$, $i = 1, 2, \ldots, n$, where $i_x = 1, 2, \ldots, n_x$ and $i_y = 1, 2, \ldots, n_y$, $n = n_x \times n_y$ and $i = (i_x - 1)n_x + i_y$. In addition, in the function it is assumed that at each time step the target moves without jumps and is able to move a single step forward, backward, left or right, or to stay in its current location. Formally, this function implements the multiplication

$$\left\{p(\vec{x}_1, t+1), \ldots, p(\vec{x}_n, t+1)\right\} = \left\{p(\vec{x}_1, t), \ldots, p(\vec{x}_n, t)\right\} \times \rho;$$

however, since following the indicated assumption the transition probability matrix ρ is sparse with nonzero elements on its diagonal and one step near the diagonal, such multiplication can be executed with less computations. The function that implements such computations is outlined as follows.

next_location_probabilities $\left(\rho, \left\{p(\vec{x}_1, t), \ldots, p(\vec{x}_n, t)\right\}\right)$: $\left\{p(\vec{x}_1, t+1), \ldots, p(\vec{x}_n, t+1)\right\}$

1) For $i_x = 1 \ldots n_x$ do:
2) For $i_y = 1, \ldots, n_y$ do:
3) Set $i = (i_x - 1)n_x + i_y$.
4) If $i_x = 1$ and $i_y = 1$, then (northwest corner of the grid X)
5) Set $R = p((i_x, i_y + 1), t)\rho(i + 1, i) + p((i_x + 1, i_y), t)\rho(i + n_y, i)$;
6) Else if $i_x = 1$ and $i_y = n_y$, then (northeast corner of the grid X)
7) Set $R = p((i_x, i_y + 1), t)\rho(i + 1, i) + p((i_x + 1, i_y), t)\rho(i + n_y, i)$;
8) Else if $i_x = n_x$ and $i_y = 1$, then (southwest corner of the grid X)
9) Set $R = p((i_x, i_y + 1), t)\rho(i + 1, i) + p((i_x - 1, i_y), t)\rho(i - n_y, i)$;
10) Else if $i_x = n_x$ and $i_y = 1$, then (southeast corner of the grid X)
11) Set $R = p((i_x, i_y - 1), t)\rho(i - 1, i) + p((i_x - 1, i_y), t)\rho(i - n_y, i)$;
12) Else if $i_x = 1$ and $1 < i_y < n_y$, then (northern bound of the grid X)
    $R = p((i_x, i_y - 1), t)\rho(i - 1, i) + p((i_x, i_y + 1), t)\rho(i + 1, i) + p((i_x + 1, i_y), t)\rho(i + n_y, i)$;
13) Else if $i_x = n_x$ and $1 < i_y < n_y$, then (southern bound of the grid X)
    $R = p((i_x, i_y - 1), t)\rho(i - 1, i) + p((i_x, i_y + 1), t)\rho(i + 1, i) + p((i_x - 1, i_y), t)\rho(i - n_y, i)$;
14) Else if $1 < i_x < n_x$ and $i_y = 1$, then (western bound of the grid X)
    $R = p((i_x, i_y + 1), t)\rho(i + 1, i) + p((i_x - 1, i_y), t)\rho(i - n_y, i) + p((i_x + 1, i_y), t)\rho(i + n_y, i)$;
15) Else if $1 < i_x < n_x$ and $i_y = n_y$, then (eastern bound of the grid X)
    $R = p((i_x, i_y - 1), t)\rho(i - 1, i) + p((i_x - 1, i_y), t)\rho(i - n_y, i) + p((i_x + 1, i_y), t)\rho(i + n_y, i)$;
16) Else (internal points of the grid X)
    $R = p((i_x, i_y + 1), t)\rho(i + 1, i) + p((i_x, i_y - 1), t)\rho(i - 1, i) + p((i_x - 1, i_y), t)\rho(i - n_y, i) + p((i_x + 1, i_y), t)\rho(i + n_y, i)$;
17) End if.

18) Set $p((i_x, i_y), t+1) = p((i_x, i_y), t)\rho(i, i) + R$.
19) End for.
20) End for.
21) Return $\{p(\vec{x}_1, t+1), \ldots, p(\vec{x}_n, t+1)\}$.

An assumption regarding the target's motion is, certainly, not crucial for the search Algorithm 7.7 and the outlined function, and is applied following practical needs only. If the considered task requires additional target motion, the function can be easily modified with respect to the abilities of the target.

Finally, let us outline the function *next_observed_area*(...), which executes the main decision-making task regarding the activity of the searcher. This function consists of two general operations: the choice of the next observed area by maximal gradient of the probability map in the searcher's current and the resolving of the uncertainties in the case, which such choice is impossible.

next_observed_area $\left(\vec{x}(t), \mathcal{D}, \rho, \{p(\vec{x}_1, t+1), \ldots, p(\vec{x}_n, t+1)\}\right) : a(t+1)$

Try the choice by maximal gradient of the map $\{p(\vec{x}_1, t+1), \ldots, p(\vec{x}_n, t+1)\}$ *in the point* $\vec{x}(t)$.

1) Set $a(t+1) = choose\_area\left(\vec{x}(t), \mathcal{D}, \{p(\vec{x}_1, t+1), \ldots, p(\vec{x}_n, t+1)\}\right)$.
2) If $p(\vec{x}, t+1) = 0$ for all $\vec{x} \in a(t+1)$ then *the choice by maximal gradient failed. Resolve uncertainty.*
3) Set $\{\tilde{p}(\vec{x}_1), \ldots, \tilde{p}(\vec{x}_n)\} = resolve\_uncertainty\left(\vec{x}(t), \mathcal{D}, \rho, \{p(\vec{x}_1, t+1), \ldots, p(\vec{x}_n, t+1)\}\right)$.
   Choose by maximal gradient of the map $\{\tilde{p}(\vec{x}_1), \ldots, \tilde{p}(\vec{x}_n)\}$ *in the point* $\vec{x}(t)$.
4) Set $a(t+1) = choose\_area\left(\vec{x}(t), \mathcal{D}, \{\tilde{p}(\vec{x}_1), \ldots, \tilde{p}(\vec{x}_n)\}\right)$.
5) Return $a(t+1)$.

The functions *choose_area*(...) and *resolve_uncertainty*(...) use several additional functions for calculations; for briefness, these functions are omitted and the calculations are included in the form of direct formulas. As above, it is assumed that the points of the grid X are $\vec{x}_i = (i_x, i_y)$, $i = 1, 2, \ldots, n$, with the same as above definition of the indices. The outlines of these functions follow the procedures presented in the books by Kagan and Ben-Gal (2013, 2015).

choose_area $\left(\vec{x}(t), \mathcal{D}, \rho, \{p(\vec{x}_1), \ldots, p(\vec{x}_n)\}\right) : a$

1) For all movements $\delta_k \in \mathcal{D}$ do:
2) Determine the candidate area $a_k \subset X$ with respect to the movement δ_k from the location $\vec{x}(t)$.
3) Set weight $w_k = \sum_{\vec{x} \in a_k} p(\vec{x})$.
4) For $i = 1 \ldots n$ do:
5) If $\vec{x}_i \in a_k$ then set $P_k(\vec{x}_i) = 1$, else set $P_k(\vec{x}_i) = p(\vec{x}_i)$.
6) End for.

7) For $i = 1 \ldots n$ do:
8) Set $\overline{P}_k(\vec{x}_i) = P_k(\vec{x}_i)/\sum_{j=1}^{n} P_k(\vec{x}_j)$.
9) End for.
10) For $i = 1 \ldots n$ do:
11) If $\vec{x}_i \notin a_k$ then set $Q_k(\vec{x}_i) = 0$, else $Q_k(\vec{x}_i) = p(\vec{x}_i)$.
12) End for.
13) For $i = 1 \ldots n$ do:
14) Set $\overline{Q}_k(\vec{x}_i) = Q_k(\vec{x}_i)/\sum_{j=1}^{n} Q_k(\vec{x}_j)$.
15) End for.
16) Set $W(a_k) = \sum_{i=1}^{n} w_k \left|\overline{P}_k(\vec{x}_i) - p(\vec{x}_i)\right| + (1 - w_k)\left|\overline{Q}_k(\vec{x}_i) - p(\vec{x}_i)\right|$.
17) End for.
18) Set $a = \text{argmax}\{W(a_k) \text{ over all } a_k\}$; ties are broken randomly.
19) Return a.

Resolving uncertainties in the choice of the next observed area is based on the informational heuristic, which, together with the distance from the current searcher's location to the estimated target's location (center of the probability map) is used for specifying the number of diffusion steps. The function is outlined as follows.

resolve_uncertainty$(\vec{x}(t), \mathcal{D}, \rho, \{p(\vec{x}_1), \ldots, p(\vec{x}_n)\})$: $\{\tilde{p}(\vec{x}_1), \ldots, \tilde{p}(\vec{x}_n)\}$

Calculate center $\vec{c}$ *of the probability map* $\{p(\vec{x}_1), \ldots, p(\vec{x}_n)\}$

1) Set $c_x = 0$ and $c_y = 0$.
2) For $i_x = 1 \ldots n_x$ and $i_y = 1, \ldots, n_y$ do:
3) Set $c_x = c_x + i_x p(\vec{x}_i)$ and $c_y = c_y + i_y p(\vec{x}_i)$, where $i = (i_x - 1)n_x + i_y$.
4) End for.
5) Set $\vec{c} = (x_c, y_c)$, where $x_c = [c_x]$, $y_c = [c_y]$.
   Calculate distance between searcher's location $\vec{x}(t) = (x(t), y(t))$ *and the maps' center* $\vec{c} = (x_c, y_c)$
6) Set $d = |x(t) - x_c| + |y(t) - y_c|$ (in the grid, X the Manhattan distance is used).
   Calculate the lack of information regarding the target's location.
7) Set $I = \log_s(n) - \sum_{i=1}^{n} p(\vec{x}_i)\log_s(1/p(\vec{x}_i))$, where $s = |a|$ is the size of the observed areas a.
   Calculate the number m *of the diffusion steps.*
8) Set $m = \lceil I \rceil \times [d]$.
   Apply m *steps of diffusion process governed by the transition matrix* ρ *to the map.* $\{p(\vec{x}_1), \ldots, p(\vec{x}_n)\}$
9) Set $\{\tilde{p}(\vec{x}_1), \ldots, \tilde{p}(\vec{x}_n)\} = \{p(\vec{x}_1), \ldots, p(\vec{x}_n)\}$.
10) Do m times:
11) Set $\{\tilde{p}(\vec{x}_1), \ldots, \tilde{p}(\vec{x}_n)\} = \textit{next\_location\_probabilities}\left(\rho, \{\tilde{p}(\vec{x}_1), \ldots, \tilde{p}(\vec{x}_n)\}\right)$.
12) Done.
13) Return $\{\tilde{p}(\vec{x}_1), \ldots, \tilde{p}(\vec{x}_n)\}$.

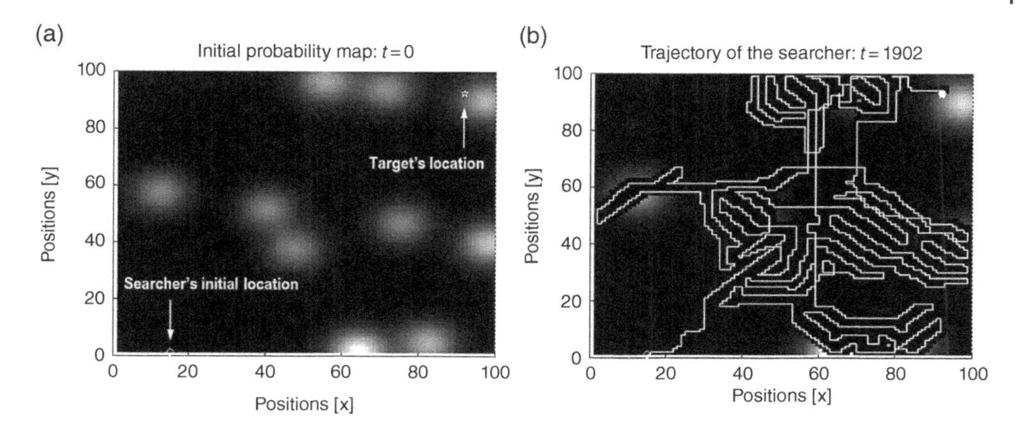

Figure 7.6 Probabilistic search according to Algorithm 7.7: (a) initial searcher's location and location of the target; and (b) trajectory of the searcher up to finding the target at time t = 1902. The search is conducted following the target's location probabilities; in the figures the higher probabilities are specified by white color and the lower probabilities by black color.

In the function (lines 5 and 8), $[a]$ stands for the integer number closest to the real number a and $\lceil a \rceil$ stands for a closest integer greater than or equal to the real number a. In the line 8, $\log_s$ stands for the logarithm base s and, following the convention, it is assumed that $0 \times \log_s(0) = 0$.

The trajectory of the searcher acting according to the Algorithm 7.7 is illustrated in Figure 7.6. In the figure, the square grid X includes $n = 100 \times 100$ points that is $n_x = 100$ and $n_y = 100$. At the time $t = 0$, the searcher starts in the point $\vec{x}(0) = (15, 1)$ and search for a target located in the unknown position. In the implemented routine, it is assumed that during the search at each time $t = 0, 1, \ldots$ the observed areas $a(t)$ are the squares of the size 3×3 around the searcher's locations $\vec{x}(t)$. Initial target's location is specified by random with respect to the target's location probabilities. The higher target's location probabilities are specified by white color and the lower probabilities by black color. Figure 7.6a shows initial searcher's location (depicted by white circle) and the target's location (depicted by a white star). Trajectory of the searcher up to detecting the target at time t = 1902 is shown in Figure 7.6b.

It is seen that according to the algorithm, if the choice of the next movement is possible, then the searcher moves according to the maximum gradient of the target's location probabilities. However, in the uncertain situation when the choice according to the maximal gradient is impossible, the searcher applies the defined above heuristics and moves in the direction of maximal probability density rated by the distance (Figure 7.7).

The presented algorithm illustrates the simplest case of motion planning in belief space, where the robot's beliefs are associated with the target's location probabilities. The target's location, in this case, defines the landmark, and the goal of the robot is to define its position in the small region close to this landmark. Notice again that in the presented algorithm, it is assumed that the observations of the environment are errorless so the uncertainties resulted by sensing inaccuracies are avoided. For additional information regarding this algorithm and its extensions for search with erroneous observations and for the search for moving target, see Kagan, Goren, and Ben-Gal (2010) and

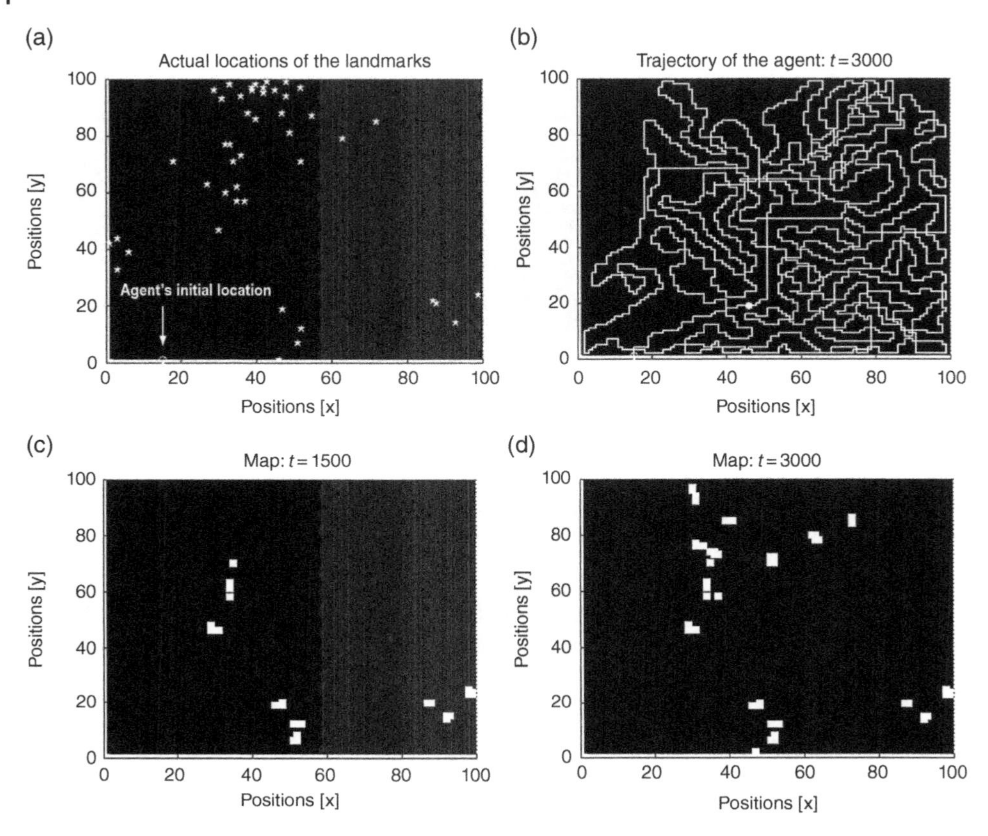

Figure 7.7 Mapping of the unknown gridded environment according to Algorithm 7.8: (a) the actual locations of the landmarks and Initial location of the agent; (b) trajectory of the agent up to termination of the process at time t = 3000; (c) map of the environment at time t = 1500; and (d) map at time t = 3000.

its continuations (Chernikhovsky et al. 2012; Kagan, Goren, and Ben-Gal 2012). In a wide framework of search problems, these methods are considered in Kagan and Ben-Gal (2013, 2015). In Chapter 10, we will consider an implementation of this algorithm for the search by several searchers. The next section presents the implementation of this algorithm for the search for multiple targets that corresponds to the mapping of the gridded environment with several identical landmarks.

7.3.2 Mapping of the Environment

Finally, let us consider an example of mapping a gridded terrain using the presented above algorithm of motion planning in belief space. Formally, such task can be reduced to the destructive probabilistic search for several targets with simultaneous indications of the target's positions. However, notice that in the algorithm should be included the mechanism of obstacle avoidance, which will guarantee that the robot will not collide with the landmarks. In Algorithm 7.7, this mechanism is omitted and will be considered in detail in Section 10.3.2, where it is used for navigation of the robot's team with obstacle and collision avoidance.

As above, assume that that the robot moves in the two-dimensional grid $X = \left\{\vec{x}_1, \vec{x}_2, \ldots, \vec{x}_n\right\}$ starting from the point $\vec{x}(0) \in X$ and at each time t = 0, 1, 2, ... it observes the area $a(t) \subset X$ around its location $\vec{x}(t)$. Similar to previous sections, denote by $\mathcal{M}$ the map of X. At the beginning of mapping, the map $\mathcal{M}$ is empty and, with the progress of the process, it will include the coordinates $\vec{m}_k$ of the landmarks k = 1, 2, ... found by the robot. It is assumed that physically, the landmarks occupy equal areas and these areas are denoted by $o_k \subset X$, k = 1, 2, ...; in Section 10.3.2 the same notation will be used for the areas of obstacles.

The motion and actions of the robot are defined similarly to Algorithm 7.7 with certain modification concerning the landmarks' detection. The robot starts with the initial beliefs $bel(\vec{x}(0)) = p(\vec{x},0)$, $\vec{x} \in X$. Notice that if the robot starts in completely unknown environment, then all points $\vec{x} \in X$ have equal probabilities $p(\vec{x},0) = 1/n$. During the motion and observation of the areas $a(t) \subset X$, $t = 0, 1, 2, \ldots$, the robot obtains observation results $z(t) \in \{0, 1\}$ that indicate detection of the landmark in the area $a(t)$. If the landmark was not detected, then the probabilities $p(\vec{x},t)$ for the points $\vec{x} \in a(t)$ are decreased to zero and the probabilities for the points in the area $X\ a(t)$ are updated. In opposite, if the landmark was detected in the area $a(t)$, then its position $\vec{m}_k$ (may be with the coordinates of the points in the landmark's area o_k), k = 1, 2, ..., is included into the map $\mathcal{M}$, and the probabilities both for the points $\vec{x} \in a(t)$ and for the points $\vec{x} \in o_k$ are decreased to zero. Then, the motion continues until some termination condition holds. The mapping algorithm is outlined as follows.

Algorithm 7.8 (probabilistic multiple targets search – mapping of the gridded terrain). Given the domain X, the initially empty map $\mathcal{M}$ and the probabilities $p(\vec{x},0)$, $\vec{x} \in X$, do:

1) Start with t = 0 and initial probabilities $p(\vec{x},t)$, $\vec{x} \in X$.
2) Choose observed area a according to the line 2 of Algorithm 7.7. If the choice is impossible, apply diffusion process to the map of the target's location probabilities (see line 3 of Algorithm 7.7).
3) For each already detected landmark k = 1, 2, ... do:
4) Set the probabilities $p(\vec{x},t) = 0$ for all points $\vec{x} \in o_k$.
5) End for.
6) Normalize the probabilities $p(\vec{x},t)$, $\vec{x} \in X$, such that $\sum_{\vec{x} \in X} p(\vec{x},t) = 1$.
7) Return to choosing the observed area in line 2.
8) Move to the chosen area a, observe it and obtain observation result $z(t)$.
9) If $z(t)$ = 1 (the landmark is detected), do:
10) Add position $\vec{m}$ of the landmark into the map $\mathcal{M}$.
11) Set the probabilities $p(\vec{x},t) = 0$ for all points $\vec{x} \in o$ in the landmarks area o.
12) End if.
13) Set the probabilities $p(\vec{x},t) = 0$ for all points $\vec{x} \in a$ in the observed area a.
14) Increase time t and continue starting from line 2 up to termination of the process.
15) Return map $\mathcal{M}$.

It is clear that the presented algorithm implements the same approach as Algorithm 7.7 of probabilistic search. The termination condition can be defined in different manners. Namely, if the number of landmarks is known, then the algorithm is terminated when the map $\mathcal{M}$ includes this number of positions, and if the time is limited by some T, then the algorithm is terminated when t reaches this limit. Finally, in the case of unknown number of landmarks and unlimited time, the process is terminated when $p(\vec{x},t) = 0$ for all points $\vec{x} \in X$ except a single point $\vec{y}$, where $p(\vec{y},t) = 1$.

It is seen that following Algorithm 7.8, which implements the probabilistic search for several targets, the agent at first detects the landmarks that are located in the areas, in which the landmarks density is higher and which are located closer to the agent's initial location. Then, it screens the remaining areas and detects the landmarks located in these areas. Notice that since the agent has no prior information regarding the areas occupied by the landmarks, it includes them into the map $\mathcal{M}$ at each time when they are detected while the agent arrives to the landmark from different direction.

The presented example finalizes the consideration of the methods of motion planning of the mobile robots in an unknown environment. Additional methods and algorithms with examples have been presented (Thrun, Burgard, and Fox 2005; Choset et al. 2005; Siegwart, Nourbakhsh, and Scaramuzza 2011) and references therein. Further development of the methods of navigation in belief space and contemporary overview of the existing techniques appear in Indelman, Carlone, and Dellaert (2013a,b). Additional useful references in the field are presented in the summary section.

7.4 Summary

In this chapter, we considered the methods of motion planning in an unknown environment. The consideration follows the approach widely known as probabilistic robotics, in which the uncertainties are resolved by means of probabilities (in opposite to, e.g., fuzzy logic methods used in fuzzy control) and motion planning is conducted in the robot's belief space. Online control of the robot in such a case is based on Markov decision-making methods.

Following this approach, the chapter considered four basic problems appearing in the motion planning in unknown environment:

1) Localization of robot based on the known map of the environment – the robot is required to detect the exact landmark observed by the robot's sensors. In general, such localization follows the Markov approach and does not assume any priory knowledge regarding robot's motion and characteristics of uncertainties (Section 7.1.1). Certain improvement of the localization results can be reached using the assumptions regarding the rules governing the robot's motion and probability distributions of the sensing and acting errors that allows prediction of the robot's locations and observation results. In particular, if the robot is governed by linear dynamical system and distributions of the errors are Gaussian, then prediction is executed using the Kalman filter methods (Section 7.1.2).
2) Mapping of the environment without localization requires creation of the environmental map during exploration of the unknown domain, but with known or estimated

positions of the robot. The map is created based on the sensorial information, using which the robot obtains positions of the landmarks with respect to its own locations. The best-known methods for solving this problem are the occupancy grid algorithms. The most known algorithm from this family is the Elfes algorithm (Section 7.2.1), which in its complete version provides both the map of the environment as it is observed by the robot's sensors and the geometry and symbolic maps, which include meaningful interpretations of the observed environment and labeling them on the map.
3) Simultaneous localization and mapping (SLAM) implies that the robot has no prior information about the landmark locations and creates the environmental map during exploration of the domain. In the probabilistic setup, the SLAM tasks are solved in the same manner as the localization tasks and usually implement the same assumptions regarding the robot's motion and observation errors that allows application of the Markov and Kalman filtering techniques (Section 7.2.1).
4) For errorless sensing, decision-making under uncertainties is usually conducted based on the methods of MDP. For erroneous sensing, it is based on the POMDP. For offline path planning, such problems are solved by widely known value iteration and policy iteration algorithms (Section 7.2.2), while for online decision making, certain heuristics are applied; Algorithm 7.7 is an example of such a heuristic.

The considered models are illustrated by running examples (Sections 7.1.1 and 7.1.2) and numerical examples, which include heuristic algorithms of motion planning (Section 7.3.1) and localization in belief space (Section 7.3.2).

References

Aoki, M. (1967). *Optimization of Stochastic Systems*. New York: Academic Press.

Astrom, K.J. (1970). *Introduction to Stochastic Control Theory*. New York: Academic Press.

Bailey, T. and Durrant-Whyte, H. (2006). Simultaneous localization and mapping (SLAM). Part II: state of the art. *IEEE Robotics and Automation Magazine* 13: 108–117.

Bertsecas, D.P. and Shreve, S.E. (1978). *Stochastic Optimal Control: Discrete TIme Case*. New York: Academic Press.

Bertsekas, D.P. (1995). *Dynamic Programming and Optimal Control*. Boston: Athena Scientific Publishers.

Borenstein, J., Everett, H. R., and Feng, L. (1996). *Where Am I? Sensors and Methods for Mobile Robot Positioning*. Retrieved December 7, 2015, from University of Michigan for ORNL D&D Program and US Department of Energy: www-personal.umich.edu/~johannb/papers/pos96rep.pdf.

Brown, S.S. (1980). Optimal search for a moving target in discrete time and space. *Operations Research* 28 (6): 1275–1289.

Castellanos, J.A. and Tardos, J.D. (1999). *Mobile Robot Localization and Map Building*. New York: Springer.

Chernikhovsky, G., Kagan, E., Goren, G., and Ben-Gal, I. (2012). Path planning for sea vessel search using wideband sonar. In: *Proc. 27-th IEEE Conv. EEEI* https://doi.org/10.1109/EEEI.2012.6377122.

Choset, H., Lynch, K., Hutchinson, S. et al. (2005). *Principles of Robot Motion: Theory, Algorithms, and Implementation*. Cambridge, MA: Bradford Books/The MIT Press.
Cuesta, F. and Ollero, A. (2005). *Intelligent Mobile Robot Navigation*. Berlin: Springer-Verlag.
DeGroot, M.H. (1970). *Optimal Statistical Decisions*. New York: McGraw-Hill.
Derman, C. (1970). *Finite State Markov Decision Processes*. New York: Academic Press.
Dudek, G. and Jenkin, M. (2010). *Computational Principles of Mobile Robotics*, 2e. New York: Cambridge University Press.
Durrant-Whyte, H. and Bailey, T. (2006). Simultaneous localization and mapping (SLAM). Part I: the essential algorithms. *IEEE Robotics and Automation Magazine* 13: 99–110.
Dynkin, E.B. and Yushkevich, A.A. (1979). *Controlled Markov Processes*. New York: Springer.
Elfes, A. (1987). Sonar-based real-world mapping and navigation. *IEEE Journal of Robotics and Automation* 249–265.
Elfes, A. (1990). Occupancy grids: a stochastic spatial representation for active robot perception. In: *Proc. 6th Conference on Uncertainty in AI*, 60–70. North-Holland.
Ephraim, Y. and Merhav, N. (2002). Hidden Markov processes. *IEEE Transactions on Information Theory* 48 (6): 1518–1569.
Everett, H.R. (1995). *Sensors for Mobile Robots: Theory and Application*. Wellesley: A K Peters.
Faragher, R. (2012). Understanding the basis of the Kalman filter via a simple and intuitive derivation. *IEEE Signal Processing Magazine* 29 (5): 128–132.
Filar, J. and Vrieze, K. (1997). *Competitive Markov Decision Processes*. New York: Springer.
Fox, D., Burgard, W., Kruppa, H., and Thrun, S. (2002). Collaborative multirobot localization. In: *Robot Teams: From Diversity to Polymorphism* (ed. T. Balch and L.E. Parker), 161–189. Natick, MA: A K Peters.
Hahnel, D., Triebel, R., Burgard, W., and Thrun, S. (2003). Map building with mobile robots in dynamic environments. In: *Proc. IEEE Conf. Robotics and Automation*, 1557–1562. Institute of Electrical and Electronics Engineers (IEEE).
Indelman, V., Carlone, L., and Dellaert, F. (2013a). Towards planning in generalized belief space with applications to mobile robotics. In: *The Israeli Conference on Robotics (ICR'13)*, –94. Tel-Aviv. The Israeli Robotics Association.
Indelman, V., Carlone, L., and Dellaert, F. (2013b). Towards planning in the generalized belief space. In: *International Symposium on Robotics Research (ISRR'13)*. Singapore. Berlin Heidelberg: Springer-Verlag.
Kaelbling, L.P., Littman, M.L., and Moore, A.W. (1996). Reinforcement learning: a survey. *Journal of Artificial Intelligence Research* 4: 237–285.
Kaelbling, L.P., Littmann, M.L., and Cassandra, A.R. (1998). Planning and acting in partially observable stachastic domains. *Artificial Intelligence* 101 (2): 99–134.
Kagan, E. and Ben-Gal, I. (2008). Application of probabilistic self-stabilization algorithms to the robot's control. In: *Proc. 15th Industrial Engineering and Management Conference IEandM'08*, 3e. Tel-Aviv, Israel. Atlantis Press.
Kagan, E. and Ben-Gal, I. (2013). *Probabilistic Search for Tracking Targets*. Chichester: Wiley.
Kagan, E. and Ben-Gal, I. (2015). *Search and Foraging. Individual Motion and Swarm Dynamics*. Baco Raton, FL: Chapman Hall/CRC/Taylor and Francis.
Kagan, E., Goren, G., and Ben-Gal, I. (2010). Probabilistic double-distance algorithm of search after static or moving target by autonomous mobile agent. In: *Proc. 26-Th IEEE Conv. EEEI*, 160–164. Institute of Electrical and Electronics Engineers (IEEE).

Kagan, E., Goren, G., and Ben-Gal, I. (2012). Algorithm of search for static or moving target by autonomous mobile agent with erroneous sensor. In: *Proc. 27-th IEEE Conv. EEEI*. https://doi.org/10.1109/EEEI.2012.6377124. Institute of Electrical and Electronics Engineers (IEEE).

Kagan, E., Rybalov, A., Sela, A. et al. (2014). Probabilistic control and swarm dynamics of mobile robots and ants. In: *Biologically-Inspired Techniques for Knowledge Discovery and Data Mining* (ed. S. Alam). Hershey, PA: IGI Global.

Kalman, R.E. (1960). A new approach to linear filtering and prediction problems. *Journal of Basic Engineering* 35–45.

Konolige, K. (1997). Improved occupancy grids for map building. *Autonomous Robots* 351–367.

Koopman, B.O. (1956–1957). The theory of search. *Operations Research* 4 (5), 324-346, 503-531, 613-626.

Leonard, J.J. and Durrant-Whyte, H.F. (1991). Simultaneous map building and localization for an autonomous mobile robot. In: *Proc. IEEE/RSJ Int. Workshop on Intelligent Robots and Systems*, 1442–1447. Osaka, Japan. Institute of Electrical and Electronics Engineers (IEEE).

Leonard, J., Durrant-Whyte, H., and Cox, I.J. (1990). Dynamic map buiding for an autonomous mobile robot. In: *Proc. IEEE Int. Workshop on Intelligent Robots and Systems*, 89–96. Institute of Electrical and Electronics Engineers (IEEE).

Luce, R.D. and Raiffa, H. (1957). *Games and Decisions*. New York: John Wiley & Sons.

Monahan, G.E. (1982). A survey of partially observable Markov decision processes: theory, models, and algorithms. *Management Science* 28 (1): 1–16.

Pearl, J. (1984). *Heuristics: Intelligent Search Strategies for Computer Problem Solving*. Reading, MA: Addison-Wesley.

Rabiner, L.R. (1989). A tutorial on hidden Markov models and selected applications in speech recognition. *Proceedings of the IEEE* 77 (2): 257–286.

Raiffa, H. (1968). *Decision Analysis: Introductory Lectures on Choices under Uncertainty*. Reading, MA: Addison-Wesley.

Ross, S.M. (2003). *Introduction to Probability Models*. San Diego, CA: Academic Press.

Roumeliotis, S.I. and Bekey, G.A. (2000). Bayesian estimation and Kalman filtering: a unified framework for mobile robot localization. In: *Proc. IEEE Conference on Robotics and Automation ICRA'00, 3*, 2985–2992. San Francisco, CA.

Siegwart, R., Nourbakhsh, I.R., and Scaramuzza, D. (2011). *Introduction to Autonomous Mobile Robots*, 2e. Cambridge, MA: The MIT Press.

Singh, S. and Krishnamurthy, V. (2003). The optimal search for a moving target when the search path is constrained: the infinite-horizon case. *IEEE Transactions on Automatic Control* 48 (3): 493–497.

Stewart, T. (1979). Search for a moving target when searcher motion is restricted. *Computers and Operations Research* 6: 129–140.

Stone, L.D. (1975). *Theory of Optimal Search*. New York: Academic Press.

Sutton, R.S. and Barto, A.G. (1998). *Reinforcement Learning: An Introduction*. Cambridge, MA: The MIT Press.

Thrun, S. (2001). A probabilistic online mapping algorithm for teams of mobile robots. *International Journal of Robotics Research* 20: 335–363.

Thrun, S. (2002). Robotic mapping: a survey. In: *Exploring Artificial Intelligence in the New Millennium* (ed. G. Lakemeyer and B. Nebel), 1–35. San Francisco, CA: Morgan Kaufmann.

Thrun, S. and Leonard, J.J. (2008). Simultaneous localization and mapping. In: *The Springer Handbook of Robotics* (ed. B. Siciliano and O. Khatib), 871–890. Berlin: Springer.

Thrun, S., Burgard, W., and Fox, D. (2005). *Probabilistic Robotics*. Cambridge, MA: The MIT Press.

Tou, J.T. and Gonzales, R.C. (1974). *Pattern Recognition Principles*. Reading, MA: Addison-Wesley.

Tsetlin, M.L. (1973). *Automaton Theory and Modeling of Biological Systems*. New York: Academic Press.

Viswanathan, G.M., da Luz, M.G., Raposo, E.P., and Stanley, H.E. (2011). *The Physics of Foraging*. Cambridge: Cambridge University Press.

Wald, A. (1950). *Statistical Decision Functions*. New York: John Wiley & Sons.

Washburn, A.R. (1980). On a search for a moving target. *Naval Research Logistics Quarterly* 27: 315–322.

Washburn, A.R. (1983). Search for a moving target: the FAB algorithm. *Operations Research* 31 (4): 739–751.

White, D.J. (1993). *Markov Decision Processes*. Chichester: John Wiley & Sons.

8

Energy Limitations and Energetic Efficiency of Mobile Robots

Michael Ben Chaim

8.1 Introduction

In the previous chapters, we considered the methods and algorithms of navigation and motion-planning of mobile robots, avoiding the fact that each mobile robot is a physical device with certain limitation on energy support. However, energy supply is one of the most important challenges for mobile robots, and most existing studies on mobile robots focus on motion planning in order to reduce the required energy. Notice that the movement and flight endurance and payload capabilities of miniature robots are strongly limited due to the unfavorable scaling effects and the lack of portable power storage devices with a sufficient power/mass ratio. These payload and flight endurance constraints strongly limit the applicability of miniature robots.

This chapter deals with the analysis of energy consumers, presents a methodology of power calculation and optimization of the power/mass ratio of miniature robots, and provides several examples that demonstrate how to reduce the energy consumption in the robots. These examples together with motion planning provide greater opportunities to achieve better energy efficiency for mobile robots.

The experimental verification of the suggested methods was performed on autonomous mobile wheeled robot, which was developed in collaboration with the students of the Department of Machines, Mechatronics, and Robotics of Ariel University. The results of suggested methods can be used to determine the setup of the motion parameters in certain motion-planning tasks and can be considered as a recommendation to the robots' producers to append new energy consumption items into driving system menu.

8.2 The Problem of Energy Limitations in Mobile Robots

In general, mobile robots are robots that can move from place to place across the ground or in the air. Mobility gives a robot much greater flexibility to perform new, complex, exciting tasks. The world does not have to be modified to bring all needed items within

Autonomous Mobile Robots and Multi-Robot Systems: Motion-Planning, Communication, and Swarming, First Edition. Edited by Eugene Kagan, Nir Shvalb and Irad Ben-Gal.

Companion website: www.wiley.com/go/kagan/robotsystems

reach of the robot. The robots can move where needed. Fewer robots can be used. Robots with mobility can perform more natural tasks, in which the environment is not designed especially for them.

A mobile robot needs locomotion mechanisms that enable it to move unbounded throughout its environment. Energy is the most important challenge for mobile robots. Power consumption is one of the major issues in robot design. Mobile robots usually have multiple components, such as motors, sensors, and microcontrollers and embedded computers. Internal combustion engine (ICE), battery and fuel cell, direct current (DC) motors are often used to drive the robots. Sensors collect data from the environment and provide information to robots. The most often used sensors are vision, infrared, sonar, and laser rangers. Many robots use embedded computers for high-level computation and microcontrollers for low-level controls. It hides the hardware details from the embedded computer, and provides an application programming interface (API) for the embedded computer. The embedded computer handles high-level computation, including motion planning, image processing, and scheduling. The separation of the microcontroller and embedded computer makes the designs more flexible.

Existing studies on energy reduction for robots focus on motion planning to reduce motion power. However, other components like sensing, control, communication, and computation also consume significant amounts of power. It is important to consider all components to achieve better energy efficiency.

This part has two major contributions. First, energy consumption calculation of a ground robots and power models from real measurement results for motion and sensing. Second, introduce methodology of dynamic energy management and real-time scheduling, to reduce the energy consumption of mobile robots. These techniques, together with motion planning, provide greater opportunities for energy-efficient designs of mobile robots.

Although this study is based on the data of ground robots, the methods can be applied to other types of robots. Robots should be able to go over long distances and operate for relatively long periods. Since robots are deployed with a finite power source, it is critical to ensure before executing the mission that the robot has sufficient energy to complete the task by predicting the energy requirements beforehand. In addition, the robot should be able to take action if its remaining energy is not enough to make a return trip to its charging station.

Robots need to manage their energy consumption to avoid running out of energy, as this would require sending personnel to retrieve or recover the robot from the potentially hazardous area where it was deployed. Also, the robot's energy autonomy should be optimized to provide more flexibility in possible interventions (in terms of distance traveled and mission time). Predicting the energy requirements of various components in a robot helps making decisions on optimizing the tasks and the mission to be accomplished with the available energy capacity. It is important to develop energy efficient.

However, in the presence of a sufficiently large number of publications, insufficiently elaborated solutions to optimize the energy consumption and the implementation of precision management coordinating level is actual. The use of the velocity profile correction of robots to solve this problem introduces additional uncertainty, worsening the accuracy of working off the required trajectory, and essentially depends on the correct selection criteria and analytical study. Although there are a sufficiently large number of publications on the optimization of the energy state approaches, there is a need for the development of optimization methods, depending on the response of the energy resources (engines, batteries, fuel cells). Therefore, this chapter discusses the velocity profile correction for energy optimization approaches in the direction of energy resources.

8.3 Review of Selected Literature on Power Management and Energy Control in Mobile Robots

The significance of power management for long-term operation of autonomous robots was discussed in the report by Deshmukh et al. (2010), with challenges in terms of battery technology, power estimation, and auto recharging. In an earlier report (Austin and Kouzoubo 2002), a robot with an auto-recharging system was proposed, with emphasis on the aspects improving the robustness (or reliability) of the system. A method for automatically recharging the batteries using the robot's built-in sensors to control docking with a recharging station was proposed by Oh and Zelinsky (2000).

Intensive research has been conducted on mobile robot motion energy optimization through motion-planning (Mei et al. 2004; Kim and Kim n.d.) and path-planning (Wei et al. 2012; Zhuang et al. 2005) techniques. Models for locomotion power and dynamics have been widely studied. In particular, it is emphasized (Yu et al. 2011) that the power models of motors are needed for the locomotion planning to complete time- or energy-constrained missions. Power models for skid steered wheeled (Yu et al. 2011; Chuy et al. 2009) and tracked (Morales et al. 2009) vehicles were proposed for various turning radius and surface conditions.

Some works analyze the energy consumption of different components in robots. The report (Liu et al. 2001) presents an energy breakdown table of a Mars rover, and the paper (Lamon 2008) estimates the energy consumption of a rover including the communication power. However, in both works the authors do not build power models for each component.

In the study (Mei et al. 2005) it is indicated that sensing, computation, and communication consume significant amounts of power compared to locomotion power. Therefore, management of all power-consuming modules is important. In another study (Mei et al. 2006), the power models were used to optimize the deployment of robots under energy and timing constraints. In the reports (Austin et al. 2001; Bonani et al. 2010) it was proposed that power models of various components allow the robot to estimate when it should exchange its battery.

A useful behavioral model for finding the optimal time at which a robot should go back to recharge is presented by Wailwa and Birk (1996). However, hostile or scientific facilities generally are not designed to accommodate mobile teleported robots. It is complicated to navigate the mobile robot to the work site due to the compact spaces and unstructured sections. In particular, for the special robot passing through the sector, special provisions were made (Hedenstrom 2003).

Because of such restrictions, auto-recharging techniques (Austin et al. 2001) cannot be used for telerobotic applications in scientific facilities emitting ionizing radiations. It can be observed that as the robot becomes heavier, the locomotion power accounts for a higher percentage of the total power because the locomotion power depends on the size and the mass of the robot, while the computers and sensors are relatively independent of the robot's size and mass. It should be noted that adding more batteries to the robot adds more mass, thereby requiring more power for locomotion.

Analysis of previous studies demonstrate that there is no common approach for creating power and energy consumption models of various components irrespective of the type of mobile robot. Hence, a generic modular approach for building power models and predicting the energy consumption of a mobile robot should be proposed. The most relevant aspect to this work is on action-selection. Perhaps the most standard and simple way to determine when it is time for the robot to recharge is to set a fixed threshold. This can either be a threshold directly on the energy supply as it was suggested by Austin et al. (2001). The latter is usually easier to implement, but it is less accurate, because one has to have some model of the energy supply. However, it was shown (Hedenstrom 2003; Wailwa and Birk 1996) that the fixed threshold policy can be improved upon. While it holds true that maximizing the energy intake rate maximizes potential work rate, it is not optimal with respect to when the work is done. It also assumes that recharging is always valuable.

And it often holds true, but not always. Also notable is that all of research just discussed refuel the robot to maximum capacity at each opportunity, and do not consider that this may not be the best policy. In the report (Litus et al. 2007), the authors consider the problem of energy efficient rendezvous as an action selection problem, and so investigate where (but not when and how long) to refuel. Work (Wailwa and Birk 1996) considers the robots living in a closed ecosystem learn to "survive." Here the robots learn to choose between recharging and fighting off competitors.

The Birk's agents (Hedenstrom 2003) value function of *survivability* is different to that considered here. The rational robot and its owner are interested in gaining maximum reward by working at the robot's task, and are indifferent to the lifespan of the robot. The survival quest of a mining robot on a distant planet contributed significantly to ideas of embodiment and whole agents (Houston and McNamara 1999; Kacelnik and Bateson 1996), but the action selection problem presented has yet to be solved in more than, the trivial way of a fixed threshold policy. In the report (Litus et al. 2007), the authors investigate refuel cycles, or as they call them the *basic cycles,* and show a simple rule based on cue and deficit. The majority of this work uses dynamic programming (DP) as a means of evaluating. In many studies in mobile robots, however, the issue of optimization of energy resources is missed. This work fills this gap and considers energy consumption and selection of the optimal mode of power plants (engines, batteries, fuel cells), while the trajectory choice is carried out (see also EG&G Technical Services, Inc. 2004).

8.4 Energetic Model of Mobile Robot

The energy needed by the robot to perform any task can be predicted with the help of energy consumption models. Energy consumption depends on power. A one-time process for generating the energy models can be developed by conducting some predetermined series of operations on the robot. The instantaneous energy consumption of the mobile robot is the sum of the energy consumed by its components or functions. The energy use of mobile robots could be divided into two categories: mobility and robotics consumption. The energy required for mobility is dependent on rover drive train, terrain type, traverse distance, and rover mass.

The total energy consumed by a robot is given as a sum of energy consumed by computer, controller, sensors, motors, and communication devices:

$$E_c = E_{comp} + E_{sens} + E_{motion} + E_{comun},$$

where E_c is the summary energy, E_{comp} is the energy for computer, E_{sens} is the energy for sensors, E_{motion} is the energy for motion and E_{comun} is the energy for communication.

Assume that all energy consumed by a robot is constant, except to motion, i.e.:

$$E_c = E_{const} + E_{motion},$$

where

$$E_{const} = E_{comp} + E_{sens} + E_{comun}.$$

The motion energy, in its turn, is defined as follows:

$$E_{motion} = E_{trac} + E_{man},$$

where E_{trac} is the traction energy and E_{man} is the maneuvering energy.

Using the corresponding forces F_{const}, F_{trac}, and F_{man} the energy E_c is represented by the following integral:

$$E_c = \int_{t_0}^{t} (F_{const} + F_{trac} + F_{man})ds.$$

The maneuvering force F_{man} can be defined in two manners. On one hand,

$$F_{man} = \frac{M_{man}}{r},$$

where M_{man} is the moment in steering systems and r is the length of the shoulder. On the other hand,

$$F_{man} = m_f g \varphi,$$

where m_f is the mass of the front park of the robot and φ is the friction coefficient.

Finally, the power P_W needed for the robot's motion can be calculated using the tracking force F_{trac} as

$$P_W = F_{trac} V,$$

where V is the velocity of the robot.

Given a rover on a certain terrain, energy required for mobility is approximately constant for a certain range, although this may vary slightly due to different steering activity. The mobility consumption does not depend on rover velocity or traveling time, since the mobility power increases with velocity, while traveling time decreases accordingly, which keeps the final mobility energy unchanged.

However, the energy for sensing, computing, and communication is expended at all-time whether moving or sitting. Thus, the energy consumption of robotic functions is mainly determined by total mission time, therefore by the rover speed in a given certain range. Some rovers do not move all the time during traverse. They have to stop intermittently for reasons such as navigation, planning, operation, data collection, and so forth. So they refer to the percentage of time that the rover is actually driving at payload operation in the total traverse time. In the time period when the rover stops and does not have mobility consumption, the robotic functions still continuously consume energy.

8.5 Mobile Robots Propulsion

The kinematic structures of wheeled mobile robots differ in the type and number of wheels and the positioning of the latter on the chassis of the robot. The driving wheels of the wheeled kinematic structures are symmetric about its principal or roll axis and rests on the ground on its contact patch. The contact patch is a small area which is in frictional contact with the ground such that the forces required to cause relative sliding between the wheel and ground are large for linear displacements and small for rotational motions. Thus, we assume that a wheel undergoing pure rolling has a contact point with no slip laterally or longitudinally, yet is free to twist about the contact point. The kinematic constraint of rolling is called a higher-pair joint. The kinematic pair has two constraints so that two degrees of freedom are lost by virtue of the rolling constraint. The main energy consumption in propulsion is the work to overcome resistance from terrain. In this section where only the mobility of robot is focused on, mobile robots are treated only as simple vehicles, not including the robotic parts. So the energy used for robot propulsion can be approximated by an ideal model that equates propulsion work to total engine energy. In this model, the propulsion work is the product of resistance force and distance. When a robot moves on paved surfaces and highways, it consumes energy to overcome the rolling resistance between the tires and the ground, as well as gravitational and inertial forces. At higher speeds, aerodynamic forces become the main mechanism of energy losses. The rolling resistance between the tire and the ground is attributed to tire slip, scrubbing in the contact patch, deflection of the road surface, and energy losses due to tire adhesion on the road and hysteresis. Rolling resistance varies with the type and material of the tire tread, the velocity of the vehicle, and environmental parameters such as temperature and humidity. For locomotion on unprepared, off-road terrain, the main mechanisms of energy losses are the wheel's compaction, bulldozing, and dragging of soil. On slopes, resistance due to the gravitational component parallel to a slope is an additional impediment to forward motion. The ability to overcome the resistance of an obstacle on a slope usually determines the extreme terrainability of the robot. The summary of such resistive forces is known as the "external motion resistance." Resistance to motion is also caused by frictional forces between drivetrain components, mechanical linkages, and hysteresis within the mechanical components of the robot. This resistance to motion is known as *internal motion resistance.* A study of the effects of motion resistance on the performance of a robot is accomplished by estimating configuration parameters that minimize the amount of energy dissipated into the terrain and the forces opposed to the motion of the wheel. Soil compaction resistance loss of soil thrust in unprepared terrains is primarily due to the compaction resistance of the soil. This form of motion resistance can be analyzed considering the mechanics of a wheel rolling into soft terrain.

8.5.1 Wheeled Mobile Robots Propulsion

In addition to soil compaction and bulldozing, motion resistance is caused by the deflection of the tire and the tread elements, wheel slip, and scrubbing at the wheel-soil interface. The combined effect of these forms of motion resistance is known as rolling resistance. The most common definition of rolling resistance is that it is the product between the vertical load applied on the wheel and an experimental coefficient. This formulation seems to deviate from the general principle of this thesis that a mathematical expression can be used as a configuration equation if it includes both configuration and performance parameters.

However, the values of both the coefficient of rolling resistance and the wheel loading depend on configuration parameters. The calculation of the coefficient of rolling resistance is a fairly involved process that considers various factors such as traveling speed, wheel slip, tire material, design, inflation pressure, temperature and loading, and the type of soil. Gravitational and inertial load distribution on the wheels depend on the geometric configuration of the chassis and mass distribution.

Consider the forces in the wheeled robot shown in Figure 8.1.

As indicated, in this robot, gravitational resistance F_w and ground slopes F_i are added with a component to the motion resistance F_f, which is proportional to the component of the total weight parallel to the slope.

Assuming the random location of a robot's center of gravity and that the robot is negotiating a combined crosshill/uphill slope, the gravitational resistance force on each wheel can be estimated, assuming that the magnitude of the gravitational load on a wheel is inversely proportional to the distance of the wheel contact from the projection of the center of gravity to the contact plane (defined by the contact points of at least three wheels). If crosshill/uphill performance is a critical design requirement, a detailed analytical investigation of the impact of the location of the center of gravity, number, and disposition of wheels on the optimal distribution of the gravitational load among the wheels is required.

When a wheeled robot is climbing an obstacle, an additional component of motion resistance is developed at the tire/obstacle interface due to the change in the normal

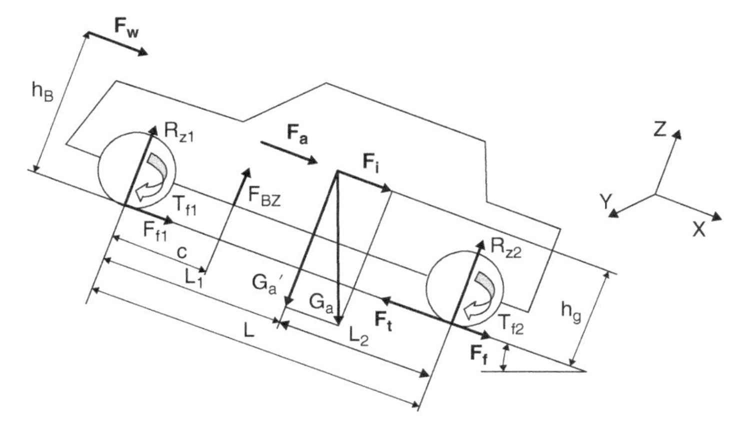

Figure 8.1 Forces acting on the wheeled robot.

contact force. In fact, as the posture of the robot changes due to obstacle climbing, so does the weight distribution over the wheels. This is a similar situation to that of a robot climbing a slope, but in this case the "grade" is determined by the angle between the line that connects the front/rear wheel contact points and the ground level. Modeling the resistance force due to obstacle climbing in soft soils and compliant obstacles is an extremely involved process that goes beyond the scope of robotic locomotion configuration. However, a manageable configuration equation of obstacle resistance can be derived from the equations of static equilibrium of a robot climbing a discrete obstacle on a hard surface or compacted soil.

Consider, for instance, the case of a four-wheel wheel-drive robot with rigid suspension climbing a square obstacle. The tracking force for the robot of the mass m can be calculated as follows:

$$F_{\text{trac}} = F_f \pm F_i \pm F_w \pm F_a,$$

where F_{trac} is the tracking force discussed in Section 8.4, $F_f = Wf = mgf$ is the rolling resistance force of the road with the rolling resistance coefficient f, $F_i = Wi = mgi$ is the grade resistance force with the scope of road $i = \sin\alpha$, $F_w = \frac{1}{2}\rho cAV^2$ is the aerodynamic resistance force while ρ is the air density, c is the coefficient of aerodynamic resistance of the robot, A is the characteristic area of the robot and V is the average speed of the robot, and $F_a = ma\gamma_m$ is the power required to overcome the resistance of the inertial acceleration while a is the acceleration of robot and γ_m is the mass factor of robot.

When the robot is skid steering, the lateral motion of the wheels causes a significant dissipation of energy due to the bulldozing and compaction of the terrain. The lower the wheel linkage, the higher the efficiency of the steering motions and the lower the power draws to complete those motions.

Consider the case of a four-wheel skid steered robot. To achieve a specified heading change, the wheels on the left and right sides of the chassis are served to different velocities. This particular steering is known as differential steering. Assuming that the robot is on level ground and that the contact pressure is uniformly distributed on each wheel, the wheels are subjected to longitudinal resistive forces F_i (primarily due to soil compaction) and lateral resistive forces (due to scrubbing on the ground or bulldozing of the soil). When the skid steering maneuver is performed at low speeds, one can describe the kinetics of the robot in steady-state terms.

8.5.2 Propulsion of Mobile Robots with Caterpillar Drive

Now consider the forces in the robot with caterpillar drive shown in Figure 8.2.

For such a robot of the mass m, the tracking force

$$F_{\text{trac}} = R_f \pm F_i \pm F_w \pm F_a,$$

where the forces F_f, F_i F_w, and F_a are defined as above, is often calculated by empiric formula (Austin and Kouzoubo 2002):

$$F_{\text{trac}} = 10^6 b \int_0^L \tau_x dx,$$

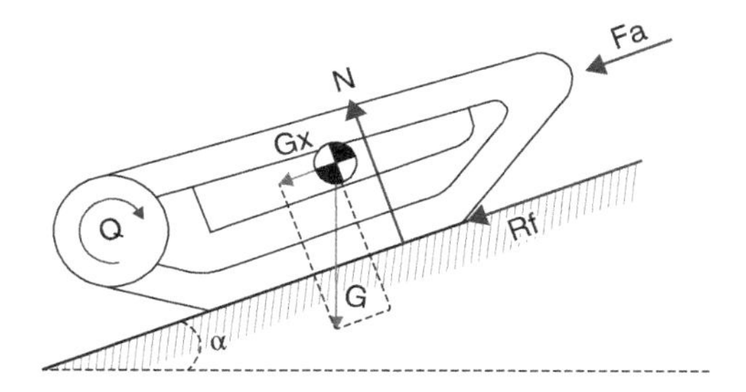

Figure 8.2 Forces acting on the robot with caterpillar drive.

where b is a track width, L is the track carrier segment length, and τ_x is a shear stress in the ground.

The maximum shear stress in the ground is defined by the Coulomb model (Austin and Kouzoubo 2002):

$$\tau_{\max} = c + \mu_0\sigma = c + \sigma\tan\rho,$$

where ρ is an internal friction angle of the ground particles, σ stands for the compressive stresses in the ground, μ_0 is the friction coefficient between the ground particles together, and c is a density of the ground.

The shear stress depends on the deformation based on the mathematical analogy between the course of the curves of shear stress and the course of the damped oscillation amplitude of the curve. They have the following form (Bekker 1956):

$$\tau_x = (c + \sigma\tan\rho)\frac{\exp\left[\left(-K_2 + \sqrt{K_2^2 - 1}\right)K_1\Delta l_x\right] - \exp\left[\left(-K_2 - \sqrt{K_2^2 - 1}\right)K_1\Delta l_x\right]}{Y_{\max}},$$

where $Y_{\max}$ is the maximal value of the expression given in the numerator fraction, Δl_x is deformation of the ground layer at the x-axis caused by slipping parallel to the ground, K_1 is the coefficient of the ground deformation during the shearing, K_2 is the coefficient characterizing the curve.

Assuming that in the course of deformation parallel to the ground it is linear, this deformation can be expressed by the formula

$$\Delta l_x = x s_b,$$

where s_b is a slip and x is the distance from the point for which the slip is calculated to the point of contact with the ground of track; the largest slip occurs for $x = L$.

On the bases of the description connected with the contact of the track with the ground, it is possible to describe the rotation of our robot in the (x, y) system of coordinates. It is necessary to assume the characteristic point C of the robot. The scheme of robot motion has been presented in the velocity components of point C. After extension by angular velocity of the frame, we receive kinematics equations in the form that allows us to solve the forward kinematics problem (Kim and Kim n.d.; Wei et al. 2012):

$$\dot{x}_C = \frac{r\dot{\alpha}_1(1 - s_1) + r\dot{\alpha}_2(1 - s_2)}{2}\cos\beta,$$

$$\dot{y}_C = \frac{r\dot{\alpha}_1(1 - s_1) + r\dot{\alpha}_2(1 - s_2)}{2}\sin\beta,$$

$$\dot{\beta} = \frac{r\dot{\alpha}_2(1-s_2) - r\dot{\alpha}_1(1-s_1)}{H},$$

where H is distance between the axes of the robot.

In order to achieve the desired trajectory, it is necessary to solve the inverse kinematics problem. The inverse kinematics equations are presented in the following form:

$$V_C = \sqrt{\dot{x}_C^2 + \dot{y}_C^2},$$

$$\dot{\alpha}_1 = \frac{V_C - 0.5\dot{\beta}H}{r(1-s_1)},$$

$$\dot{\alpha}_2 = \frac{V_C + +0.5\dot{\beta}H}{r(1-s_2)}.$$

These kinematic equations allow for controlling the position and orientation of the robot and tracing the desired trajectory.

Calculation of energy consumption provided according following scenes:

1) Free path at a predetermined speed
2) The trajectory of a given point
3) The free trajectory on the given point
4) The trajectory at a predetermined power
5) Given point in free or predetermined trajectory

for a variety of energy resources:

- ICE
- Electric batteries
- Fuel cells

In the next section, we consider the energetic model of the robot with different energy resources.

8.6 Energetic Model of Mechanical Energies Sources

This section considers the energetic model that analyzes energy utilization in mobile robot traverse and estimates maximum range achievable by wheeled and caterpillar mobile robots operating on a single battery discharge. After taking into account different energy utilizations, such as propulsion and steering, the model indicates that, as already noted, the most energy-consuming part of a mobile robot is its robotics functions, such as computing, sensing, and communication. Based on this, it points out ways to improve maximum robot traverse range: increasing rover velocity, driving duty cycle (ratio of driving time to total mission time), and decreasing robotics functions power.

Considering the significant energy proportion of robotics functions, the leftover propulsive consumptions are analyzed, which directly determines the maximum range using a classic terramechanics model. The proportions of energy expended in wheel and caterpillar drive robot. It is known that the energetic efficiency of lithium batteries depends on the rate of discharge.

8.6.1 Internal Combustion Engines

The energy required for mobility is dependent on mode of robot motion:

$$E_s = E_1 + E_2 + E_3 + E_4,$$

where E_1 is the energy required for overcoming the forces of resistance at constant speed (cruise speed) mode, E_2 is the energy required for overcoming the forces of resistance at constant acceleration mode, E_3 is the energy required for overcoming the forces of resistance at constant deceleration mode, and E_4 is the energy required for idle mode.

Energy E_1 can then be calculated as follows:

$$E_1 = \frac{1}{\eta_t}\sum_{i=1}^{k}\frac{E_{cr_i}}{\eta_{Pn_i}}, \quad E_{cr_i} = \left(mgc_r + \frac{1}{2}\rho c A_f V_{a_i}^2\right)S_i,$$

where m is the mass of the robot, g is the acceleration of gravity, c_r is the rolling resistance coefficient, A_f is the characteristic area of the robot, V_{a_i} is the cruise speed of the vehicle, S_i is the distance at average speed intervals, η_t is the efficiency of the transmission, and η_{Pn} is the varying efficiency of the engine, which depends on the coefficient of the degree of power utilization and the coefficient of the engine speed mode (Ben Haim and Shmerling 2013; Ben Haim et al. 2013):

$$\eta_{Pn} = \eta_e \mu_P \mu_n,$$

where η_e is the effective (maximal) efficiency of the engine, $\mu_P = f\left(\frac{P_i}{P_e}\right)$ is the coefficient where P_i is the instantaneous required engine power, P_e is the instantaneous actual engine power, through which the influence of the tractive efficiency on the effective efficiency of the engine is expressed, and μ_n is the coefficient through which the influence of engine speed mode on the effective efficiency of the engine is expressed. The value of the parameters μ_P, μ_n, P_i and P_e are determined by the methodology presented (Ben Haim and Shmerling 2013; Ben Haim et al. 2013). Notice that from the viewpoint of minimizing energy consumption, the speed of the robot should be chosen in such a manner that the coefficient η_{Pn} reaches its maximum.

The energy required for motion with accelerations E_2 is calculated as follows:

$$E_2 = \frac{1}{\eta_t}\sum_{j=1}^{q}\frac{E_{ac_j}}{\eta_{Pn_j}},$$

where q is the number of acceleration intervals (different for urban and nonurban traffic). The energy E_{ac_j} required for motion with accelerations E_2 is calculated via the following formula:

$$E_{ac_i} = \left(mgc_r + KA_f V_{av_j}^2 + m_a a_j \gamma_{m_j}\right)S_j,$$

where V_{av_j} is the average speed in interval of the accelerations, γ_{m_j} is the mass factor of the vehicle, a_j is the acceleration of the vehicle, and S_j is the acceleration distance of the vehicle.

The mass factor of the vehicle is calculated by empiric equation (Ben Haim and Shmerling 2013; Ben Haim et al. 2013):

$$\gamma_{m_j} = 1 + \tau \xi_k^2,$$

where τ is a coefficient of the empirical equation that for wheeled vehicles is $\tau = 0.05 \div 0.07$ and ξ_k is the gear box ratio for a definite average speed of vehicle.

Energy E_3 is equivalent to fuel consumption at movement with deceleration, and energy E_4 is equivalent to the energy of fuel consumption expended at idling mode.

At the deceleration mode, in actual, power from the engine is not required. Therefore, E_3 is the energy expended when the vehicle is moving with deceleration mode. Part of the energy, and therefore, the amount of fuel consumption during deceleration mode, has been determined analytically under a new control technology of gasoline (also of diesel) engines with fuel injection – common rail (Ben Haim et al. 2015; Ben Haim and Leybovich 2014). By this control technology, fuel supply is stopped after $2.0 \div 3.0$ seconds from the beginning of the deceleration (braking) and restored when vehicle reaches the desired final speed or the idling mode, i.e. when $n_e \rightarrow n_{im}$, where n_{im} is the engine speed at time of idling mode.

As the deceleration mode always follows after the constant speed mode, one can use for calculation the same expression as for constant average speed:

$$E_3 = \frac{1}{\eta_t} \sum_{i=1}^{k} \frac{V_i}{\eta_{Pn_i}} \left(m_a g c_r + K A_f V_{a_i}^2 \right) t_{d_i},$$

where V_{a_i} is the cruise speed of the vehicle before deceleration and t_{d_i} is a time allocated for engine control ($2 \div 3$ seconds) (Ben Haim and Leybovich 2014).

Finally, energy E_4 is defined as follows (Ben Haim et al. 2013):

$$E_4 = \sum_{z=1}^{r} \frac{n_{im}}{120} \frac{1}{14.7\lambda} V_h \eta_v \frac{\rho_a}{\rho_f} H_l t_z,$$

where n_{im} is an engine speed at idling mode, λ is the fuel-to-air equivalence ratio, V_h is the engine displacement, ρ_f is the fuel density, and $\rho_a = \frac{P_s}{RT_s}$ is the air (air-fuel mixture) density at time of idling mode, where P_s is the mixture pressure at time of idling mode, R is the gas constant, T_s is the mixture temperature at idling mode, η_v is the volumetric efficiency at idling mode, and t_z is the time of idling mode.

8.6.2 Lithium Electric Batteries

The battery-powered robots without replenishment represent the generality of the most mobile robots and provide the simplicity to focus on a constant energy amount while examining energy utilization. Total energy for driving is primarily independent on speed since driving energy is primarily related to rover mass, gravity, distance traveled, and road resistance. By comparison, the energy for robotics functions (e.g., sensing, computing, communication) utilizes considerable power, whether driving or sitting.

It is the optimal temperature range for LIBs is 15 °C–35 °C. Hence, we can design the system so that there is a good air circulation around the battery while driving and, if necessary, we can design the battery cooling ribs to keep it as low a temperature as possible. In addition, literature given us the optimal discharge rate system loses energy. Minimum battery discharge is a discharge rate of between 5% and 10% when the battery is charged. This discharge rate discharge efficiency is 99%.

In view of the above the best battery power source lithium batteries for optimal discharge rate.

8.7 Summary

In the chapter, we addressed the problem of energy limitations of mobile robots during their motion and considered the energetic model for mobile robot (Section 8.4). The total energy consumed by the robot includes the energy required by computer, controller, sensors, motors, and communication devices. However, since most of the energy is used for supplying the robot's mobility, the model is concentrated on the energy required for the robots' motion:

1) In this chapter, we considered the two main types of robots: the wheeled robots (Section 8.5.1) and the robots with caterpillar drive (Section 8.5.2). For both types of drives, we considered the forces acting on the robots and defined the parameters required for definition of energy consumption.
2) The energetic model of the robot (Section 8.6) considers the energies required for overcoming the resistance forces at the cruise mode, at the acceleration and deceleration modes, and idle mode. In addition to the main consideration of internal combustion engine, the section included certain remarks on the usage of lithium batteries and corresponding velocity modes.

In addition, the chapter provided a detailed review (Section 8.3) of the contemporary sources in the field of power management and energy control in mobile robots.

References

Austin, D. and Kouzoubo, K. (2002). Robust, long term navigation of a mobile robot. In: *Proc. IARP/IEE-RAS Joint Workshop on Technical Challenges for Dependable Robots in Human Environments*, 67–72. Institute of Engineering and Technology (IET).

Austin, D., Fletcher, L., and Zelinsky, A. (2001). Mobile robotics in the long term – exploring the fourth dimension. In: *Proc. IEEE/RSJ Int. Conf. Intelligent Robotics and Systems*, 613–618. Institute of Electrical and Electronics Engineers (IEEE).

Bekker, M.G. (1956). *Theory of Land Locomotion: The Mechanics of Vehicle Mobility*. University of Michigan Press.

Ben Haim, M. and Leybovich, E. (2014). Fuel consumption at conditions of the Israel's higways driving cycle. In: *Proc. Int. Conf. Energy, Environment, Development and Economics*, 92–94. Santorini Island, Greece. World Energy and Environment Technology.

Ben Haim, M. and Shmerling, E. (2013). A model of vehicle fuel consumption at conditiond of the URBAN. *International Journal of Mechanics* 7: 10–17.

Ben Haim, M., Shmerling, E., and Kuperman, A. (2013). Analytic modeling of vehicle fuel consumption. *Energies* 6 (1): 117–127.

Ben Haim, M., Avrahami, I., and Sapir, I. (2015). Analytical determination of fuel consumption the car with diesel engine at urban driving cycle. In: *Int. Conf. Energy and Environment Engineering and Management*, 110–112. Paris, France. World Energy and Environment Technology.

Bonani, M., Longchamp, V., Magnenat, S. et al. (2010). The MarxBot, a miniature mobile robot opening new perspectives for the collective robotic research. In: *Proc. Int. Conf. Intelligent Robotis and Systems*. Taiwan. Institute of Electrical and Electronics Engineers (IEEE).

Chuy, O., Collins, E., Yu, W., and Ordonez, C. (2009). Power modeling of a skid steered wheeled robotic ground vehicle. In: *Proc. IEEE Int. Conf. Robotics and Automation*, 4118–4123. Institute of Electrical and Electronics Engineers (IEEE).

Deshmukh, A., Vargas, P., Aylett, R., and Brown, K. (2010). Towards socially constrained power management for long-term operation of mobile robots. In: *Proc. Towards Autonomous Robotic Systems TAROS'10*. Springer Verlag.

EG&G Technical Services, Inc. (2004). *Fuel Cell Handbook*. Morgantown, WV: US Department of Energy Office of Fossil Energy, National Energy Technology Laboratory.

Hedenstrom, A. (2003). Optimal migration strategies in animals that rum: a range equation and its consequences. *Animal Behavior* 66: 631–636.

Houston, A.I. and McNamara, J.M. (1999). *Models of Adaptive Behavior*. Cambridge: Cambridge University Press.

Kacelnik, A. and Bateson, M. (1996). Risky theories: the effects of varience on foraging decisions. *American Zoologist* 36: 402–434.

Kim, C. and Kim, B. (2008). Minimum-energy motion planning for differential driven wheeled mobile robots. In: *Mobile Robots Motion Planning: New Challenges* (ed. X.-J. Jing), 193–226. IntechOpen.

Lamon, P. (2008). The Solero rover. 3D-position tracking and control for all-terrain robots. In: *Springer Tracts in Advanced Robotics*, vol. 43, 7–19. Berlin, Heidelberg: Springer.

Litus, Y., Vaughan, R.T., and Zebrowski, P. (2007). The frugal feeding problem: energy-efficient, multi-robot, multi-place rendezvous. In: *Proc. IEEE Int. Conf. Robotics and Automation*, 27–32. Roma, Italy. Institute of Electrical and Electronics Engineers (IEEE).

Liu, J., Chou, P., Bagherzadeh, N., and Kurdahi, F. (2001). Power-aware scheduling under timing constraints for mission-critical embeded systems. In: *Proc. Conf. Design and Automation*, 840–845. Institute of Electrical and Electronics Engineers (IEEE).

Mei, Y., Lu, Y.-H., Hu, Y., and Lee, C.S. (2004). Energy-efficient motion planning for mobile robots. In: *Proc. IEEE International Conference on Robotics and Automation*, vol. 5, 4344–4349. Institute of Electrical and Electronics Engineers (IEEE).

Mei, Y., Yung-Hsiang, Y., Hu, L., and Lee, C. (2005). A case study of mobile robot's energy consumption and conservation techniques. In: *Proc. 12th Int. Conf. Advanced Robotics*, 492–497. Institute of Electrical and Electronics Engineers (IEEE).

Mei, Y., Lu, Y.-H., Hu, Y., and Collins, E. (2006). Deployment of mobile robots with energy and timing constraints. *IEEE Transactions on Robotics* 22 (3): 507–522.

Morales, J., Martinez, J., Mandow, A. et al. (2009). Power consumption modeling of skid-steer tracked mobile robots on rigid terrain. *IEEE Transactions on Robotics* 25 (5): 1098–1108.

Oh, S. and Zelinsky, A. (2000). Autonomous battery recharging for indoor mobile robots. In: *Proc. Australian Conference on Robotics and Automation*. Australian Robotics and Automation Association.

Wailwa, H. and Birk, A. (1996). Learning to survive. In: *Proc. 5th Europ. Workshop Learning Robots*, 1–8. Bari, Italy. Springer Verlag.

Wei, H., Wang, B., Wang, Y. et al. (2012). Staying-alive path planning with energy optimization for mobile robots. *Expert Systems with Applications* 29 (3): 3559–3571.

Yu, W., Collins, E., and Chuy, O. (2011). Dynamic modeling and power modeling of robotic skid-steered wheeled vechicles. In: *Mobile Robots: Current Trends* (ed. D.Z. Gacovski), 291–318. IntechOpen.

Zhuang, H.-Z., Du, S.-X., and Wu, T.-J. (2005). Energy-efficient motion planning for mobile robots. In: *Proc. Int. Conf. Machine Learning and Cybernetics*. Springer Verlag.

9

Multi-Robot Systems and Swarming

Eugene Kagan, Nir Shvalb, Shlomi Hacohen, and Alexander Novoselsky

This chapter provides an introduction into collective behavior in multi-robot systems. In particular, the chapter considers different types of communication and control in multi-agent systems, basic flocking rules, and general methods of aggregation and collision avoidance. In addition, the chapter considers the corresponding networking techniques, cybernetic concepts, and positioning problems.

9.1 Multi-Agent Systems and Swarm Robotics

In previous chapters, we considered the motion of a single robot in static and dynamic environments and the methods of motion planning with complete and incomplete information. This section starts the consideration of multi-robot systems, in which the robots act in the teams and collaborate for executing definite common mission. It is necessary to stress that in the multi-robot systems the mission and the goal are defined for the team as a unit (Balch and Parker 2002), while the missions and the goals of individual robots of the team can strongly differ both from the team's mission and from the mission and the goal of each other robot.

The activity of multi-robot systems in general and of the teams of mobile robots can be considered from different points of view. For example, in the case of a manufacturing line that includes arm manipulators, conveyers, machines with computer numerical control (CNC) and automatic storages, positions of all the devices in global coordinates of the manufacturing line and available trajectories are known beforehand. Then, the activity of the line is defined according to the production needs, and the problem of motion planning is reduced to the problem of synchronization of the robotic manipulators and other equipment. Communication between the devices is conducted via the central unit that controls the activity of the line. The similar situation appears in the team of mobile robots, in which each robot has access to a global positioning system (GPS) or similar device and communication and control are conducted using a central unit (see Section 10.1).

Another situation appears in the multi-robot systems, which consist of mobile robots acting without complete access to global coordinates and/or with limited

Autonomous Mobile Robots and Multi-Robot Systems: Motion-Planning, Communication, and Swarming, First Edition. Edited by Eugene Kagan, Nir Shvalb and Irad Ben-Gal.

Companion website: www.wiley.com/go/kagan/robotsystems

communicating and sensing abilities. In this case, each mobile robot acts under uncertainties led by the environment, sensors, actuators, models, and computation abilities (see Section 7.1) and by the dynamics of the other robots in the system. Thus, the motion of each robot is conducted in the dynamical partially observed environment and should be planned using the appropriate techniques, which can strongly differ from the methods used for motion planning in static environments. Chapter 7 considers the probabilistic methods of motion planning under uncertainties for a single robot; in this section, these methods are extended for navigation of the robot's teams.

9.1.1 Principles of Multi-Agent Systems

The most general approach to the studies of the teams of mobile robots acting under uncertainties follows the philosophy of multi-agent systems (Shoham and Leyton-Brown 2008; Weiss 1999; Wooldridge 2009). According to Wooldridge (2009, p. xiii):

> *multi-agent systems are systems composed of multiple interacting computing elements, known as* agents. *Agents are computer systems...* [such that] *they are at least to some extent capable of* autonomous action – *of deciding* for themselves *what they need to do...* [and] *are capable of interacting with other agents – not simply by exchanging data, but by engaging in analogous of the kind of social activity...*

The indicated properties of the agents are also known as (Wooldridge 1999, p. 32):

- *Reactivity* is the ability of perceiving the agent's environment and responding to its changes.
- *Pro-activeness* is the goal-directed behavior satisfying the desired objectives.
- *Social ability* is the ability to interact with other agents.

It is clear that the last property is crucial and forms a basis for consideration of the set of individual agents as a multi-agent system, while the first two properties are necessary for the intelligent behavior that implies decision-making abilities. The agents that demonstrate the decision-making abilities are often considered as *intelligent agents*, and the decision-making conducted by such agents acting in groups is considered as distributed decision-making.

There are several approaches to decision-making in multi-agent systems. For example, following the "social choice" methods, decision-making is considered as a voting process, in which each agent declares its individual decision and collective decision is obtained by certain average or majority rule (Shoham and Leyton-Brown 2008; Wooldridge 2009). Such an approach is essentially based on the game theory, especially when implementing the methods of nonantagonistic games.

The other approach follows general methods of Markov decision processes (MDPs) (White 1993; Ross 2003) that in the case of uncertain observations are considered as a partially observable Markov decision process (POMDP) (Aoki 1967; Monahan 1982) or hidden Markov models (HMMs) (Rabiner 1989). Algorithms that implement these methods for navigation of a single robot were considered in Section 7.2.2; an extension of these algorithms for decision-making in multi-agent systems is presented in Section 9.2.2.

In spite of different directions in the considerations of multi-agent systems, both game theory and MDP models imply that even the agents acting under uncertainties are completely informed about their goals, the goal of the system, and goals of the other agents. In addition, these models assume that the agents act rationally in such a sense that they always tend to maximize actual or expected reward (or minimize actual or expected payoff).

However, notice that the success of a certain agent resulted by maximization of its own reward (as well as minimization of its own payoff) does not lead to the failure of other agents. In other words, the agents are considered as "egoistic" or self-interested in dealing with their own incomes without taking into account the payoffs and rewards of other agents. Certainly, it does not mean that the agents care only about themselves but that each agent defines how it wants to see the world and acts in such a manner that changes the world according to the agent's beliefs.

The well-known approach for representing such rational behavior of the agents is based on the utility theory (Friedman and Savage 1948), which provides quantitative representation of the alternatives' desirability and considers the alternating options both in uncertain and risky conditions. Such representation implements the game theory approach (von Neumann and Morgenstern 1944) and applies the *utility function* that maps the states of the world or the agent's preferences and risk tolerance to real numbers. Formally, it is defined as follows (Luce and Raiffa 1957; Shoham and Leyton-Brown 2008); the consideration below partially follows the report by Sery and Feldman (2016).

Let $\mathcal{A} = \{a_1, a_2, \ldots, a_n\}$ be a finite set of all possible alternatives of the agent; often, the alternatives are interpreted as the outcomes obtained as a result of choosing certain alternative $a \in \mathcal{A}$ or as a result of the action represented by this alternative. Assume that for every pair of the alternatives $a_i, a_j \in \mathcal{A}$, $i, j = 1, 2, \ldots, n$, there are defined the following relations:

- *Preference "$\succ$" (often called strong preference).* For the alternatives a_i and a_j, relation $a_i \succ a_j$ means that for the agent the alternative a_i is preferable to the alternative a_j.
- *Indifference "$\sim$".* For the alternatives a_i and a_j, relation $a_i \sim a_j$ means that the agent has no preferences while considering the alternatives a_i and a_j.

The relation between the alternatives a_i and a_j such that the agent prefers a_i to a_j or indifferent regarding these alternatives is denoted by $a_i \succcurlyeq a_j$ and is often called weak preference. Below, both strong and weak preferences are indicated by the term *preference* with corresponding notation "$\succ$" or "$\succcurlyeq$," if it is needed.

Denote by $\ell = [p(a_1), p(a_2), \ldots, p(a_n)]$ the probability distribution over the set $\mathcal{A}$ that defines the objective probabilities $p(a_i)$ of obtaining the alternatives $a_i \in \mathcal{A}$, $i = 1, 2, \ldots, n$. It is assumed that the probabilities $p(a_i)$ do not depend on the agent's relation to the corresponding alternatives a_i and are defined by external objective factors. For example, if the alternatives $a_i \in \mathcal{A}$ are the results of throwing a dice, then of the probabilities

$p(a_1) = p(a_2) = \ldots = p(a_6) = 1/6$ with no respect to the agent's preferences and expected results of the throwing in the particular game. The probability distribution ℓ with such a property is called lottery.

In order to obtain the quantitative representation of the agent's relation to the alternatives it is assumed that the relations $\succ$ and $\sim$ (and consequently the relation $\succcurlyeq$) meet the following axioms (Shoham and Leyton-Brown 2008):

1) *Completeness.* For each pair $a_i, a_j \in \mathcal{A}$ of alternatives it holds true that either $a_i \succ a_j$ or $a_j \succ a_i$ or $a_i \sim a_j$. In other words, the preference relation is defined over all pairs of the alternatives.
2) *Transitivity.* For all triples $a_i, a_j, a_k \in \mathcal{A}$ of the alternatives it hold true that if $a_i \succcurlyeq a_j$ and $a_j \succcurlyeq a_k$, then $a_i \succcurlyeq a_k$. Such a property guarantees that the preferences follow a linear structure and do not introduce cyclic relations between the alternatives.
3) *Decomposability.* For each alternative $a_i \in \mathcal{A}$ and all pairs of the lotteries ℓ' and ℓ'' it holds true that if $p_{\ell'}(a_i) = p_{\ell''}(a_i)$, then $\ell' \sim \ell''$, where $p_\ell(a_i)$ stands for the probability of obtaining the alternative a_i in the lottery ℓ. In other words, the agent is indifferent between the lotteries that provide certain alternative with the same probabilities.
4) *Monotonicity.* If for the alternatives $a_i, a_j \in \mathcal{A}$ it holds true that $a_i \succ a_j$ and $p(a_i) > q(a_j)$, where p and q are two probability distributions defined over $\mathcal{A}$, then for the lotteries $\ell' = [p(a_i), 1 - p(a_j)]$ and $\ell'' = [q(a_i), 1 - q(a_j)]$ it holds true that $\ell' \succ \ell''$. It means that the agent prefers the lottery, which provides the preferable alternative with greater probability.

The presented axioms lead to the following lemma that states an existence of the probability distributions, which completely define the relations between the alternatives.

Lemma 9.1 Let the preference relation "$\succ$" satisfy the indicated axioms 1–4. If for the alternatives $a_i, a_j, a_k \in \mathcal{A}$ it holds true that $a_i \succ a_j$ and $a_j \succ a_k$, then there exists some probability distribution p where for all probability distribution p' such that $p'(a_i) < p(a_i)$ it holds true that $a_j \succ [p'(a_i), 1 - p'(a_k)]$, and for all probability distribution p'' such that $p''(a_i) > p(a_i)$ it holds true that $a_j \prec [p''(a_i), 1 - p''(a_k)]$. This lemma states the possibility to formulate the preference relation between the alternatives in the terms of probabilities of obtaining these alternatives in certain lotteries.

Finally, for existence of the utility function, it is assumed that the preference relation meets an additional axiom (Shoham and Leyton-Brown 2008):

5) *Continuity.* If for the alternatives $a_i, a_j, a_k \in \mathcal{A}$ it holds true that $a_i \succ a_j$ and $a_j \succ a_k$, then there exists a probability distribution p such that $a_j \sim [p(a_i), 1 - p(a_k)]$. In other words, it is possible to define the lottery with probabilities of the alternatives a_i and a_k such that the agent is indifferent between obtaining the alternative a_j with certainty and participating in this lottery.

The existence of the utility function is stated by the following theorem by von Neumann and Morgenstern.

Theorem 9.1 (von Neumann and Morgenstern 1944). If the preference relation "$\succcurlyeq$" satisfies axioms 1–5, then there exists a real-valued function u with the values from the interval [0, 1] such that:

1) $u(a_i) \geq u(a_j)$ if and only if $a_i \succcurlyeq a_j$, $a_i, a_j \in \mathcal{A}$;
2) $u(p(a_1), p(a_2), \ldots, p(a_n)) = \sum_{i=1}^{n} p(a_i) u(a_i)$.

As indicated above, the utility function u forms a basis for the utility theory that represents the decision-making with rational judgments.

Notice that in addition to the indicated axioms, it is usually assumed that the preference relation satisfies the axiom.

6) *Substitutability*. If for the alternatives $a_i, a_j \in \mathcal{A}$ it holds true that $a_i \sim a_j$, then for all sequences $[a_1, a_2, \ldots, a_{i-1}, a_{i+1}, \ldots, a_{j-1}, a_{j+1}, \ldots, a_n]$ of the alternatives, the probability distributions such that $p + \sum_{k=1, k \neq i, k \neq j}^{n} p(a_k) = 1$ and the lotteries

$$\ell_{-i} = [p(a_1), p(a_2), \ldots, p(a_{i-1}), p(a_{i+1}), \ldots, p(a_{j-1}), p(a_j), p(a_{j+1}), \ldots, a_n]$$

and

$$\ell_{-j} = [p(a_1), p(a_2), \ldots, p(a_{i-1}), p(a_i), p(a_{i+1}), \ldots, p(a_{j-1}), p(a_{j+1}), \ldots, a_n]$$

it holds true that $\ell_{-i} \sim \ell_{-j}$.

Following this axiom, if the agent is indifferent regarding the alternatives, then it is indifferent regarding the lotteries, in which the alternatives appear with the same probabilities and all other alternatives and their probabilities are also the same.

The possibility of probabilistic representation of the relations between the agents' preferences forms a basis for the noticed above Markov models of decision-making, where the agents' rewards are defined by the means of their utility functions. In particular, if the alternatives $a \in \mathcal{A}$ are interpreted as the actions $\mathfrak{a} \in \mathfrak{A}$ of the agent, then the activity of a single agent is defined by the well-known value iteration and policy iteration algorithms (see Section 7.2.2), in which the agent follows such a policy that maximizes its expected utility:

$$\mathcal{E}\left[\sum_{t=0}^{T} \gamma^t u(\mathfrak{s}(t), \mathfrak{a}(t))\right], \tag{9.1}$$

where $u(\mathfrak{s}(t), \mathfrak{a}(t)) \geq 0$ is an immediate utility obtained by the agent while being in the state $\mathfrak{s}(t) \in \mathfrak{S}$ at time t the agent executes the action $\mathfrak{a}(t) \in \mathfrak{A}$; as in the Section 7.1, $\mathfrak{S}$ stands for the set of states, $\mathfrak{A}$ stands for the set of the available actions and $\gamma \in (0, 1]$ is a discount factor. In the finite horizon setup, it is assumed that the time is $T < \infty$ and the discount factor is $\gamma = 1$, and for the infinite horizon setup $T = \infty$ and $\gamma < 1$. Notice that in the Markov decision models considered in Section 7.2.2, the agent's utility u is directly interpreted as a reward r obtained by the agent after execution of the chosen action.

In the case of multi-agent systems, the activities of the agents are usually described in the game theory terms (von Neumann and Morgenstern 1944; Luce and Raiffa 1957;

Shoham and Leyton-Brown 2008) (for introduction into the games theory, see, e.g., Osborne and Rubinstein (1994)). In the normal form, the game of m agents is defined over the Cartesian product $\mathfrak{A}^m = \mathfrak{A}_1 \times \mathfrak{A}_2 \times \ldots \times \mathfrak{A}_m$ of the agents' actions, where $\mathfrak{A}_i$, $i = 1, \ldots, m$, is a set of actions available to the agent i, and common utility $\vec{u} = (u_1, u_2, \ldots, u_m)$, where $u_i : \mathfrak{A}^m \to \mathbb{R}$ is a utility (or payoff) of the agent i, which defines its reward (or payoff) in the game. In game theory, the vector $\vec{\alpha} = (\alpha_1, \alpha_2, \ldots, \alpha_m) \in \mathfrak{A}^m$, where α_i, $i = 1, \ldots, m$, is an action chosen by the ith agent is called the *action profile* (Shoham and Leyton-Brown 2008), and in the artificial intelligence researches and the studies of multi-robot systems it is called *joint action* (Seuken and Zilberstein 2008).

As above, the goal of each agent is to find such a policy that maximizes its own reward (or minimizes its own payoff). However, notice that for each agent its utility is defined over the product set $\mathfrak{A}^m$ and, consequently, the reward (or payoff) of each agent depends on the actions of all m agents. To represent such a situation, the solution of the game that specifies the optimal policies of the agents is defined in the terms of equilibrium, which provides the best responses of the agents to the actions of the other agents.

Denote by $P_i = P(\mathfrak{A}_i)$ the set of all probability distributions over the set $\mathfrak{A}_i$ of actions of the ith agent, $i = 1, \ldots, m$, and by $p_i \in P_i$ the particular probability distribution from this set. The probability distribution p_i is called the mixed strategy of the agent i and, consequently, the set P_i contains all mixed strategies of this agent. The value $p_i(\alpha_i)$ is interpreted as a probability that the ith agent will choose the action α_i under its mixed strategy p_i. It is clear that for each ith agent, such probabilities of choosing the actions $\alpha_i \in \mathfrak{A}_i$ are literally the same as the previously discussed probabilities $p(a)$ of choosing the alternatives $a \in \mathcal{A}$ under the certain lottery ℓ.

Finally, denote by $\vec{p} = (p_1, p_2, \ldots, p_m) \in P_1 \times P_2 \times \ldots \times P_m$ the vector of the agents' mixed strategies that represents the mixed strategies profile. Then, the expected utility $u_i\left(\vec{p}\right)$ of the ith agent, $i = 1, \ldots, m$, in the game is defined as follows (Shoham and Leyton-Brown 2008):

$$u_i\left(\vec{p}\right) = \sum_{\vec{\alpha} \in \mathfrak{A}^m} u_i\left(\vec{\alpha}\right) \prod_{j=1}^{m} p_j(\alpha_j). \tag{9.2}$$

In other words, in the game in which the agents 1, 2, …, m choose their actions α_1, α_2, …, α_m according to their probability distributions p_1, p_2, …, p_m, the expected utility of the ith agent is defined as a sum of the expected utilities obtained for each common action $\vec{\alpha} = (\alpha_1, \alpha_2, \ldots, \alpha_m)$, and the probability of choosing this action is specified by the product of the probabilities $p_j(\alpha_j)$ that the agents $j = 1, 2, \ldots, m$ will chose their individual actions α_j.

Such a definition assumes that the agents' activity is considered from the external point of view, where the observer analyses the already-conducted choices of all agents. However, from the point of view of each individual agent, its choice is conducted as a response to the choices of the other agents. To define such a situation, denote by $\vec{p}_{-i} = (p_1, p_2, \ldots, p_{i-1}, p_{i+1}, \ldots, p_m)$ the mixed strategy profile of the agents excluding the agent i (cf. the notation appearing in the axiom 6 above), $i = 1, \ldots, m$; consequently, the mixed-strategy profile for all m agents is $\vec{p} = \left(p_i, \vec{p}_{-i}\right)$. Then, the idea of maximizing the expected utility of the ith agent is formally represented by task of finding the best response strategy $p_i^* \in P_i$ such that it results in maximal expected utility $u_i\left(p_i^*, \vec{p}_{-i}\right) \geq u_i\left(p_i, \vec{p}_{-i}\right)$. In other words, the best response strategy p_i^* of the agent i is such a mixed strategy that:

$$p_i^* = \arg\max_{p_i \in P_i} u_i\left(p_i, \vec{p}_{-i}\right). \tag{9.3}$$

The mixed-strategy profile $\vec{p}^* = \left(p_1^*, p_2^*, \ldots, p_m^*\right)$ such that for each agent i, $i = 1, \ldots, m$, the mixed strategy p_i^* is the best response strategy is called the Nash equilibrium. Intuitively, this equilibrium means that in the noncooperative game of m agents any change of the strategy of the agent will lead to less utility obtained by this agent; hence, in such a game, no agent needs to change its strategy, given the strategies of the other agents.

An existence of the Nash equilibrium is stated by the following Nash theorem.

Theorem 9.2 (proven by Nash in 1951). Every noncooperative game with a finite number of agents and a finite number of actions has at least one Nash equilibrium in mixed strategies.

The presented theorems by von Neumann and Morgenstern (Theorem 9.1) and by Nash (Theorem 9.2) form a basis for the probabilistic planning the activity both of the single agent (see Section 7.1) and of the multi-agent systems (see Section 9.2.2). However, notice that in the case of multi-agent systems, it is assumed that each agent has complete information about the probability distributions used by the other agents and is aware that the other agents have the same information about itself. Such assumption allows an application of the centralized control of the group activity, as is stated by the Aumann "common knowledge" theorem (Aumann 1976). The meaning of this theorem is the following.

Let $(\Omega, \mathcal{B}, p)$ be a probability space, where Ω is a set of atomic events ω, $\mathcal{B}$ is a Boreal algebra on Ω that supports formal operations between the subsets of Ω and an existence of the empty set and p is a probability function, which defines the probabilities $p(\omega) \in [0, 1]$ of the events $\omega \in \Omega$. In addition, let $U \in \mathcal{B}$ be an event over Ω.

Consider the system, which includes two agents, agent 1 and agent 2. The "common knowledge" of these agents is defined as follows. Assume that both agent 1 and agent 2 complete prior information about the event U, and assume that the agent 1 knows that the agent 2 is aware of U and vice versa. In addition, assume that the agent 2 knows that the agent 1 knows that the agent 2 is aware of U and the same is true regarding the agent 1, which knows that the agent 2 knows that the agent 1 knows that the agent 2 is aware of U, and so far. Then, according to the Aumann theorem (Aumann 1976), the posterior information of the agents 1 and 2 about the event U is equal by necessity and the agents "cannot agree to disagree."

Formally, this fact is represented as follows. Denote by $\mathcal{P}_1$ and $\mathcal{P}_2$ the partitions of the set Ω associated with the knowledge of the agent 1 and the agent 2, respectively. These partitions are called information partitions; in game theory, such partitions define the split of the positions into information sets (Osborne and Rubinstein 1994). Since both agents deal with the same probability space $(\Omega, \mathcal{B}, p)$, the probability p is their common prior probability.

Let ω be a real true state of the agents and the environment. Then, the ith agent, $i = 1, 2$, is aware of the event $P_i(\omega) \in \mathcal{P}_i$, which includes the event ω. In other words, instead of knowing the exact event $\omega \in \Omega$, each agent i is informed about some subset $P_i(\omega) \subset \Omega$, which contains this event. The common knowledge of the agents 1 and 2 about the event $\omega \in \Omega$ is represented by the event $P_{com} \in \mathcal{P}_1 \vee \mathcal{P}_2$, where $\mathcal{P}_1 \vee \mathcal{P}_2 = \{P_1 \cap P_2 \mid P_1 \in \mathcal{P}_1, P_2 \in \mathcal{P}_2\}$ stands for the partition, which consists of all possible intersections of the events form the agents' partitions $\mathcal{P}_1$ and $\mathcal{P}_2$.

Now, let U be an event and $p_i(U) = p(U \mid \mathcal{P}_i)$, $i = 1, 2$, are the posterior probabilities of U given the agents' information partitions $\mathcal{P}_i$. Then, if $\omega \in \Omega$ is a real true event, then $p_i(\omega) = p(U \cap P_i(\omega))/p(P_i(\omega))$.

Theorem 9.3 (Aumann 1976). Let $\omega \in \Omega$ be an elementary event, and let q_1 and q_2 be two real numbers. If at the event ω there is common knowledge that $p_1(\omega) = q_1$ and $p_2(\omega) = q_2$, then $q_1 = q_2$.

In other words, if the agents start with the same prior information and utilize their common knowledge, then from the probabilistic point of view their posterior probabilities are equal and they "cannot agree to disagree" in the knowledge regarding these posteriors.

As indicated above, this result allows an extension of the value iteration presented in Section 7.2.2 and policy iteration algorithms to the multi-agent systems. The resulting (near)-optimal joint actions of the agents provide maximum of the expected utility of the system and the maximum of the expected utility of each agent. That means that the agents acting both in the group and individually obtain rational decisions, which provide their maximal utility. In practice, however, the common knowledge assumption does not always holds and in order to maximize utility, the agents should apply the judgments that appear to be irrational.

The simplest way to define the irrationality of the agents' judgments implements such a utility function u that represents the relation of the agent to the risky alternatives, where the risk is considered as an amount of discomfort both by the chosen alternative and measured in the payoff units (Friedman and Savage 1948). In the other approach (Kahneman and Tversky 1979), the risk is considered in terms of uncertainty, and it was demonstrated that the agents prefer such alternatives that provide certain results rather than the alternatives with greater but uncertain expected rewards. In particular, it means that the indifference $a_j \sim [p(a_i), 1 - p(a_k)]$ appearing in the indicated above continuity axiom 5 holds true only for trivial probability distributions p, which reduce the lottery $[p(a_i), 1 - p(a_k)]$ to the certain choice of either the alternative a_i or the alternative a_k, while for the other distributions the agents prefer the alternative a_j chosen with certainty.

Such an approach to the irrationality gave a rise to the prospect theory (Kahneman and Tversky 1979) that substitutes the utility theory in the considerations of decision-making under uncertainties. While analyzing the decisions made by humans, in this theory there are emphasized several effects, in particular, the empirical tests resulted in the following:

- *Certainty effect.* Most people prefer an alternative that leads to less reward with certainty rather that an alternative that leads to greater reward with some probability and remains the chance of obtaining zero reward.

- *Reflection effect.* If the alternative includes some lost, then most people change their priorities; in some cases, people demonstrate risk indifference between losing options, while in the other cases they demonstrate clear risk preference or aversion.
- *Isolation effect.* Comparison between a number of alternatives is conducted on the basis of certain feature that distinguish these alternatives, while different combinations of the features can result in different priorities.

In other words, the decisions obtained under uncertainties depend not only on the expected reward but also on the expected information gain and the expected payoff provided by the decision. Moreover, additional features included into the alternatives can change the preferences of the decision maker.

In the multi-agent systems, the situation is more complex since each agent has to take into account the decisions and actions of the other agents and the goal of the system considered as a unit. In this case, the activity of each individual agent can appear as irrational, but, in fact, such irrationality is resulted by the dependence of the agent's decisions on the activity and abilities of the group. For example (Rubin and Shmilovitz 2016), in the multi-pursuer pursuit evasion game there exist such positions of the pursuers and of the evader that some of the pursuers have to obtain irrational decision to terminate the pursuit, otherwise the evader will not be caught by any of pursuers from their team.

Formal description of such irrationality, where the agents' decisions are not necessary lead to maximum of the expected reward or minimum of the expected payoff, can follow different approaches; in particular, using the indicated above methods of MDPs it can be represented as follows.

Assume that the individual decision making by the agent is governed by Eq. (9.1). As indicated above, such decision making with rational judgments is based on the assumption that the decision maker is able to predict its future behavior and its current decision is influenced by the expected future decisions with certain degree that monotonously decreases in time. In the case of irrational judgments, such assumption fails and the expected future decisions can influence to the current decision in different and monotonic decreasing of the discount factor not always holds.

In order to represent such a situation, in the pricing theory (Cochrane 2001; Smith and Wickens 2002) it was suggested that the stochastic discount factor be applied to represent the idea that the decrease of the asset's price in time with respect to the asset's payoffs is influenced by some random factors. Then, the constant discount factor $\gamma \in [0, 1]$ is substituted by the random variable $\tilde{\gamma}$ drawn from the interval [0, 1] according to certain distribution. Then, instead of Eq. (9.1) the expected utility of the agent is defined as follows:

$$\mathcal{E}\left[\sum_{t=0}^{T} \tilde{\gamma}^{t} u(\mathfrak{s}(t), \mathfrak{a}(t))\right], \tag{9.4}$$

where $\tilde{\gamma}$ is a random variable, $0 \le \tilde{\gamma} \le 1$. Similarly, for the multi-agent system including m agents, the expected total utility is defined as a sum of the utilities of individual agents:

$$\mathcal{E}\left[\sum_{t=0}^{T} \sum_{i=1}^{m} \tilde{\gamma}_i^t u_i\left(\mathfrak{s}(t), \vec{\mathfrak{a}}(t)\right)\right], \tag{9.5}$$

where $\tilde{\gamma}_i$, $0 \le \tilde{\gamma}_i \le 1$, $i = 1, 2, \dots, m$, is a random discount factor of the ith agent, $\mathfrak{s}(t)$ is a state of the system at time t and $\vec{\mathfrak{a}}(t)$ is a joint action or action profile of the system at this time. As above, the time $t = 0, 1, 2, \ldots, T$, where for finite horizon $T < \infty$ and for infinite horizon $T = \infty$.

Since in the framework of the Markov decision model, the discount factor defines the relation of the decision maker to the expected results of future decisions, the randomness of the discount factor represents possible variations in the decision maker's relation to the information regarding expected results of chosen actions. Notice that the game theoretic definition of the expected utility given by Eq. (9.2), follows the same principles, where the randomness of the agent's decisions is refined with respect to the probabilities of the decisions obtained by the other agents. The algorithmic definition of the decision-making processes with stochastic discount factor implements the conventional value iteration or policy iteration algorithms (see Section 7.2.2).

Finally, let us notice two different implementations of the decision-making process in the multi-agent systems: in the systems with centralized control of simple agents and in the systems with peer-to-peer communication between the agents with enough computational and communication abilities (see Section 10.1). In the first case, the decision-making process is conducted by the central unit, which holds a central database, transmits the commands to the agents, and perceives the results of the executed actions. In the second case, decision-making is conducted by each agent on the base of the information perceived from the other agents using the methods of so-called social choice, in which the decisions are obtained by some voting procedures (Shoham and Leyton-Brown 2008; Wooldridge 2009). These implementations are briefly considered in Section 9.2.1.

9.1.2 Basic Flocking and Methods of Aggregation and Collision Avoidance

The presented principles of rational judgments and corresponding decision-making techniques provide a general framework for consideration of the collaborative activity of the agents of different nature. In the case of the teams of mobile robots, the behavior of the agents is restricted by certain physical limitations (e.g., possible agents' positions, velocities, available energy and similar, limited communication and computation abilities) and by the need of preserving the group of the agents as a unit rather than a collection of separate agents. For the mobile robots, the last task is usually considered as a problem of flocking or swarming and is reduced to the problem of preserving a geometrically distinguished cluster formed by the agents in the Euclidian space.

Formally, the basic swarming (without addressing the mission of the agents) is defined by the Reynolds rules (Reynolds 1987), which were suggested for simulations of the behavior of the birds' flocks. In the modern terms, these rules are defined as follows (Gazi and Passino 2011); for convenience the original Reynolds terms (see Introduction) are given in brackets.

1) *Separation* (collision avoidance) is required to preserve certain minimal distances between the agents.
2) *Alignment* (velocity matching) prescribed to each agent preserves its velocity, including the value and the direction, as similar as possible to the velocities of the agent's neighbors.
3) *Cohesion* (centering) requires each agent to stay as closer as possible to the agent's neighbors.

In spite of the simplicity of these rules, they provide a complete system of necessary conditions for preserving the swarm. An application of these rules in different multi-robot systems of mobile robots can result in rather different phenomena and gives rise to different problems; a list of some problems appearing in this framework and detailing consideration of the swarms' objectives and possible theoretical solutions are presented in Gazi and Passino (2011).

However, from a practical point of view, an implementation of the Reynolds rules implies that each agent is able to recognize its neighbors in the domain and to measure the distances between itself and the other agents and the velocities of the other agents. Such abilities of the agents are easily provided by some GPS and centralized communication and control (see Section 10.1), and then the decision-making regarding the direction of each agent follows the techniques of rational judgments with common knowledge. In contrast, if the agents are not aware of their exact coordinates and about the coordinates and velocities of the other agents in the team, then the implementation of the Reynolds rules requires both additional sensing abilities of each agent and more complex techniques of collective decision-making. We will discuss only the simple Cucker-Smale model (Cucker and Smale 2007a,b) and the basic methods of aggregation and collision avoidance; more complex methods of decision-making that follow from the considered above rational and irrational judgments are considered in Section 9.2.1.

Let us consider the swarm of $m \geq 1$ point-like agents moving in two-dimensional domain $X \subset \mathbb{R}^2$ with the points $\vec{x} = (x, y)$. The coordinate of each jth agent, $j = 1, 2, \ldots, m$, agent at time $t \in \mathbb{R}^+$ is denoted by $\vec{x}_j(t)$. Then, the Cucker-Smale model of the agents' behavior is defined as follows (Cucker and Smale 2007a,b):

$$\frac{d}{dt}\vec{x}_j(t) = \vec{v}_j(t), \quad \frac{d}{dt}\vec{v}_j(t) = \frac{\alpha}{m}\sum\nolimits_{i=1}^{m} \psi\left(\vec{x}_i(t), \vec{x}_j(t)\right)\left(\vec{v}_i(t) - \vec{v}_j(t)\right), \tag{9.6}$$

where $\alpha \geq 0$ is a coupling strength and $\psi : X \times X \to \mathbb{R}^+$ is a communication rate function such that for all $i, j = 1, 2, \ldots, m$ the symmetry condition $\psi\left(\vec{x}_i, \vec{x}_j\right) = \psi\left(\vec{x}_j, \vec{x}_i\right)$, and the translation invariance $\psi\left(\vec{x}_i, \vec{x}_j\right) = \psi\left(\vec{x}_i + \vec{c}, \vec{x}_j + \vec{c}\right)$, $\vec{c} \in \mathbb{R}^2$, are satisfied. In the basic model, this communication rate function ψ is defined as follows (cf. navigation function $\varphi_\kappa\left(\vec{x}\right)$ considered in Section 4.5.1):

$$\psi\left(\vec{x}_i, \vec{x}_j\right) = \frac{b}{\left(1 + \left\|\vec{x}_i - \vec{x}_j\right\|^2\right)^\kappa}, \tag{9.7}$$

with some fixed constants $b > 0$ and $\kappa \geq 0$.

It was demonstrated (Cucker and Smale 2007a,b) that if the agents allow long-range communication (the values $\psi\left(\vec{x}_i, \vec{x}_j\right)$ decrease slowly with the distances $\left\|\vec{x}_i - \vec{x}_j\right\|$), then their velocities converge in time to the same asymptotic velocity for any compact set of initial conditions. In contrast, if the agents allow only short-range communication (the values $\psi\left(\vec{x}_i, \vec{x}_j\right)$ becomes close to zero already for small distances $\left\|\vec{x}_i - \vec{x}_j\right\|$), then the swarming occurs only for a limited class of initial agents' positions.

Notice that the system (9.6) can be written in the following form:

$$\frac{d}{dt}\vec{x}_j(t) = \vec{v}_j(t), \quad \frac{d}{dt}\vec{v}_j(t) = -\gamma_j\left(\vec{x}_j\right)\vec{v}_j(t) - \nabla\mathcal{U}_j\left(\vec{x}_j, t\right), \tag{9.8}$$

where both friction coefficient

$$\gamma_j\left(\vec{x}_j\right) = \frac{\alpha}{m}\vec{v}_j(t)\sum_{i=1, i\neq j}^{m}\psi\left(\vec{x}_i(t), \vec{x}_j(t)\right)$$

and the gradient of the potential field

$$\nabla\mathcal{U}_j\left(\vec{x}_j, t\right) = -\frac{\alpha}{m}\sum_{i=1, i\neq j}^{m}\psi\left(\vec{x}_i(t), \vec{x}_j(t)\right)\vec{v}_i(t)$$

are specified by the interactions with the neighbors of the jth agent. Such representation immediately relates to the Cucker-Smale model (9.6) with the model based on active Brownian particles (Schweitzer 2003; Romanczuk et al. 2012) that is briefly considered in Section 10.1.1 (see also Erban, Haskovec, and Sun (2016)). The next example illustrates motion of the agents following this model.

Example 9.1 (Ben-Haim and Elkayam 2017). Consider the motion of $m = 25$ point-like agents acting in the square-gridded domain X of the size $n = 100 \times 100$ with definite external activity potential field $\mathcal{U}^{act}$ (see Section 10.1.2), which specifies the topography of the domain, and the agents' potentials $\mathcal{U}_j$, $j = 1, 2, \ldots, m$, defined according to the Cucker-Smale model written in the form of Eq. (9.8).

The initial positions of the agents and their trajectories in the homogeneous domain with $\mathcal{U}^{act} \equiv 0$ are shown in Figure 9.1. In the figure, the parameters of the communication rate function ψ are $b = 1$ and $\kappa = 2$.

It is seen that after the choice of the neighbors, the agents are moving toward these chosen agents. Notice that since the initial positions form a regular lattice, some of the agents cannot choose the preferable neighbor and stay in their positions without movements.

In contrast, if the agents act in the heterogeneous domain with nonzero external activity potential field $\mathcal{U}^{act} \neq 0$, then they tend to follow toward the regions with lower potential. An example of the agents' trajectories in this scenario is shown in Figure 9.2. As above, in this figure, the parameters of the function ψ are $b = 1$ and $\kappa = 2$.

As expected, in this case most of the agents that initially were located in the regions with high gradient of the potential field $\mathcal{U}^{act}$ rapidly move to the regions with lower potential and then continue they motion in these regions according to the Cucker-Smale model with the communication rate function ψ. However, the agents that start their motion in the regions with lower gradient of the potential field $\mathcal{U}^{act}$ (the brightest regions at the bottom and right sides of the figure) stay moving in these regions according to the Cucker-Smale model without following to the regions with lower potential.

(a)

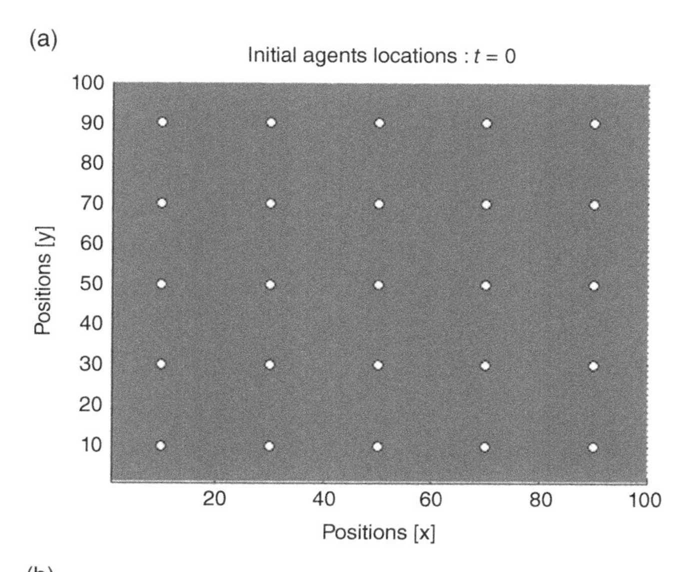

(b)

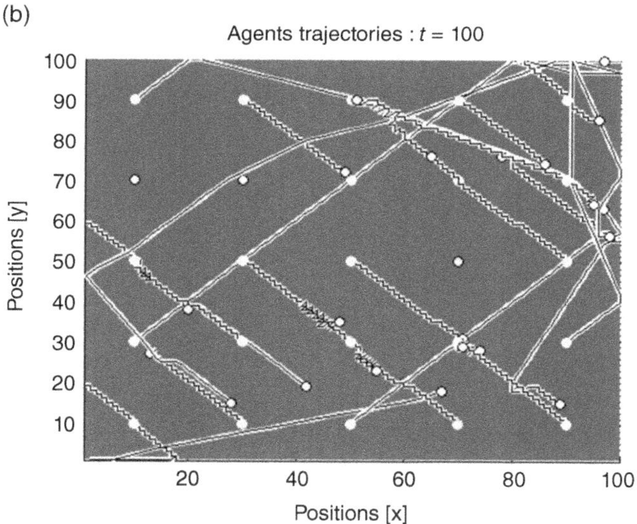

Figure 9.1 The $m = 25$ agents acting in the homogeneous gridded domain of the size $n = 100 \times 100$ according to the Cucker-Smale model in the form of Eq. (9.8): (a) initial positions of the agents; (b) trajectories of the agents in time $t = 0, 1, 2, \ldots, 100$.

The presented examples demonstrate the reaction of the agents to the neighbors with respect to the communication rate function ψ and to the external potential field. In the absence of the external potential as well as at the lower distances between the agents, the agents move with respect to the positions of the neighbors, while in presence of external potential field, they are governed by this field. Certainly, the influence of the neighbors and of the external potential field depends on the parameters and, more generally, on the form of the function ψ and on the form of the potential field $\mathcal{U}^{act}$. ∎

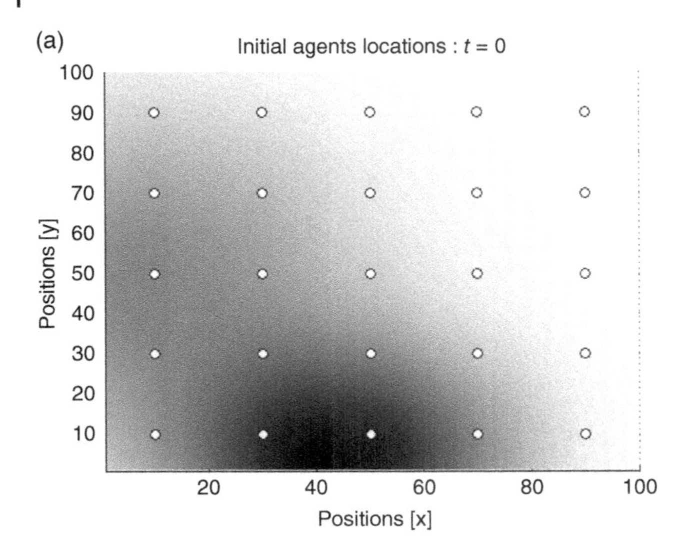

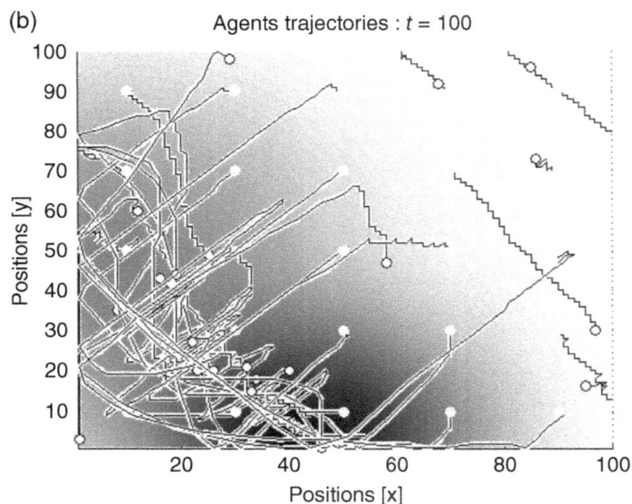

Figure 9.2 The $m = 25$ agents acting in the heterogeneous gridded domain of the size $n = 100 \times 100$ according to the Cucker-Smale model in the form of Eq. (9.8): (a) initial positions of the agents; (b) trajectories of the agents in time $t = 0, 1, 2, \ldots, 100$. In both figures, the regions with higher potential are marked by the brighter gray color and the regions with lower potential are marked by darker gray color.

In addition to the original papers by Cucker and Smale (2007a,b), the model (9.6) was intensively studied in its different setups. In particular, Ha, Lee, and Levy (2009) and then Ton, Linh, and Yagi (2015) studied this system with stochastic interactions with the environment (cf. the model of active Brownian motion in heterogeneous environment considered in Section 10.2.1) and Shen (2007) and then Li and Ha (2013) considered it with different abilities of the agents such that some of the agents act as leaders of the swarm.

The similar model of swarming with and without leadership was independently suggested in 2005 by Couzin et al. (2005), who considered the activity of the groups of animals. In this model, it is assumed that the agents move in discrete time and the alignment rule prescribes to adjust the direction of the agent's motion to the average direction of the swarm. Namely, if $\vec{x}_j(t)$ is a position of the jth agent at time t and $\vec{\phi}_j(t)$ is its direction vector (normalized velocity per step of the agent) at this time, then the direction vector $\vec{\phi}_j(t+1)$ at the next time $t+1$ is defined as follows (Couzin et al. 2005; Qu 2009):

$$\vec{\phi}_j(t+1) = -\sum_{i=1,i\neq j}^{N_j^{short}(t)} \frac{\vec{x}_i(t)-\vec{x}_j(t)}{\left\|\vec{x}_i(t)-\vec{x}_j(t)\right\|}, \tag{9.9}$$

where the sum is taken over the set $N_j^{short}(t) = \left\{i \mid 1 \le i \le m, \left\|\vec{x}_i(t)-\vec{x}_j(t)\right\| \le d_{short}\right\}$ of the neighbors of the jth agent that are detected by short-distance sensors. If the agent does not detect any agent in its close neighborhood, then the direction $\vec{\phi}_j(t+1)$ is calculated based on the set $N_j^{long}(t) = \left\{i \mid 1 \le i \le m, \left\|\vec{x}_i(t)-\vec{x}_j(t)\right\| \le d_{long}\right\}$ of neighbors as follows:

$$\vec{\phi}_j(t+1) = \sum_{i=1,i\neq j}^{N_j^{long}(t)} \frac{\vec{x}_i(t)-\vec{x}_j(t)}{\left\|\vec{x}_i(t)-\vec{x}_j(t)\right\|} + \sum_{i=1}^{N_j^{long}(t)} \frac{\vec{\phi}_i(t)}{\left\|\vec{\phi}_i(t)\right\|}, \tag{9.10}$$

where $\left\|\vec{\phi}_i(t)\right\|$ stands for the norm of the velocity vector $\vec{\phi}_i(t)$. These two rules guarantee that the robots preserve the completeness of the group and the swarm acts as a unit. The remaining issue is the control of the group activity.

Then, the alignment rule specifies that if $\vec{v}_j^{desired}$ is a predefined desired direction of the jth agent, and its real direction $\vec{v}_j(t)$ is defined as

$$\vec{v}_j(t+1) = \frac{\vec{\phi}_j(t+1) + w_j\,\vec{v}_j^{desired}}{\left\|\vec{\phi}_j(t+1) + w_j\,\vec{v}_j^{desired}\right\|}, \tag{9.11}$$

where $0 \le w_j \le 1$ is a weighting coefficient. This approach guarantees that the agents preserve the completeness of the swarm and its activity as a unit, while the control is implemented by specifying the desired directions $\vec{v}_j^{desired}$ and the coefficients w_j, $j = 1, 2, \dots, m$; such an interpretation of the model was formulated in collaboration with Levy (2016).

It is clear that in the homogeneous swarm with the weighting coefficients $w_j = 0$ for all agents $j = 1, 2, \dots, m$, the agents follow the alignment rule with no influence of the desired direction $\vec{v}_j^{desired}$. If the weighting coefficients are $w_j = 1$ for all agents $j = 1, 2, \dots, m$, then the agents tend to follow the desired direction $\vec{v}_j^{desired}$ with preserving the swarm as a geometrical cluster.

In the heterogeneous swarm with the leading agents, which are defined as informed agents with $w_j > 0$, in contrast to the naïve agents, for which $w_j = 0$, it was found (Couzin et al. 2005) that for a given swarm size m the accuracy of swarm motion in the desired direction increases with the number of the informed agents. Moreover, for the larger numbers m of the agents, the accuracy of motion in the desired direction is reached for the smaller proportion of the informed agents. In other words, for the following predefined direction, the smaller swarms require larger proportion of the leaders, while the larger swarms can be led by a small number of the leaders.

The considered Cucker-Smale and Couzin et al. models address the alignment rule and implement it in the simple forms defined by the functions (9.7) and (9.11), correspondingly. The decision-making regarding the direction is then reduced to the deterministic relations that prescribe to match the velocity of each agent to the velocities of its neighbors without alternating possibilities. In more complex models, the agents are able to decide whether to match its velocity to the velocities of its neighbors or to follow some additional criterions and chose the velocity with respect to its own preferences as it is described in Section 9.1.1. In particular, such criterions are specified by the separation and cohesion rules, which, even in the deterministic case, strongly influence the directions chosen by the agents according to the rule of alignment.

The widely accepted manner of defining cohesion and separation is based on the aggregation potential functions $\mathcal{U}_j^{agr}$, $j = 1, 2, \ldots, m$, that define the forces with which the agents attract each other at long distances – which results in swarm centering – and that repulse each other at short distances – which provides collision avoidance. Usually, such aggregation functions are combined from two separate functions – attraction function $\mathcal{U}_j^{atr}$ and repulsion function $\mathcal{U}_j^{rep}$ – and are specified as follows (Gazi and Passino 2011):

$$\mathcal{U}_j^{agr}(\vec{x}_j,t) = -\mathrm{r}(\vec{x}_j,t)\left(\mathcal{U}_j^{atr}(\mathrm{r}(\vec{x}_j,t)) - \mathcal{U}_j^{rep}(\mathrm{r}(\vec{x}_j,t))\right), \tag{9.12}$$

where $\mathrm{r}(\vec{x}_j,t) = \sum_{i=1}^{m} \left\| \vec{x}_i(t) - \vec{x}_j(t) \right\|$ is a distance between agent j and the other agents of the swarm. In addition, regarding the function $\mathcal{U}_j^{atr}$ and $\mathcal{U}_j^{rep}$ it is assumed that for each agent $j = 1, 2, \ldots, m$ there exists a unique equilibrium distance $\mathrm{r}^*(\vec{x}_j,t)$ such that

$$\mathcal{U}_j^{atr}(\mathrm{r}^*(\vec{x}_j,t)) = \mathcal{U}_j^{rep}(\mathrm{r}^*(\vec{x}_j,t)),$$

and it holds true that

$$\mathcal{U}_j^{atr}(\mathrm{r}(\vec{x}_j,t)) > \mathcal{U}_j^{rep}(\mathrm{r}(\vec{x}_j,t)) \text{ for } \mathrm{r}(\vec{x}_j,t) > \mathrm{r}^*(\vec{x}_j,t)$$

and

$$\mathcal{U}_j^{atr}(\mathrm{r}(\vec{x}_j,t)) < \mathcal{U}_j^{rep}(\mathrm{r}(\vec{x}_j,t)) \text{ for } \mathrm{r}(\vec{x}_j,t) < \mathrm{r}^*(\vec{x}_j,t).$$

In most cases in which attraction and repulsion forces are based on the Euclidian distances between the agents, the aggregation potential function is defined as

$$\mathcal{U}_j^{agr}(\vec{x}_j,t) = -\mathrm{r}(\vec{x}_j,t)\left(\alpha_a - \alpha_r \exp\left[-\frac{1}{\beta_r}\left[\mathrm{r}(\vec{x}_j,t)\right]^2\right]\right), \tag{9.13}$$

that implements the domination of attraction at long distances and of repulsion at short distances. The equilibrium distance for such aggregation is $r^*(\vec{x}_j,t) = \sqrt{\beta_r \ln \frac{\alpha_r}{\alpha_a}}$, $\alpha_a > 0$. Practical implementation of such aggregation functions in the swarms of mobile robots is based on sonar or similar distance sensors that are applied in the same manner as in the Elfes mapping scheme (Elfes 1987, 1990) (see Section 7.2.1).

In the other situations, for example, in the problems of probabilistic search (Kagan and Ben-Gal 2013, 2015), the aggregation can be defined using the other distance measures, which represent the knowledge of the agents about the target or, more generally, about the goal of the swarm's mission. In particular, assume that the agents act in the gridded domain $X = \{\vec{x}_1, \vec{x}_2, \dots, \vec{x}_n\} \subset \mathbb{R}^2$ with the probability mass function $p : X \times [0,\ T) \to [0,\ 1]$, which defines the probability of finding the target in the points $\vec{x}$ of the domain X at time t and such that for any $t \in [0,\ T)$ it holds true that $\sum_{i=1}^{n} p(\vec{x}_i,t) = 1$. Then for each agent j, $j = 1, 2, \dots, m$, the attraction potential $\mathcal{U}_j^{agr}(\vec{x}_j,t)$ in the point $\vec{x}_j$ of the agent's location at time t is defined proportionally to the probabilities available to the agents:

$$\mathcal{U}_j^{atr}\left(\vec{x}_j,t\right) \sim \left| \sum_{\vec{x} \in a_i(t)} p(\vec{x},t) - \sum_{\vec{x} \in a_j(t)} p(\vec{x},t) \right|, \tag{9.14}$$

where $a_i(t) \subset X$ and $a_j(t) \subset X$ are the areas observed by the agents i and j at time t, respectively. Such type attraction is implemented in the method of collective search considered in Section 10.3.1 and in the search with biosignaling considered in Section 11.3.2.

One of the implementations of such attraction function is based on the informational distance measures (Cover and Thomas 1991) that apply the Shannon entropy; for detailed consideration of this approach, see Kagan and Ben-Gal (2013, 2015). In particular, let the probabilities available for the ith and jth agents, $i, j = 1, 2, \dots, m$, be defined as follows:

$$p_i(\vec{x},t)\Big|_{\vec{x} \in X} = \begin{cases} p(\vec{x},t) \text{ if } \vec{x} \in a_i(t) \\ 0 \quad \text{otherwise} \end{cases} \text{ and } p_j(\vec{x},t)\Big|_{\vec{x} \in X} = \begin{cases} p(\vec{x},t) \text{ if } \vec{x} \in a_j(t) \\ 0 \quad \text{otherwise} \end{cases}.$$

Using these probabilities, the Kullback-Leibler distance between the agents i and j at time t is

$$D_i(\vec{x}_i,\vec{x}_j,t) = D(\vec{x}_i \| \vec{x}_j,t) = \sum_{\vec{x} \in X} p_i(\vec{x},t) \log \frac{p_i(\vec{x},t)}{p_j(\vec{x},t)},$$

and the Kullback-Leibler distance between the agents j and i at time t is

$$D_j(\vec{x}_i,\vec{x}_j,t) = D(\vec{x}_j \| \vec{x}_i,t) = \sum_{\vec{x} \in X} p_j(\vec{x},t) \log \frac{p_j(\vec{x},t)}{p_i(\vec{x},t)}.$$

Then, the attracting potential functions for the agents i and j are defined as follows (such an approach was developed in collaboration with Levy (2016)):

$$\mathcal{U}_i^{atr}(\vec{x}_i,t) = \sum_{j=1, j \neq i}^{m} D_j(\vec{x}_i,\vec{x}_j,t) \text{ and } \mathcal{U}_j^{atr}(\vec{x}_j,t) = \sum_{i=1, i \neq j}^{m} D_i(\vec{x}_i,\vec{x}_j,t). \tag{9.15}$$

Notice that in general $D_i(\vec{x}_i, \vec{x}_j, t) \neq D_j(\vec{x}_i, \vec{x}_j, t)$ and, consequently, $\mathcal{U}_i^{atr}(\vec{x}_i, t) \neq \mathcal{U}_j^{atr}(\vec{x}_j, t)$. The difference between aggregation with attraction based on probabilistic and informational distances (Eqs. (9.14) and (9.15), respectively) is illustrated by the following example.

Example 9.2 Assume that $m = 25$ agents act in the squire gridded domain X of the size $n = 100 \times 100$ with definite topography and probability distribution that follows this topography (Levy 2016). The topography over the domain and the probability distribution are shown in Figure 9.3a,b, correspondingly. Initial agents' positions are shown in Figure 9.3c, where the lower regions (and, that is the same, the regions with lower probabilities) are depicted by black color and the higher regions (and the regions with higher probabilities) are depicted by white color.

The trajectories of the agents driven during the period $T = 100$ by topography and by probabilistic and informational attraction potential functions are shown in Figure 9.4.

It is seen that following the topography of the domain, the agents tend to move to the lower regions (Figure 9.4a), while following the probabilistic (Figure 9.4b) and informational (Figure 9.4c) attraction functions the agents tend to move in the regions with

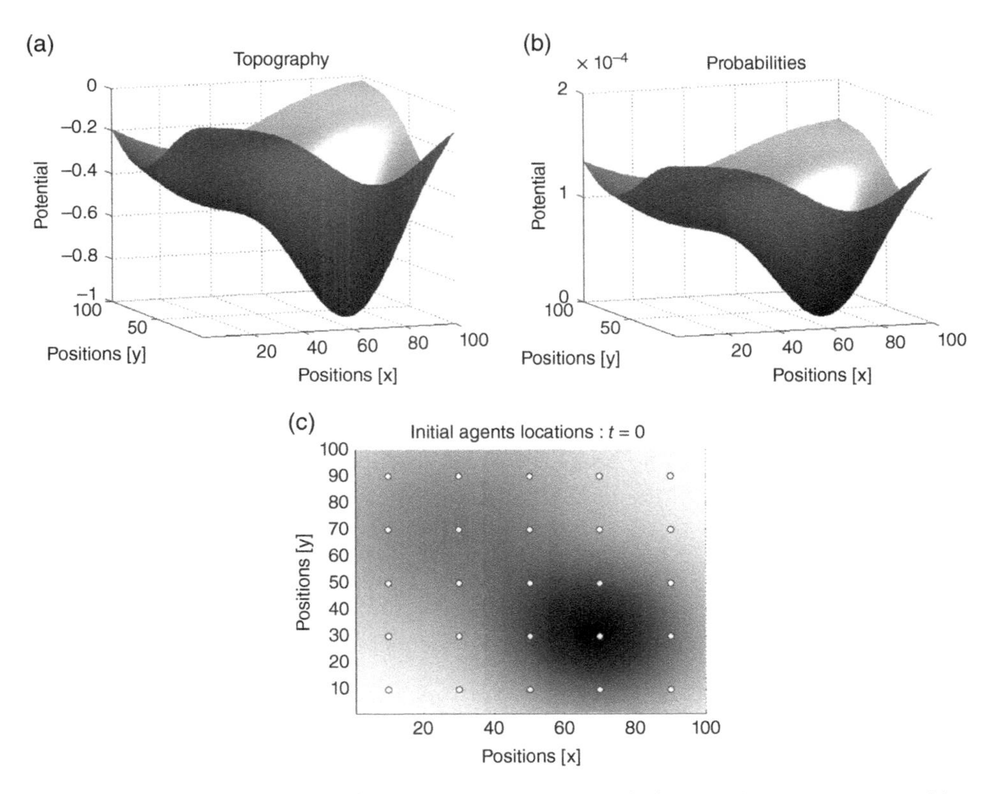

Figure 9.3 The gridded domain of the size $n = 100 \times 100$, in which are acting $m = 25$ agents: (a) topography of the domain; (b) probability distribution; and (c) initial positions of the agents. The lower regions (that are the regions with lower probabilities) are depicted by black color and the higher regions (that are the regions with higher probabilities) are depicted by white color.

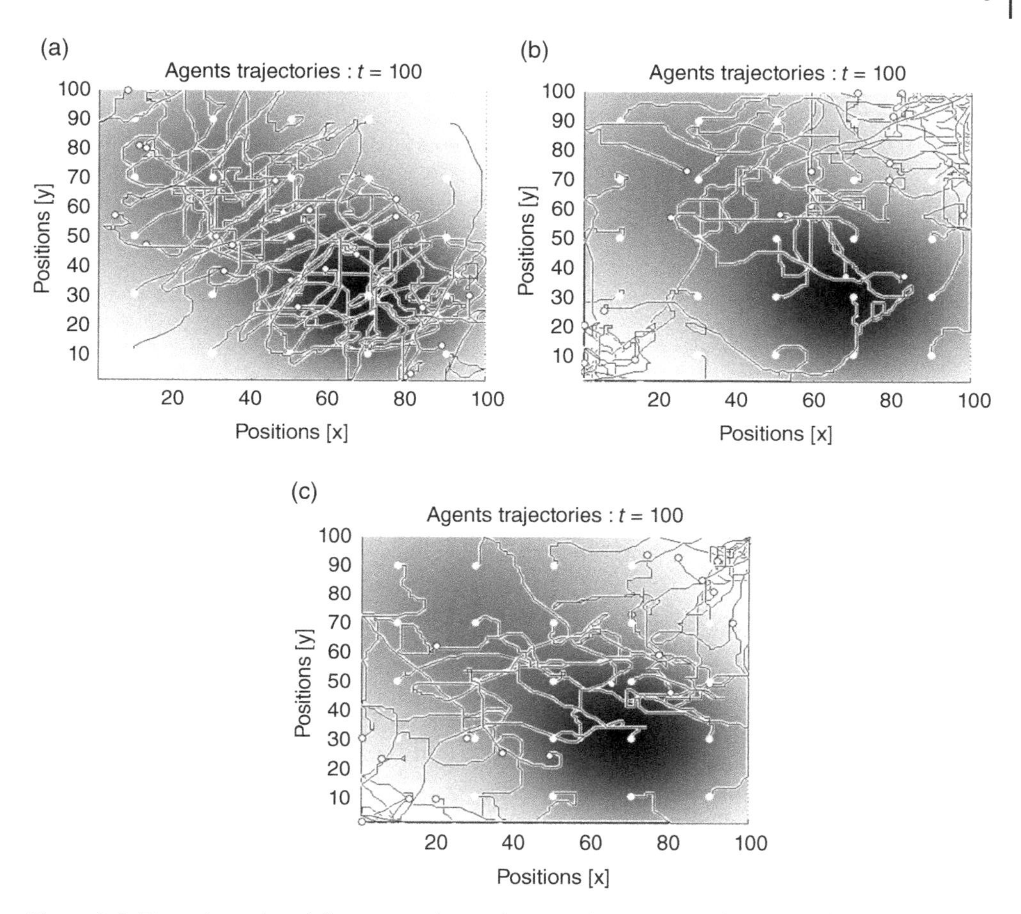

Figure 9.4 The trajectories of the agents during the period $T = 100$: (a) the agents follow the topography of the domain; (b) the agents are attracted by the probabilistic attraction potential function (9.14); (c) the agents are attracted by the informational attraction potential function (9.15). White points mark the initial positions of the agents.

higher probabilities; however, the tendency of such motion is different. As it is expected, because of logarithmic scale of the informational attraction in this case the agents move toward the regions with higher probabilities faster than in the case of probabilistic attraction, in which the velocity is a linear function of the probabilities. ■

Since the Kullback-Leibler distance is not a metric, the attraction potentials are, in general, also not symmetric and the greater attraction is provided by the agents with the greater Shannon information defined by the probabilities of the observed area. Such a property completely coincides with the Couzin et al. approach (Couzin et al. 2005), where the leaders are defined as more informed agents, and allows to use the informational distances for definition of the dynamics of heterogeneous swarms with leading agents (Levy 2016).

Finally, notice that, similarly to attraction, the methods of defining the collision avoidance strongly depend on the nature of the agents. For the agents defined as computational units acting in some information network, collision avoidance can be defined using the similar to presented above techniques of information theory. However, for the mobile robots moving in the two- or three-dimensional space with Euclidian metric, collision avoidance is reduced to preserving certain physical distance between the robots and is provided by the repulsion function, which defined the repulsion force as greater as the robots becomes closer one to another. In Eq. (9.12) such repulsion function is defined in the quadratic form, while in the other situations the dependence between the distance and the repulsion can be specified similarly to the attraction in the exponential form (Kagan and Ben-Gal 2015).

9.2 Control of the Agents and Positioning of Swarms

In the previous section, we considered the main principles of swarming and the methods of their implementation based on the attraction and repulsion potential functions that define the relations between the agents and result in the activity of the group of agents as a swarm executing a certain mission. However, definition of the mission and corresponding control of the swarm require additional techniques and includes the following issues (Fornara 2003):

- Managing relations between the agents with synchronization of the agents, which execute different tasks and use the results obtained by the other agents.
- Managing global constraints in time, space, resources, and information in order to receive joint reward of the swarm.
- Managing shared resources and information and transmission of the required resources and information from one agent to another.

The implementation of these issues strongly depends on the level of coordination between the agents and the type of the swarm control. The Iocchi, Nardi, and Salerno taxonomy (Iocchi, Nardi, and Salerno 2001) (see also Kagan and Ben-Gal (2015)) applied to the multi-agent systems with respect to the coordination level, as shown in Figure 9.5.

Following this taxonomy, the awareness level specifies that if the agents are not aware of the other swarm members, then their actions are considered separately and the swarm is divided into independent agents acting in parallel.

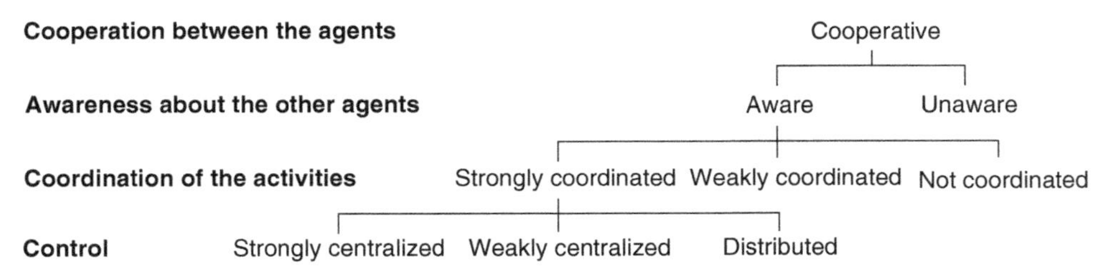

Figure 9.5 Taxonomy of the multi-agent systems with respect to the level of coordination between the agents.

On the coordination level, it is assumed that if the agents are not coordinated, they also execute their missions in parallel, but the results of the activity of one agent can depend on the results of the activities of the other agent. The weak coordination specifies that the agents act in parallel with certain corrections of the behavior with respect to the other agents, which are still considered as the elements of external environment. In contrast, strong coordination implies that the agents support the coordination protocol and make decisions regarding the chosen actions, taking into account their influence on the other agents.

Finally, the control level considers the roles of the agents in the decision-making process. In strongly centralized systems, the decision-making is conducted by a single constant leading agent, which obtains information about the other agents and prescribes their behavior. Weak centralization also assumes that the leading agent controls the activity of all the agents in swarm, but allows changing the leader during the mission. In contrast, in distributed systems, the agents make their decisions autonomously according to the activities of the other agents, and, consequently, distributed decision-making as well as decentralized control are considered as methods of choosing actions by the autonomous agents acting in teams in order to execute common task or reaching the common goal.

While considering the multi-agent systems consisting of the mobile robots, the main problem is the problem of navigation of the robots such that it provides a desired motion the swarm. This problem can be considered from different points of view (Bouffanais 2016). In particular, the desired motion of swarm can be obtained as a result of the planned motions of the robots, or, by contrast, the motion of each swarm mate is obtained autonomously with respect to the planned motion of the swarm. Below, we consider these possibilities using the examples of the centralized control of the swarm considered as a dynamical system and the decentralized control with the highest level of the agents' autonomy with the individual decision-making following the considered in Section 9.1.1 utility and prospect theories and social choice techniques. Finally, we present general probabilistic models of swarms and the corresponding extensions of the path-planning algorithms applied in Section 7.2.2 for navigation of a single agent.

9.2.1 Agent-Based Models

In studies of mobile robots, the robot is considered as an autonomous computer system interacting with its environment and executing computational and physical actions and movements with the aim of reaching a definite goal. In the simplest case, the objective of the robot's activity is to plan its path from the initial position to the static or moving target position, while in general, such objective can change during the motion and the robot should react to the changes of the environment. The basis of the judgments and decision-making regarding the actions of the robot is presented in Section 9.1.1; an extension of this approach to multi-agent decision-making appears in Section 9.2.1.

In such a framework, control of the robots acting both individually and in swarms is conducted in two steps:

1) Define the robots' reactions to the changes of the environment, including artificial environment that is formed by the robot itself, the other robots, and the control unit.

2) Vary the available artificial environment by the robot itself, by the other robots, and by the control unit in such a manner that the robot and the swarm follow the defined goal.

It is clear that the use of the navigation function (see Chapters 4 and 10 that consider individual and collective motion in artificial potential field, respectively) directly implements this control scheme.

With respect to the manner of implementation of the control steps, the robots are considered as reactive agents, planning agents, and autonomous agents. Following Fornara (2003, pp. 9–10):

- Reactive agents… consist only of a program that maps each possible perception or percept sequence in the corresponding action that the agent has to carry out. They need a *built-in knowledge*, which univocally determines their behavior.
- Planning agents… have more complex built-in knowledge about the set of actions that can be performed. This means that they know the preconditions and the effects of their actions on the environment… [and] choose which plan to execute among all the combinatorial combinations of their allowed actions.
- A truly autonomous artificial agent may be obtained providing it with the built-in knowledge… and with the powerful capability to *learn*. In this way, its behavior is actually determined by its own experiences.

According to this classification, most robots considered in the book are either reactive or planning agents, while autonomous agents are presented by the mobile robots, which move in unknown environment and deal with the problems of localization and mapping (see Chapter 7).

In spite of the difference between the "intelligent" abilities of the agents of the indicated agents, the common property of reactive, planning, and learning agents is their ability to react to the states and changes of the environment and to their states in the environment. Such a property is considered as a feedback control. A general scheme of the feedback control is shown in Figure 9.6 (Aoki 1967).

Following this scheme, the agent observes the environmental states and reacts to the obtained information by changing its internal state or by acting in a way that results in changing the state of the environment. Figure 9.6 takes into account that both observations and actions can be imperfect (see Section 7.1); in the models such imperfectness is usually described by the use of additional noise.

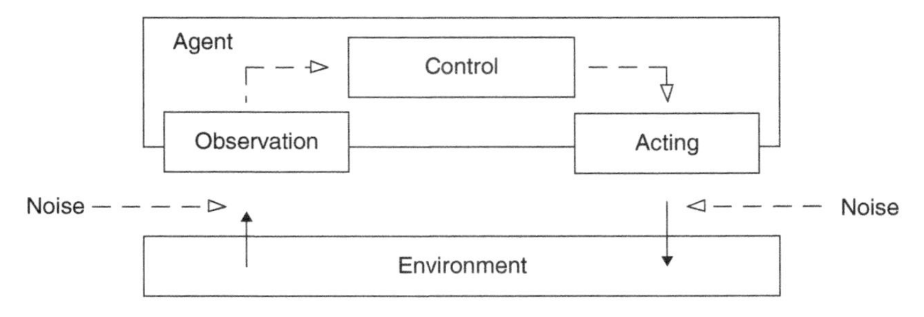

Figure 9.6 Scheme of the feedback control with imperfect observations and actions.

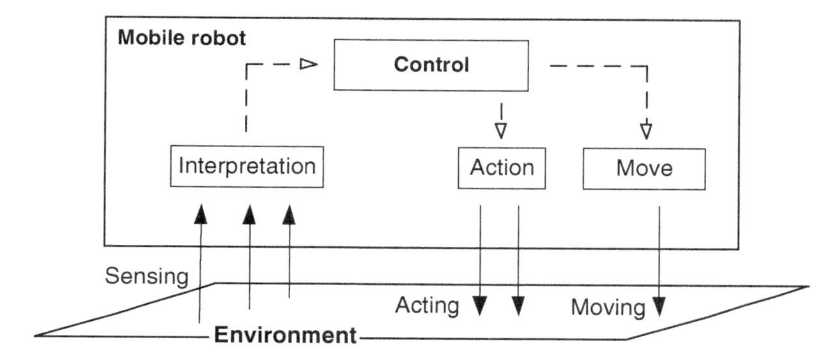

Figure 9.7 General scheme of the feedback control of mobile robot.

In the case of mobile robots moving in certain terrain, the obtained information includes physical conditions around the robot, relative locations of the other robots, and the data and commands from the other robots and from the central unit if it exists. In the details, such control follows the Elfes scheme (Elfes 1987, 1990) shown in Figure 7.4; in more general form, it is shown in Figure 9.7 (Siegwart, Nourbakhsh, and Scaramuzza 2011; Kagan et al. 2014; c.f. the scheme of the robot architecture shown in Figure 2 in the Introduction).

Following the presented scheme, the actions of mobile robot are divided into actions that lead to changing the environment and to the movements of the robot. The movements, in turn, require specific control including localization, path planning and navigation, and the resulting changes of the robot's locations and consequent changes in the sensed data close the loop of the feedback control scheme.

In the case of multi-agent system, which includes a number of agents, the feedback control is organized following several schemes. The simplest scheme implements the indirect communication scheme (see Section 11.1.2), in which the agents change the environment and communicate via environmental changes. In this case, the control of each agent follows a general feedback control scheme shown in Figure 9.6 and in Figure 9.7, while the actions include changing the environmental states such that these changes are considered by the other agents as meaningful signs or even symbols. Such communication, however, implies that the agents are intelligent enough and are able to recognize and interpret these symbols. In more complicated communication schemes, the swarm members are aware of each other and communicate using certain network with definite communication protocol. The scheme of feedback control in this case is shown in Figure 9.8.

Certainly, in the multi-agent systems that consist of intelligent agents with strong communication and computation abilities can implement both communication schemes and control is related both to each agent and to the groups of the agents considered as a unit. In this case, controller is responsible on the stable functionality of each agent and on the coordination between the agents' control. Although, in complex self-organizing systems, controllers are considered as elements of the system such that one agent is included into the feedback control loop and so is considered as a controller of the other agents and vice versa (Giovanini 2007). The swarms of living organisms are also provide an example of

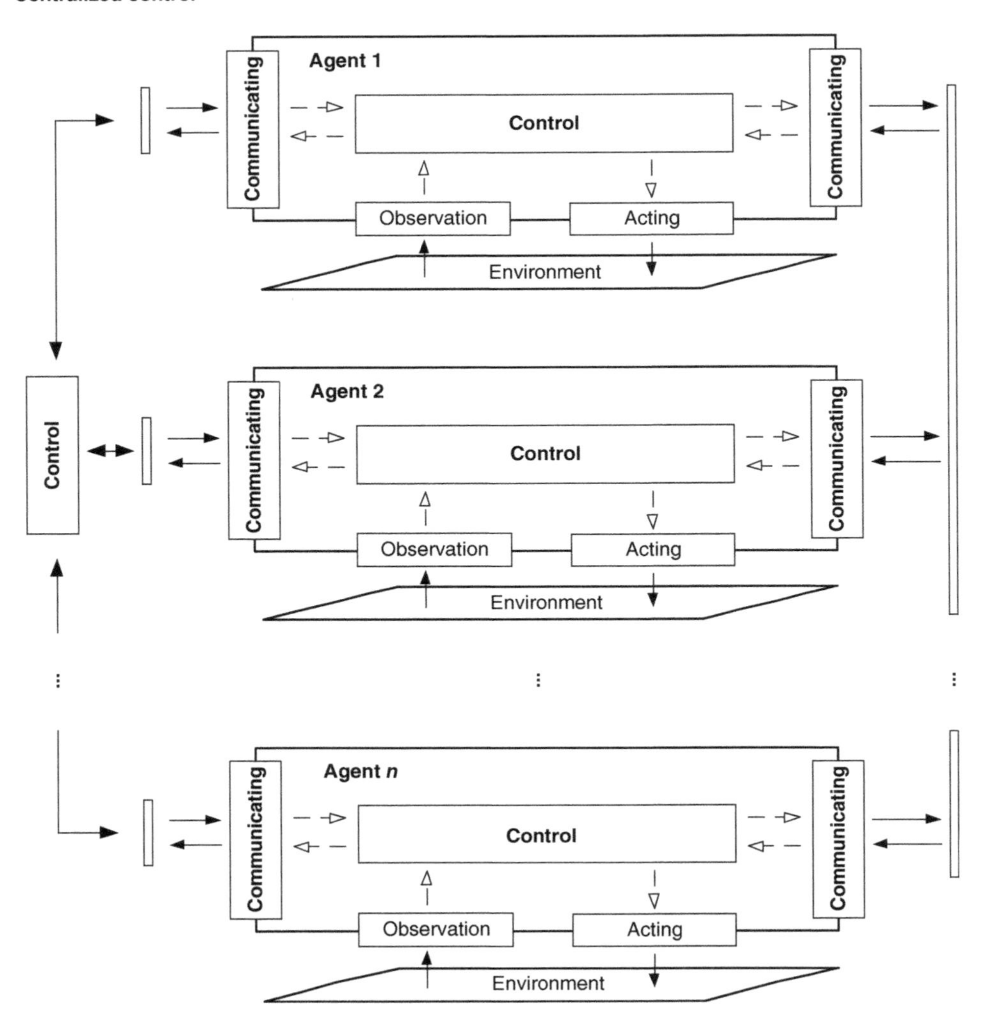

Figure 9.8 Scheme of the centralized and distributed feedback control with direct communication between the agents.

such systems and in many cases are used as a prototype for artificial systems and the teams of mobile robots. In the next chapters, we present several examples of implementation of the indicated schemes by the teams of mobile robots; in particular, in Chapter 10 we consider general models of the robots' motion with shared information about environment and in Chapter 11 the models of motion with indirect and direct communication. Below, we present the direct implementation of the feedback control scheme for control of the team of mobile robots based on the conventional methods of control of dynamical systems; the discourse partially follows the report by Cohen and Levy (2016).

As already indicated, behavior of the multi-agent system, especially of the system acting under uncertainties, can be defined by different methods, e.g. using discrete or continuous Markov models, by formal specification of the data and control flows, or by the methods of dynamical systems theory. In the last case, the dynamics of the system is represented in the form of ordinary or partial differential equations such that their solution defines the trajectories of the agents or their distributions over the domain in their states space.

Assume that the swarm consists of m agents. Then, in general, the state of each agent $j = 1, 2, \dots, m$ is defined as a vector of parameters that unambiguously specifies the position of the agent with respect to its previous positions and the environmental states. In particular, for the system of dimensionless particles moving in Euclidian space, the state of each particle is defined as a pair of variables $(\vec{x}_j(t), \vec{v}_j(t))$, $j = 1, 2, \dots, m$, that are the coordinate $\vec{x}_j(t)$ of the particle and its velocity $\vec{v}_j(t)$ at time t. Then, the motion of jth particle is specified by the system of ordinary differential equations

$$\begin{cases} \frac{d}{dt}\vec{x}_j(t) = f(\vec{x}_j, \vec{v}_j, t, u(\vec{x}_1, \dots, \vec{x}_m, \vec{v}_1, \dots, \vec{v}_m)), \\ \frac{d}{dt}\vec{v}_j(t) = g(\vec{x}_j, \vec{v}_j, t, u(\vec{x}_1, \dots, \vec{x}_m, \vec{v}_1, \dots, \vec{v}_m)) \end{cases}, \tag{9.16}$$

where f and g are some functions that correspondingly define the position and velocity of the particle and to the value $u(\vec{x}_1, \dots, \vec{x}_m, \vec{v}_1, \dots, \vec{v}_m)$ of the control function u specified with respect to the positions and velocities of all the particles in the system. Notice that the motion of each agent in the presented above Cucker-Smale model (9.6) and in the Langevin model (10.1) of active Brownian particles is defined by the system of such kind.

Let us consider the set of all possible states of each agent $j = 1, 2, \dots, m$ called the phase space. Then, the behavior of the agent is represented by the phase trajectory that specifies the changes of the agent's states in time. In other words, the behavior of the agent is defined by certain topological mapping of its phase space to itself, and the set of derivatives of this mapping by time forms the vector field that defines all possible trajectories of the agent. Vector field examples are shown in Figure 9.9 (cf. artificial potential fields shown in Figure 4.4). For detailed consideration of the indicated terms in the context of dynamical systems theory see, e.g., books by Meiss (2007) and by Brin and Stuck (2002); classical introduction into the fields of dynamical systems is presented in the book by Guckenheimer and Homes (2002).

Following the approach suggested by Ceccarelli et al. (2008), consider the motion of the system of m mobile robots on the plane and associate jth robot, $j = 1, 2, \dots, m$, with the velocity vector starting in the current location of the robot. Then, the paths of the robots are associated with the phase trajectories of certain dynamical system and the problem of control of the robots' swarm is reduced to the problem of control of the dynamical system defined by Eq. (9.16).

According to the canonical Hamiltonian formalism (Zaslavsky 2007), control of such system is defined as follows. Denote by H the Hamilton function of the system with coordinates $\vec{x}_1, \dots, \vec{x}_m$ and velocities $\vec{v}_1, \dots, \vec{v}_m$ that define the total energy $H(\vec{x}_1, \dots, \vec{x}_m, \vec{v}_1, \dots, \vec{v}_m)$ of the system, which also includes the control value $u_j = u(\vec{x}_1, \dots, \vec{x}_m, \vec{v}_1, \dots, \vec{v}_m)$ for the jth agent.

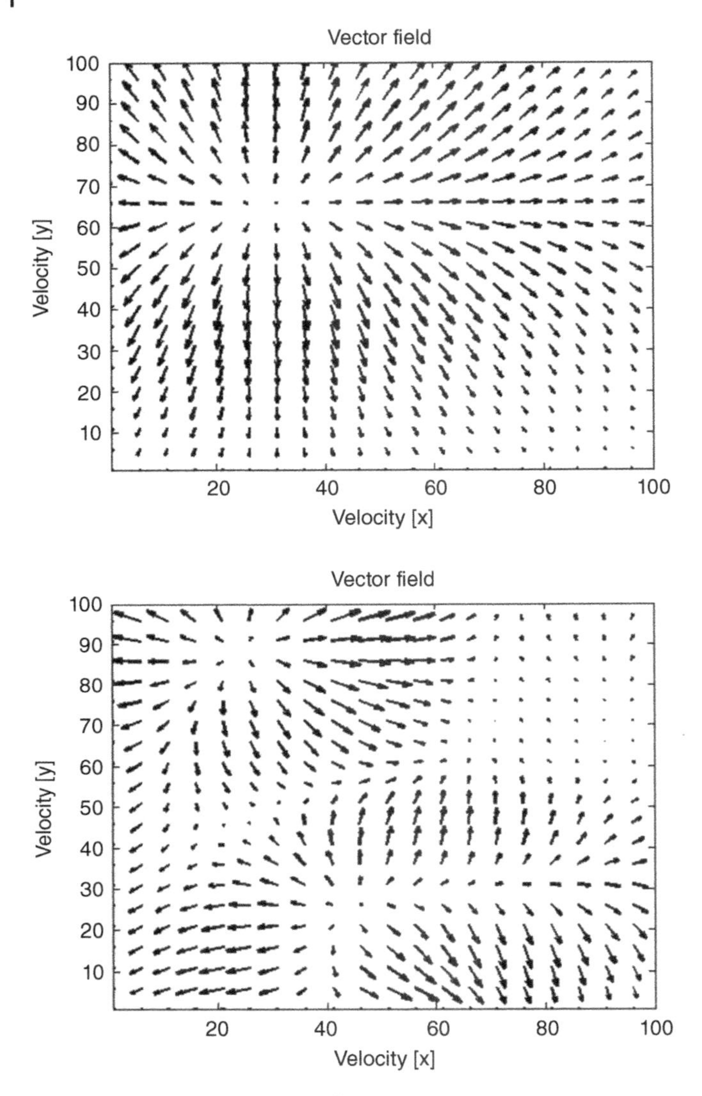

Figure 9.9 Examples of vector fields.

Then, the dynamics of the system are defined as follows:

$$\begin{cases} \dfrac{d}{dt}\vec{x}_j(t) = \dfrac{\partial}{\partial \vec{v}_j} H\left(\vec{x}_1, \dots, \vec{x}_m, \vec{v}_1, \dots, \vec{v}_m, u_j\right), \\ \dfrac{d}{dt}\vec{v}_j(t) = -\dfrac{\partial}{\partial \vec{x}_j} H\left(\vec{x}_1, \dots, \vec{x}_m, \vec{v}_1, \dots, \vec{v}_m, u_j\right), \end{cases} \tag{9.17}$$

For the limited motion, the action of the agent j is given by the integral

$$I_j = \frac{1}{2\pi}\oint \vec{v}_j d\vec{x}_j, \tag{9.18}$$

and the limited action by the integral

$$S_j = \oint^{\vec{x}_j} \vec{v}_j d\vec{x}_j. \tag{9.19}$$

Then, the angle that specifies direction of the motion of the agent j is given by the derivative of the limited action S_j by the action I_j:

$$\theta_j = \frac{\partial}{\partial I_j} S_j. \tag{9.20}$$

Given the action I_j and the angle θ_j of motion of each agent j, $j = 1, 2, \ldots, m$, the dynamics of the system of m agents can be defined in the action-angle coordinates:

$$\begin{cases} \frac{d}{dt} I_j(t) = -\frac{\partial}{\partial \theta_j} H\left(\vec{x}_1, \ldots, \vec{x}_m, \vec{v}_1, \ldots, \vec{v}_m, u_j\right), \\ \frac{d}{dt} \theta_j(t) = \frac{\partial}{\partial I_j} H\left(\vec{x}_1, \ldots, \vec{x}_m, \vec{v}_1, \ldots, \vec{v}_m, u_j\right), \end{cases} \tag{9.21}$$

which is certainly equivalent to the system (9.17) but instead of coordinates and velocities of the robots defines the lengths and directions of their steps.

Assume that the robots move with constant velocities and so make equal steps each time and that they preserve their energy. Then the system (9.21) is reduced to the system (Ceccarelli et al. 2008)

$$\begin{cases} \frac{d}{dt} I_j(t) = 0, \\ \frac{d}{dt} \theta_j(t) = u\left(\vec{x}_1, \ldots, \vec{x}_m, \vec{v}_1, \ldots, \vec{v}_m\right), \end{cases} \tag{9.22}$$

which represents the fact that the direction of the motion of each agent in the swarm depends on the positions and velocities of the other agents. It is clear that if in addition to the dependences of the agents' on each other, the swarm follows toward some goal, then the control function u includes the terms that represent such a goal and, in general, defines the topology of the space over which the vector field is defined.

On the other hand, if the swarm is controlled by a central unit and the interactions between the robots are not allowed, then the robots move with respect to the phase portrait of the dynamical system implemented in the central controller. Benedettelli et al. (2010) (see also (Ceccarelli et al. 2008)) implemented such a system for the swarm of Lego RCX robots controlled by the dynamical system of the second order that converges to the limit circle. The system, which mimics this system based on the Lego NXT robots with Bluetooth communication and OpenCV image processing, is shown in Figure 9.10 (following the scheme presented by Cohen and Levy (2016)).

The system was implemented using the Lego NXT robots programmed in the NXC language (Benedettelli 2007; Hansen 2010). Control of the robots was conducted via Bluetooth and programmed using the C/C++ libraries for the NXT robots (Monobrick 2012). Positions of the robots were captured by Microsoft digital camera and video stream was processed using the OpenCV libraries for image processing (OpenCV 2016) and OpenCV Object Tracking library by Hounslow (2013). In addition to positions, the current directions of the robots in each frame were recognized as well.

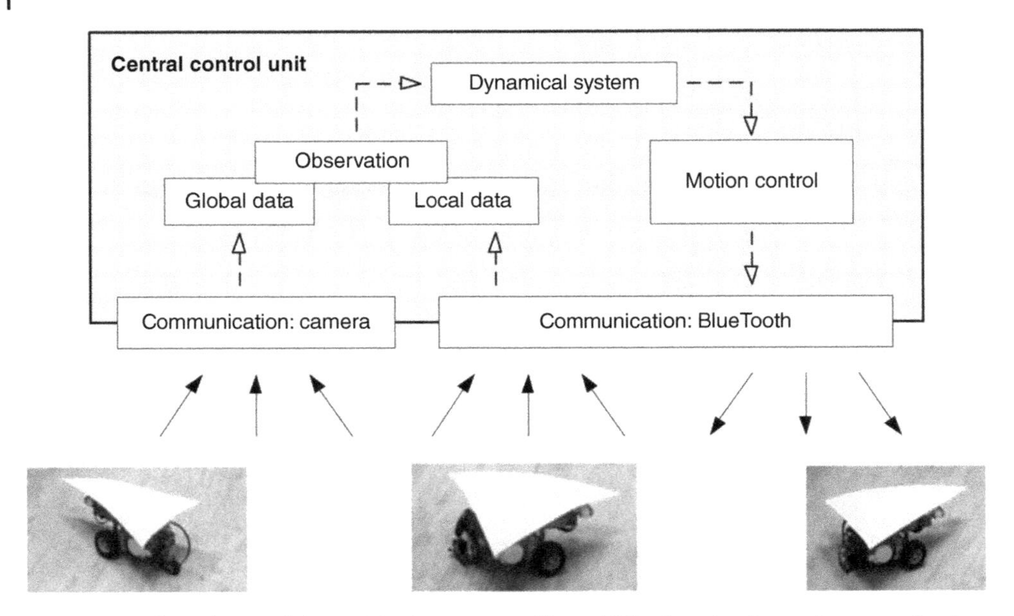

Figure 9.10 The scheme of the multi-robot system of Lego NXT robots implementing centralized feedback control.

In the control, it was assumed that the robots make equivalent steps of exact lengths each time and the dynamical system was defined following Eq. (9.22) in the action-angle coordinates.

In the considered system with centralized control, the central unit completely defines the activity of the robots, which move following the phase trajectories of the implemented dynamical system and do not allow independent decisions. In contrast, in the decentralized multi-agent systems the agents make independent decisions with respect to the observed environmental states and to the information obtained from the other agents. However, if the agents have complete information about each other, the behavior of the swarm can be defined using their joint actions and corresponding joint probability distributions of the agents' states (see Section 9.1.1). Thus, the control of such a swarm also can be conducted using the scheme of centralized control, and it is a sufficient tool for conducting the basic task of preserving the swarm that is achieved by implementing the Reynolds rules (see Section 9.1.2). For conducting the joint action and moving toward a joint goal by making autonomous decentralized decisions, the agents are required to apply additional techniques of coordination of their actions. One of the well-known techniques of such coordination is *social choice*.

The social choice method of coordination extends the alternatives and preferences and for the group of m agents is defined as follows (Shoham and Leyton-Brown 2008). Similar to Section 9.1.1, let $\mathcal{A} = \{a_1, a_2, \ldots, a_n\}$ be a set of possible alternatives such that at the current step each agent j chooses the alternative $a \in \mathcal{A}$ with respect to its preferences. Notice that here it is assumed that all the alternatives are available for all agents, although following the MDP approach (Dynkin and Yushkevich 1979; White 1993; Ross 2003) it can be defined in such a manner that the jth agent, $j = 1, 2, \ldots, m$, chooses its alternatives from the set $\mathcal{A}_j \subset \mathcal{A}$, which can change in time.

As above, it is assumed that the alternatives from the set $\mathcal{A}$ are partially ordered with respect to the preference relation $\succcurlyeq$. Since in the choice of the alternatives each jth agent follows its individual preferences, the preference of the jth agent, $j = 1, 2, \dots, m$, is denoted by $\succcurlyeq_j$ and its indifference is denoted by $\sim_j$; if it is needed, we also distinguish between strict preference $\succ_j$ and weak preference $\succcurlyeq_j$. Similarly to the actions and strategies profiles defined in Section 9.1.1, the preference profile that represents the preferences of the agents' group is denoted by $\vec{\succcurlyeq} = (\succcurlyeq_1, \succcurlyeq_2, \dots, \succcurlyeq_m)$.

Certainly, the preference relation $\succcurlyeq_j$ of each jth agent can be considered separately, as it is presented in Section 9.1.1. In particular, assume that for the jth agent, the probability distribution p_j or the lottery $\ell_j = [p_j(a_1), p_j(a_2), \dots, p_j(a_n)]$ over $\mathcal{A}$ can be defined, and assume that the relation $\succcurlyeq_j$ meets that indicated in Section 9.1.1 axioms of completeness, transitivity, decomposability, monotonicity, continuity, and substitutability. Then, following the von Neumann and Morgenstern theorem (Theorem 9.1), the ordering of the set $\mathcal{A}$ of alternatives defined by the preference relation $\succcurlyeq_j$ can be specified in the terms of the agent's utility function $u_j : \mathcal{A} \to [0, 1]$ such that for any pair of alternatives $a_i, a_k \in \mathcal{A}$ it holds true that $a_i \succcurlyeq a_k$ if and only if $u_j(a_i) \geq u_j(a_k)$.

Denote by $\mathcal{L}$ the set of all possible orders of the set $\mathcal{A}$ of alternatives. Since for each agent j its preference relation $\succcurlyeq_j$ unambiguously defines an ordering of $\mathcal{A}$, it is convenient to write that $\succcurlyeq_j$ is an element of $\mathcal{L}$, that is $\succcurlyeq_j \in \mathcal{L}$. Similarly, for the preference profile $\vec{\succcurlyeq}$ of the group of m agents it is convenient to write $\vec{\succcurlyeq} \in \mathcal{L}^m$, where $\mathcal{L}^m = \underbrace{\mathcal{L} \times \mathcal{L} \times \dots \times \mathcal{L}}_{m \text{ times}}$ stands for m Cartesian products of the set $\mathcal{L}$.

The methods of social choice are aimed to specify the decision of the group of m agents such that it coordinates the decisions, which are made by the agents with respect to their own preferences and utilities, and forms the joint decision related to group considered as a unit. In other words, given a group of m agents with their separate preferences $\succcurlyeq_j$, $j = 1, 2, \dots, m$, over the set $\mathcal{A}$ of alternatives it is required to specify a social choice function $c : \mathcal{L}^m \to \mathcal{A}$ that selects an alternative $a \in \mathcal{A}$ with respect to the preference profile $\vec{\succcurlyeq} \in \mathcal{L}^m$ of the group. Since construction of the group decision is often conducted sequentially such that the process starts with a set of candidate alternatives and enhances them up to obtaining a single decision if it exists, it is also used a set valued function $C : \mathcal{L}^m \to 2^{\mathcal{A}}$ called the social choice correspondence, which in contrast to the function c returns a set $A \subset \mathcal{A}$ of candidate alternatives.

The social choice function c and the social choice correspondence C can be constructed in different manners such that, given the same agents' preferences they provide different group decisions. Nevertheless, in order to meet the intuitive awareness on the collective non-regulated choice, usually these functions are defined on the base of voting techniques such that the winning alternative is specified as an alternative, which is chosen by the majority of the agents. Such definition of the winning alternative is widely known as the Condorcet condition (Shoham and Leyton-Brown 2008), which states that an alternative $a^* \in \mathcal{A}$ is the Condorcet winner if for any alternative $a \in \mathcal{A}$ it holds true that $(a^* \succ a) \geq (a \succ a^*)$, where $(a' \succ a'')$ stands for the number of agents that prefer alternative a' to alternative a''.

In spite of intuitive clearance of the Condorcet condition, there exist the preference profiles such that over a given set of alternatives the Condorcet winner does not exist. The simplest example is provided by the antagonistic game between $m = 2$ players over the set $\mathcal{A} = \{a_1, a_2\}$ of $n = 2$ opposite alternatives. In such a game, $a_1 \succ_1 a_2$ and $a_2 \succ_2 a_1$; thus $\#(a_1 \succ a_2) = \#(a_2 \succ a_1)$, and both alternatives are the Condorcet winners that is meaningless. The other example of nonexistence of the Condorcet winner (Shoham and Leyton-Brown 2008) considers the set $\mathcal{A} = \{a_1, a_2, a_3\}$ of three alternatives and the group of $m = 3$ agents such that their preferences are defined as follows:

$$a_1 \succ_1 a_2 \succ_1 a_3, \quad a_2 \succ_2 a_3 \succ_2 a_1, \quad a_3 \succ_3 a_1 \succ_3 a_2.$$

Thus, $\#(a_1 \succ a_2) = \#(a_2 \succ a_3) = \#(a_3 \succ a_1) = 2$, and the Condorcet winner also does not exist.

In order to overcome this problem, there were suggested several different voting techniques for choosing the winning alternative with taking into account the alternating preferences of the agents. At the same time, it was demonstrated that these techniques allow inconsistent situations, in which the candidate alternative that initially preferred by the minor number of agents is chosen as a final winner. The detailed consideration of the voting techniques and corresponding problems is presented in several sources such as in Shoham and Leyton-Brown (2008) and Nurmi (1999, 2002). Below, we restrict the discussion with the famous Arrow's impossibility theorem that is closely related to the Aumann's theorem on common knowledge (Theorem 9.3).

Given the set $\mathcal{A}$ of alternatives and the set $\mathcal{L}$ of all possible orders of $\mathcal{A}$, denote by $w : \mathcal{L}^m \to \mathcal{L}$ the function, which maps the preference profile $\vec{\succcurlyeq}$ of the group of m agents to a certain order of the set $\mathcal{A}$. In other words, the function w combines or aggregates the preferences $\succcurlyeq_j$ of the agents, $j = 1, 2, \dots, m$, and produces their joint preference relation $\succcurlyeq_w$ such that it unambiguously prescribes the alternatives, which are preferred by the agents considered as a team. Such function w is known as a social welfare function.

It is clear that the social welfare function w can be constructed in different manners and can support different approaches to the collective decision-making by the group of autonomous agents. Nevertheless, it is meaningful to assume that in contrast to the centralized control methods, it should meet at least three requirements that represent general ideology of decentralized and independent choice of the alternatives, formulated as follows (Shoham and Leyton-Brown 2008):

1) *Pareto efficiency*. Let $a_i, a_k \in \mathcal{A}$ be a pair of alternatives, $i, k = 1, 2, \dots, n$, and $\vec{\succ} \in \mathcal{L}^m$ be a preference profile of the group of m agents. Then if for the preference relation $\succ_j \in \mathcal{L}$ of each agent $j = 1, 2, \dots, m$ it holds true that $a_i \succ_j a_k$, then it also holds true that $a_i \succ_w a_k$, where w stands for the social welfare function defined with respect to the preference profile $\vec{\succ}$. In other words, if all m agents agree that the alternative a_i is preferable to the alternative a_j, then the alternative a_i is also preferable to a_j according to the social welfare function w aggregating the preferences of all the agents.
2) *Independence on irrelevant alternatives*. Let $a_i, a_k \in \mathcal{A}$ be a pair of alternatives, $i, k = 1, 2, \dots, n$, and $\vec{\succ}', \vec{\succ}'' \in \mathcal{L}^m$ be a pair of preference profiles of the group of m agents. Then if for the preference relations $\succ_j'$ and $\succ_j''$ of each agent $j = 1, 2, \dots, m$ it holds true that $a_i \succ_j' a_k$ and $a_i \succ_j'' a_k$, then it also holds true that $a_i \succ_{w'} a_k$ and $a_i \succ_{w''} a_k$, where w' and w'' stand for the social welfare functions defined with respect to the preference profiles $\vec{\succ}'$ and $\vec{\succ}''$, respectively. It means that the ordering of the alternatives defined by the social welfare function depends only on the orderings defined by the agents and

does not depend on the alternatives, which are not included into the agents' orderings.

3) *Nondictatorship*. Let $a_i, a_k \in \mathcal{A}$ be a pair of alternatives, $i, k = 1, 2, \ldots, n$, and $\vec{\succ} \in \mathcal{L}^m$ be a preference profile of the group of m agents. Then in the group does not include an agent $j^* \in \{1, 2, \ldots, m\}$ with the preference relation $\succ_{j^*}$ such that the preference $a_i \succ_w a_k$ specified by the social welfare function w with respect to the group preference profile $\vec{\succ}$ with necessity follows from preference $a_i \succ_{j^*} a_k$ of the agent j^*. In other words, there is no agent for which its preferences prescribe the preferences of the group.

The first requirement guarantees the consistency between individual and joint decisions, the second provides an independence of the decisions on external alternatives, and the third guarantees the parity of the agents.

In spite of the simplicity and clearance of these requirements, it was demonstrated a nonexistence of the social welfare function that meets all three requirements. More precisely, in 1951 Arrow proved his famous theorem that states the following (cf. the Aumann Theorem 9.3).

Theorem 9.4 For any number $m \geq 2$ of agents, if the number of alternatives is $n \geq 3$, then any social welfare function that is Pareto efficient and independent of irrelevant alternatives is dictatorial (Arrow 1951).

As a result of this theorem, several voting schemes were suggested to aim to overcome this paradoxical situation. However, after demonstrating in 1977 by Muller and Satterhwaite that even the weaker requirement of efficiency also leads to the dictatorial social welfare function, the main attention was concentrated on implementing different ranking voting schemes, in which the agents and their coalitions have different influence on the joint decision of the group.

In particular, in the framework of game theory, such a scheme is implemented in the form of the weighted voting game. Following, we consider an implementation of this game for forming subgroups in the mobile robots team and navigating the swarm toward the target position. In its basic formulation, the weighted voting game is defined as follows (Shoham and Leyton-Brown 2008; Cheng and Dashgupta 2010).

Denote by $\mathcal{J} = \{1, 2, \ldots, m\}$ the team of m agents, and let $u : 2^{\mathcal{J}} \to \mathbb{R}$ be a utility function such that for each subgroup $J \subset \mathcal{J}$ of the agents it specifies a real-valued utility (or payoff) $u(J)$. As above, for a singular agent $j \in J = \{j\}$, this utility is denoted by $u_j = u(\{j\})$. In the framework of game theory, the subgroups $J \subset \mathcal{J}$ are called coalitions, and it is assumed that the coalition J is winning if $u(J) = 1$ and it is losing if $u(J) = 0$.

The utilities $u(J)$ of the coalitions J in the weighted voting games are usually defined as follows. Let $w(j) \geq 0$ be a real value that defines the contribution of the agent j into the game, and let $w^* > 0$ be a certain threshold value. Then

$$\begin{cases} u(J) = 1, \text{ if } \sum_{j \in J} w(j) > w^*, \\ u(J) = 0, \text{ otherwise.} \end{cases} \tag{9.23}$$

The weights of the agents and the utilities of the coalitions can be defined by different manners. In particular, games known as potential games are defined using the potential function $\mathcal{P}$ (or, more precisely, the weighted potential function (Monderer and Shapley 1996)), which for any weighted coalitional game of the group $\mathcal{J}$ of agents specifies the real value $\mathcal{P}(\mathcal{J})$ such that (Petrosian, Zenkevich, and Shevkoplias 2012):

$$w(j \mid \mathcal{J}) = \mathcal{P}(\mathcal{J}) - \mathcal{P}(\mathcal{J}\ \{j\}), \quad \sum_{j \in \mathcal{J}} w(j \mid \mathcal{J}) = u(\mathcal{J}), \quad \mathcal{P}(\emptyset) = 0. \tag{9.24}$$

In other words, this function defines a trend of changing the agents' strategies that leads to success or failure of the coalition.

Following these definitions, it is possible to specify the games of different types of agents. The next example demonstrates such a game of the Tsetlin automata (Tsetlin 1973) acting in the stochastic environment.

Example 9.3 Let $\mathcal{J} = \{1, 2, \dots, m\}$ be the team of m Tsetlin automata acting in stochastic environment $C = \{c_1, c_2\}$, $|c_1|$, $|c_2| \le 1$ (Nikoy and Buskila 2017). At the times t = 0, 1, 2, ..., each jth automaton, $j \in \mathcal{J}$, receives an input value $\xi_j(t) \in \{0, 1\}$ with the probabilities

$$p(t) = Pr\{\xi_j(t) = 0\} = \frac{1 + c(t)}{2}, \quad q(t) = Pr\{\xi_j(t) = 1\} = \frac{1 - c(t)}{2},$$

defined with respect to the observed environmental state $c(t) \in \{c_1, c_2\}$. For the Tsetlin automaton, the transitions of the states $S = \{0, 1\}$ are defined with respect to the received input values by the following transition matrices:

$$\rho(\xi_j(t) = 0) = \begin{Vmatrix} 1 & 0 \\ 0 & 1 \end{Vmatrix}, \quad \rho(\xi_j(t) = 1) = \begin{Vmatrix} 0 & 1 \\ 1 & 0 \end{Vmatrix}.$$

Finally, the output $\zeta_j(t) \in \{0, 1\}$ of jth automaton is defined equally to its state $\zeta_j(t) = s_j(t)$.

Notice that following the definition of the Tsetlin automata, if the jth automaton receives an input $\xi_j(t) = 0$, then it stays in its current state, and if it receives an input $\xi_j(t) = 1$, then it changes its state. Following this property, in the game of the Tsetlin automata, the utility $u(\{j\})$ of the jth automaton is defined as

$$u(\{j\}) = (1 - \xi_j(t)).$$

In other words, if the jth automaton is led to change its state, then its utility is $u(\{j\}) = (1 - \xi_j(t)) = 1 - 1 = 0$, and if it remains in its current state, then its utility is $u(\{j\}) = (1 - \xi_j(t)) = 1 - 0 = 1$.

Notice that if the group includes one automaton, that is $\mathcal{J} = \{j\}$, then $w(j \mid \{j\}) = \mathcal{P}(\{j\}) - \mathcal{P}(\{j\}\ \{j\}) = \mathcal{P}(\{j\})$ and $u(\{j\}) = w(j \mid \{j\})$. Then the utility of two groups $\mathcal{J}_1 = \{1\}$ and $\mathcal{J}_2 = \{2\}$ of one automaton each is

$$u(\mathcal{J}_1 \cup \mathcal{J}_2) = u(\{1\}) + u(\{2\}) = \mathcal{P}(\{1\}) + \mathcal{P}(\{2\}).$$

However, for the group $\mathcal{J} = \{1, 2\}$ of two automata, the weights are

$$w(1 \mid \{1,2\}) = \mathcal{P}(\{1,2\}) - \mathcal{P}(\{2\}), \quad w(2 \mid \{1,2\}) = \mathcal{P}(\{1,2\}) - \mathcal{P}(\{1\})$$

and the utility $u(\{1, 2\})$ of the group $\mathcal{J} = \{1,2\}$ is

$$u(\mathcal{J}) = u(\{1,2\}) = 2\mathcal{P}(\{1,2\}) - \mathcal{P}(\{1\}) - \mathcal{P}(\{2\}).$$

Even in such a simple game, the automata demonstrate some basic learning properties and the behavior mimics the behavior of intelligent agents. In particular, in the group $\mathcal{J}$ of 100 automata divided by random into 10 coalitions $\{J_1, J_2, \dots, J_{10}\}$ it was observed that if half of the coalitions are successful and half are failing, then the average in time utility of the group $\mathcal{J}$ increases and deviates around the stable value. However, if 20% of the coalitions are successful and the other are failing, then averaged in time, utility of the group $\mathcal{J}$ deviates around its initial low value. The graphs of the utilities are shown in Figure 9.11.

Certainly, the presented game can be directly extended to the game of general finite automata with more wide sets of input, output, and state values and arbitrary transition

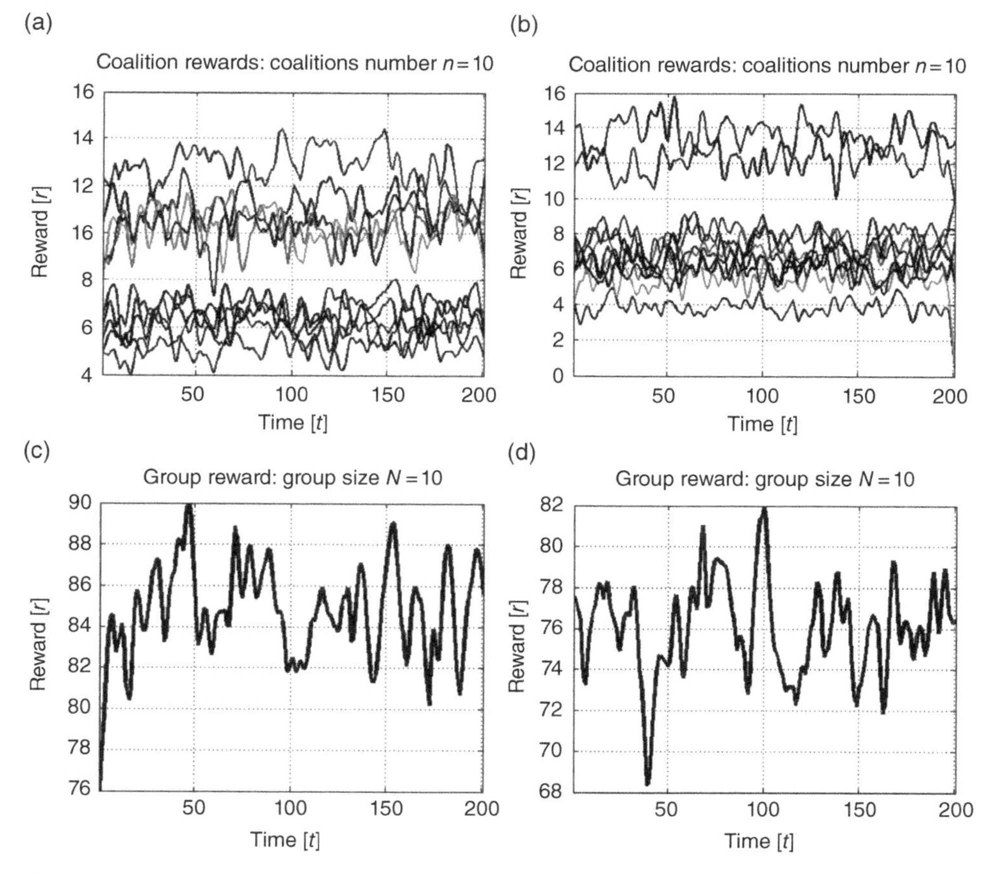

Figure 9.11 Utilities of the group of 100 Tsetlin automata for different percentage of successful and failing coalitions: figures (a, b) show the utilities of 10 coalitions with 50% and 20% successful coalitions, respectively, and figures (c, d) show the corresponding utilities of the group.

functions. Similarly, the game is not restricted by the defined utility function, and can be extended for any reasonable utilities and potential functions according to the considered problem. In particular, if the Tsetlin automata are used for control of the robots' movements in four definite direction (north, south, east, and west) (see Section 11.3.1), then the game should be defined for each of two automata of the controller. ■

In the considered model, the agents are grouped into coalitions based on the joint utility function. In other words, the agents decide either to join the coalition or not, using some external criterion that defines their preferences. Another option is the voting model, in which the alternatives are associated with the agents themselves such that the agents vote either to invite a candidate agent to the coalition or not.

Formally, the last approach is specified as follows. As above, let $\mathcal{A} = \{a_1, a_2, \dots, a_n\}$ be a set of alternatives and $\mathcal{J} = \{1, 2, \dots, m\}$ be the team of m agents. The preferences of the jth agent, $j = 1, 2, \dots, m$, regarding the alternatives are defined by its preference relation $\succcurlyeq_j$, and the preference profile that represents the preferences of the coalition $J \subset \mathcal{J}$ of the size $m_J \leq m$ is denoted by $\vec{\succcurlyeq}_J = (\succcurlyeq_1, \succcurlyeq_2, \dots, \succcurlyeq_{m_J})$.

The association of the alternatives with the agents means that while the jth agent chooses the alternative $a_i \in \mathcal{A}$, it, in fact, chooses the agent $i \in \mathcal{J}$ that is preferred to be included into the coalition according to the preferences $\succcurlyeq_j$ of the jth agent. For convenience, a single agent is considered as a coalition that includes only one agent, and including the agent i into a coalition of a single agent j means creating the coalition $J = \{i, j\}$ of two agents. Certainly, here we assume that the number n of alternatives is equivalent to the number m of agents.

Finally, in each coalition $J \subset \mathcal{J}$ with the preferences profile $\vec{\succcurlyeq}_J$ the joint decision whether to include the candidate agent $i \in \mathcal{J}$ into a coalition or not is obtained by a voting process that specifies the winning agent. The next example illustrates such an approach.

Example 9.4 Assume that the team $\mathcal{J} = \{1, 2, \dots, m\}$ of $m = 10$ agents is divided by random to $k = 5$ coalitions (including empty coalitions), and assume that the agents are required to form new division to the same number $k = 5$ coalitions with respect to their preferences (Raybi and Levy 2017).

The preferences of the agents are specified as shown in Figure 9.12.

According to the figure, first agent prefers the 9th agent to the 5th agent, the 5th agent to the 8th agent, and so on. Similarly, for example, the 7th agent prefers the 6th agent to the 1st agent, the 1st agent to the 9th agent, and so on. Notice that the agents are assumed to be "objective" and not always prefer themselves.

Agent 1 : 9, 5, 8, 10, 4, 2, 6, 3, 1, 7.
Agent 2 : 10, 8, 2, 7, 9, 4, 3, 5, 1, 6.
Agent 3 : 2, 1, 9, 5, 7, 3, 10, 8, 4, 6.
Agent 4 : 3, 10, 2, 8, 4, 5, 6, 1, 9, 7.
Agent 5 : 9, 3, 2, 10, 5, 4, 1, 6, 7, 8.
Agent 6 : 6, 7, 4, 9, 8, 2, 5, 10, 1, 3.
Agent 7 : 6, 1, 9, 10, 3, 2, 5, 7, 8, 4.
Agent 8 : 7, 4, 10, 8, 1, 2, 9, 3, 6, 5.
Agent 9 : 8, 1, 5, 4, 7, 6, 10, 9, 3, 2.
Agent 10 : 10, 4, 1, 9, 5, 6, 3, 2, 7, 8.

Figure 9.12 Preferences of $m = 10$ agents regarding the agents, which should be included into the coalitions.

Figure 9.13 Initial division of m = 10 agents to k = 5 coalitions.

Coalition: 1:
u = 1
Agent 2: w = 0.62
Agent 4: w = 0.24
Agent 5: w = 0.33
Agent 8: w = 0.72

Coalition: 2:
u = 1
Agent 1: w = 0.24
Agent 6: w = 0.59
Agent 7: w = 0.07
Agent 9: w = 0.38
Agent 10: w = 0.92

Coalition: 3:
u = 0
Agent 3: w = 0.89

Coalition: 4:
u = 0

Coalition: 5:
u = 0

In addition, following the direction of the coalitional games, assume that the contributions of the agents $j \in \mathcal{J}$ to the coalition are specified by their weights $w(j) \geq 0$ and that the utility of the coalition is defined by the formula (9.23) using the threshold value $w^* > 0$. Notice that the preferences of the agents do not depend on the weights $w(j)$, and the weights are used only for specifying the winning coalitions. In the following consideration, the weights are chosen randomly and independently on the preferences. In the considered simulation, the used threshold weight is $w^* = 1.25$.

The initial division to the $k = 5$ is conducted at random; this division is shown in Figure 9.13.

In the initial division there are two winning coalitions – coalition $J_1 = \{2, 4, 5, 8\}$ and coalition $J_2 = \{1, 6, 7, 9, 10\}$, one losing coalition $J_3 = \{3\}$ and two empty coalitions that are $J_4 = \emptyset$ and $J_5 = \emptyset$. In other words, at the beginning there are three nonempty coalitions, two of which are winning.

After voting with respect to the agents' preferences and the preference profiles of the coalitions, the agents are reorganized and form the other coalitions. The coalitions after voting are shown in Figure 9.14.

It is seen that after the voting, agent 2 left the first coalition J_1 and joined the third coalition J_3, which became the winning coalition. Nevertheless, since agent 10 left the second coalition J_2 and joined the first coalition J_1, this coalition is still winning. In addition, the second coalition lost two agents – agent 1 and agent 9 – and became the losing

Figure 9.14 Division of m = 10 agents to k = 5 coalitions after voting.

Coalition: 1:
u = 1
Agent 4: w = 0.24
Agent 5: w = 0.33
Agent 8: w = 0.72
Agent 10: w = 0.92

Coalition: 2:
u = 0
Agent 6: w = 0.59
Agent 7: w = 0.07

Coalition: 3:
u = 1
Agent 2: w = 0.62
Agent 3: w = 0.89

Coalition: 4:
u = 0

Coalition: 5:
u = 0
Agent 1: w = 0.24
Agent 9: w = 0.38

coalition. These two agents joined the fifth coalition J_5, which previously was empty; however, since the sum of the weights of these agents is less than the threshold weight, this coalition is losing. In other words, the voting resulted in division of the agents into four nonempty coalitions, $J_1 = \{4, 5, 8, 10\}$, $J_2 = \{6, 7\}$, $J_3 = \{2, 3\}$ and $J_5 = \{1, 9\}$, two of which are winning and two are losing. ■

It is clear that the presented examples do not enhance all the variety of the agent-based models of swarm dynamics and the possible implementations of the game theory models, and are used for illustration of the main ideas of such approach. The other approach applies the probabilistic models and corresponding optimization techniques mentioned in Section 9.1.1; in the next section, we consider these models in detail.

9.2.2 Probabilistic Models of Swarm Dynamics

In the deterministic case, where the agents are acting in known environment, the centralized control choice of the actions is conducted only with respect to the actions that are chosen and executed by the other agents. In the probabilistic setup, in contrast, the centralized control can be applied only to the probability distributions. One effective method of such control is the use of probability navigation function (PNF).

In our previous research (Hacohen, Shoval, and Shvalb 2017), the PNF is implemented for a single robot and a single target motion planning in an uncertain dynamic environment. For the multi-agent multi-target (MAMT) problem, we have to rebuild the function foundations, which consist of *distance to target* and *collision avoidance* functions.

The *collision avoidance* function does not change and is given by

$$\beta^j(x) = (\Delta_0 - p_0(x)) \prod_{i=0, i \neq j}^{N_a + N_o} (\Delta_i - p_i(x))$$

where Δ_i and $p_i(x)$ are the Δ and $p_{tot}(x)$ for the ith object, respectively, N_a and N_o are the number of agents and obstacles, respectively; see Figure 9.15. The quantities Δ_0 and $p_0(x)$ refer to the permitted area boundary. $p_0(x)$ is the $p_{tot}(x)$ where the geometry function gets

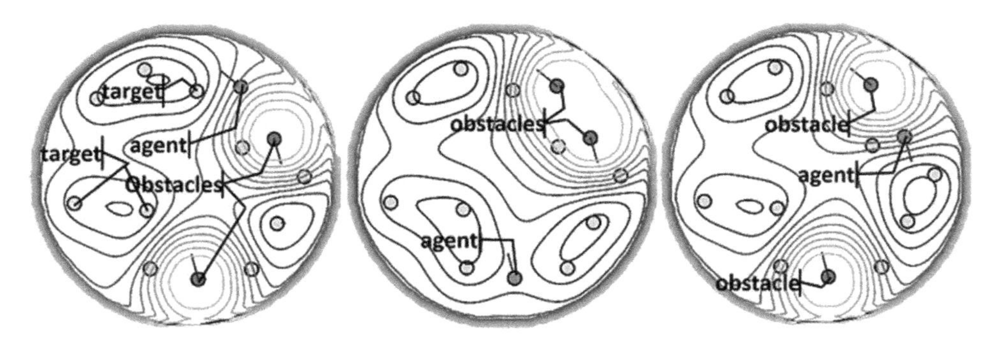

Figure 9.15 A scenario with three agents: every agent compute his own PNF in which the other agents are treated as obstacles.

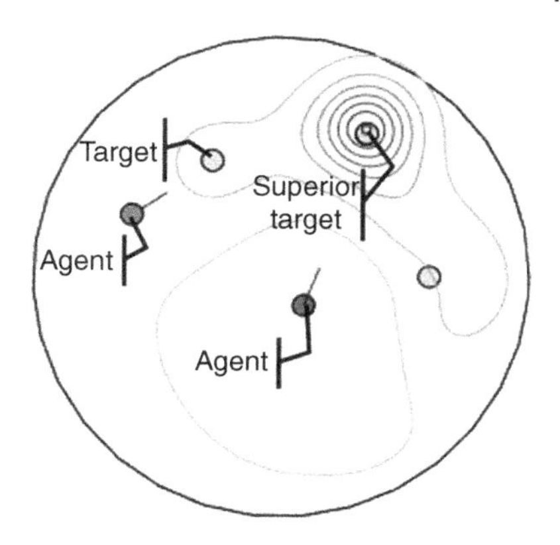

Figure 9.16 Scenario with two agents and three targets. The middle target has three times higher priority than the others. Although the right target is closer to the right agent than the middle one, the PNF gradient leads to the superior one.

a unit value outside the permitted area and it computed according to the probability density function (PDF) of the current agent only.

The *targets function*, on the other hand, is changed in two aspects. First, focus shifts to the probability of finding the target instead of the original distance to the target. Second, now there is more than one target. We defined the targets function as:

$$\lambda(x) = \sum_{j=0}^{N_t} \left(1 - p_j(x)\right)$$

where N_t is the number of unintercepted targets and $p_i(x)$ is the value of $p_{tot}(x)$ for the jth target. If the targets are prioritized, a priority value is added to the respective target. We mark the priority value of the ith target by α_i. Now, the targets function becomes (see Figure 9.16):

$$\lambda(x) = \sum_{j=0}^{N_t} \left(1 - p_j(x)\right)^{\alpha_i}$$

The direction of the agent maneuver is given by the gradient descent of the PNF:

$$\nabla\varphi(x) = \frac{\nabla\lambda\left(\lambda^K + \beta^j\right)^{\frac{1}{K}} - \frac{\lambda}{K}\left(K\nabla\lambda\lambda^{K-1} + \nabla\beta^j\right)\left(\lambda^K + \beta^j\right)^{\frac{1}{K}}}{\left(\lambda^K + \beta^j\right)^{2/K}}$$

Now we consider MAMT mission with dynamic objects (e.g., other agents, targets, or obstacles). The main principal is as follows (for more details see Hacohen, Shoval, and Shvalb (2017)). At every time step, a path for the N_{fwd} steps ahead is computed $Q = \{q_1, q_2, \ldots, q_{N_{fwd}}\}$. A transition from q_k to $q_{k+1}$is according the PNF (φ_k), which is computed using the suitable kth predicted PDFs of the objects. The entire path is improved by an optimization process using simulated annulling method. The cost function for the optimization is the sum of the PNF values that corresponds to the present node of the path, and is given by

$$E_0 = \sum_{k=1}^{N_{fwd}} \varphi_k(q_k)$$

Since the uncertainty increases at every step of the prediction, the length of the time horizon is limited, as after a finite number of time steps the probability of finding an object is almost uniform in all C.

Finally, let us demonstrate the proposed approach by numerical simulations. For all the scenarios used in the simulation, we take a disk-shaped objects with two length units radii. The agents' velocities are limited to two length units per time step and the targets and obstacles velocities are taken as one length unit per time step. One of the main issues when implementing multi-agent missions is the data transmission between the team members. Here, the only data that the agents can share is the objects' current location PDFs. All other data, such as future motions of the other agents, targets, and obstacles, are unknowns. The process and measurement noises taken here have normal distribution with zero mean and variance of one length unit.

A known problem when considering targets with normal distribution of their location is that not far from the targets the Gaussian function has negligible values. Such values cause the gradient to vanish, resulting in undefined motion direction. To avoid this, we extend the Gaussian distribution by powering it by $c << 1$ constant.

Finally, the K constant in the equation of gradient descent has a significant effect on the trajectory as it sets the priority between collision avoidance and reaching the targets. In addition, a correct selection of K guarantees convergence. In the given simulation, we select $K = \frac{N_t}{4}$ (N_t is the number of targets). A demonstration of the approach is presented in Figure 9.17.

In the figure, three agents are required to intercept 15 targets. The agents do not collide; they maintain a "safe" distance from each other, in a similar way as if an algorithm for distribution of tasks had been used. In this scenario, it takes 115 time steps to complete the entire mission.

Finally, Figure 9.18 illustrates the effect of the targets' priorities on the motion planning for single agents.

In the presented scenario, at the agent's initial position the ranges to the targets are equal. However, the function gradient leads the agent to the third target first since it has higher priority. After the interception of the third target, the agent moves toward the first since the distance to this target is much smaller than the distance to the second one.

9.3 Summary

In the chapter, we presented the concepts and methods of collective behavior in multi-robot systems. In particular, we considered the basic ideas used in the studies of multi-agent systems, the rules of flocking and aggregation, and the methods of collision avoidance. The material is organized as follows:

1) The chapter opened (Section 9.1.1) with the consideration of the main principles of multi-agent systems, and addresses the relations between the agents and their preferences, and the methods of rational and irrational decision-making under uncertainty, including voting procedures and corresponding game-theoretical facts.
2) The methods considered in Section 9.1.1 dealt with groups of abstract agents and assume that the groups are already exist. The next section (Section 9.1.2) covered

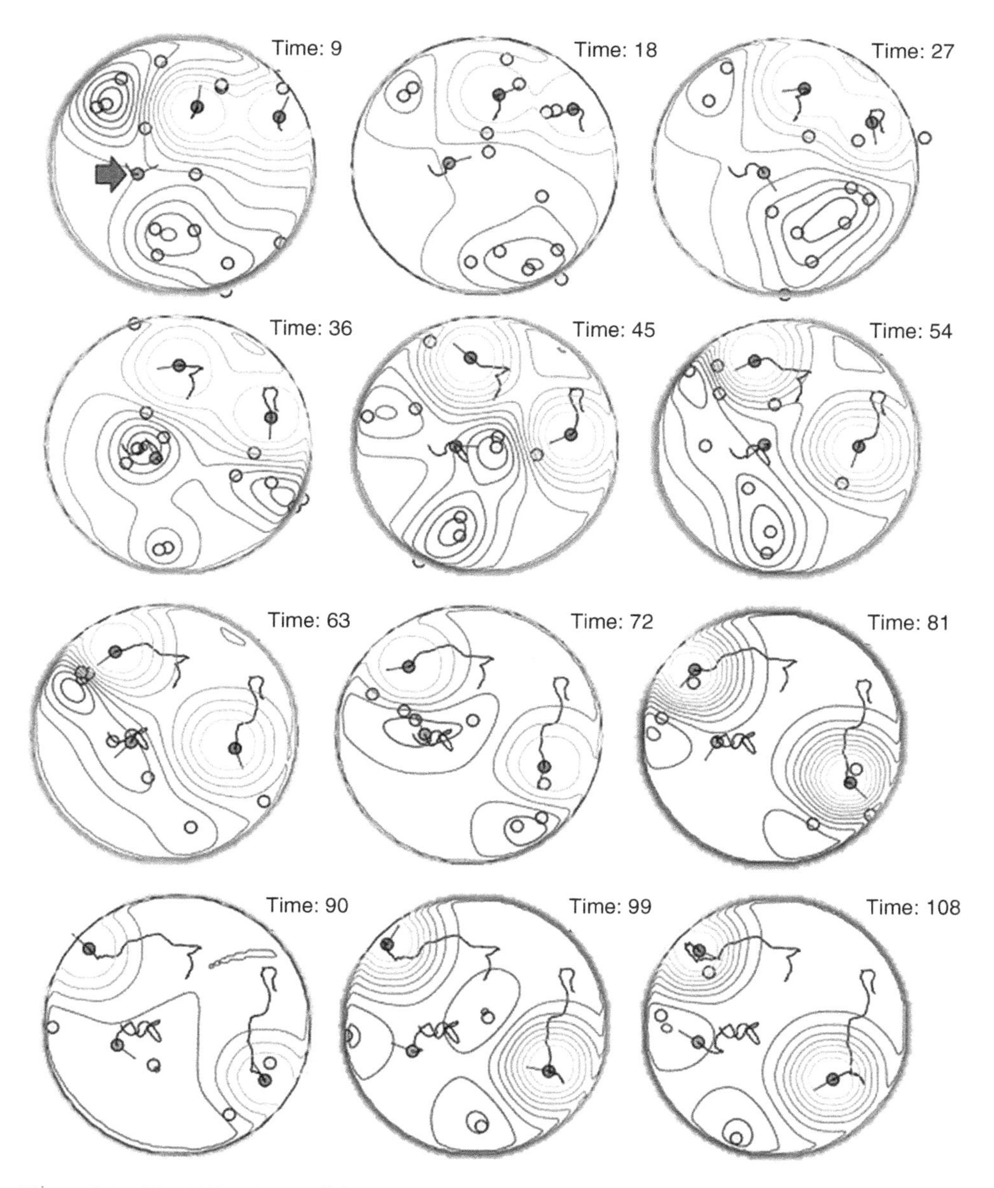

Figure 9.17 The PNFs of one of the three agents (marked in the first frame). The dark lines are the agents' motion, the dark disks are the agents and the lights discs are the targets.

concrete mobile agents acting in physical space, and addressed the basic methods of group formation – the flocking rules, and their implementation in the form of aggregation and collision avoidance techniques.

The indicated discourse provides basic ideology and philosophy of multi-robot systems that is considered in details in the next chapters of the book. As an introduction to this consideration, the second part of this chapter included some methods of

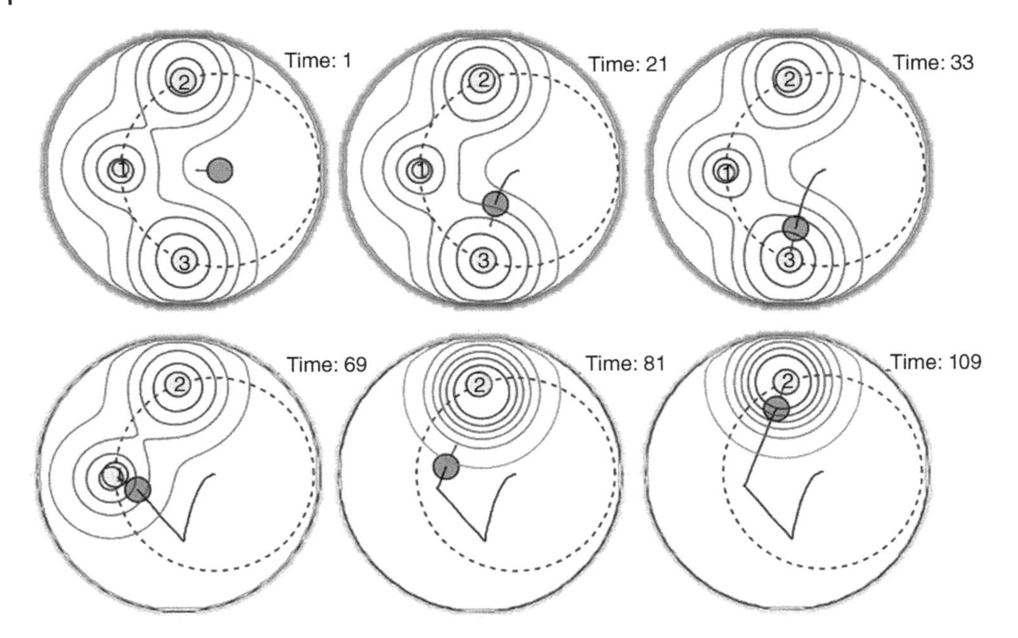

Figure 9.18 Illustration of one agent (dark disk) and three targets (light disks) which marked by their priorities. The PNF lead first toward the fourth target even though the distance to it is equal to the unity valued target.

control and positioning of the agents and their groups. In particular, it discuss the following issues.

3) (Section 9.2.1) examined the models of control of the agents and their groups, including the feedback control and relation between the agent and its neighborhood and between the agents and the environment. Such models can be implemented using different techniques; in this section, we considered their implementation in the form of controlled dynamical systems and in the form of interconnected Tsetlin automata.
4) Finally, the chapter presented some recent results in probabilistic control of the groups of mobile robots using probabilistic navigation function (Section 9.2.2). This consideration continues the material presented in Chapter 4 and extends the methods of probabilistic navigation function to multi-robot systems.

The considered models and methods were illustrated by running examples (Examples 9.1–9.4) that address the methods of collective behavior of the swarms and numerical simulations (see figures in the text), especially the examples in Section 9.2.2 that considers control of the swarm by PNF.

References

Aoki, M. (1967). *Optimization of Stochastic Systems*. New York: Academic Press.

Aumann, R.J. (1976). Agreeing to disagree. *Annals of Statistics* 4 (6): 1236–1239.

Arrow, K.J. (1951). *Social Choice and Individual Values*. (first edition) New York: Wiley.

Balch, T. and Parker, L.E. (eds.) (2002). *Robot Teams; From Diversity to Polymorphism.* Natick, MA: A K Peters.

Benedettelli, D. (2007, Jan). *NXC Bluetooth Library.* Retrieved from http://robotics/benedettelli.com/bt_nxc.htm (currently it is a part of the Lego firmware and supported by the bricxcc studio: http://bricxcc.sourceforge.net)

Benedettelli, D., Ceccarelli, N., Garulli, A., and Giannitrapani, A. (2010). Experimental validation of collective circular motion for nonholonomic multi-vehicle systems. *Robotics and Autonomous Systems* 58: 1028–1036.

Ben-Haim, S. and Elkayam, S. (2017). *The Cucker-Smale Model for the Swarm of Active Brownian Particles. B.Sc. Project.* Ariel: Ariel University.

Bouffanais, R. (2016). *Design and Control of Swarm Dynamics.* Singapore: Springer.

Brin, M. and Stuck, G. (2002). *Introduction to Dynamical Systems.* New York: Cambridge University Press.

Ceccarelli, N., Di Marco, M., Garulli, A., and Giannitrapani, A. (2008). Collective circular motion of multi-vehicle systems. *Automatica* 44: 3025–3035.

Cheng, K. and Dashgupta, P. (2010). Weighted voting game based on multi-robot team formation for distributed area coverage. In: *Proc. 3rd Int. Symp. Practical Cognitive Agents and Robots PCAR'10*, 9–15. Association for Computing Machinery (ACM Digital Library).

Cochrane, J.H. (2001). *Asset Pricing.* Princeton, NJ: Princeton University Press.

Cohen, S. and Levy, A. (2016). *Dynamical System Control of Collective Behavior. B.Sc. Project.* Ariel: Ariel University.

Couzin, I.D., Krause, J., Franks, N.R., and Levin, S.A. (2005). Effective leadership and decision-making in animal groups on the move. *Nature* 433: 513–516.

Cover, T.M. and Thomas, J.A. (1991). *Elements of Information Theory.* New York: Wiley.

Cucker, F. and Smale, S. (2007a). Emergent behavior of flocks. *IEEE Transactions on Automation and Control* 52: 852–862.

Cucker, F. and Smale, S. (2007b). On mathematics of emergence. *Japanese Journal of Mathematics* 2: 197–227.

Dynkin, E.B. and Yushkevich, A.A. (1979). *Controlled Markov Processes.* New York: Springer.

Elfes, A. (1987). Sonar-based real-world mapping and navigation. *IEEE Journal on Robotics and Automation* 3: 249–265.

Elfes, A. (1990). Occupancy grids: a stochastic spatial representation for active robot perception. In: *Proc. 6th Conference on Uncertainty in AI*, 60–70. New York, NY: Elsevier Science Inc.

Erban, R., Haskovec, J., and Sun, Y. (2016). A Cucker-Smale model with noise and delay. *SIAM Journal on Applied Mathematics* 76 (4): 1535–1557.

Fornara, N. (2003). *Interaction and Communication Among Autonomous Agents in Multiagent Systems.* Lugano, Switzerland: Universita della Svizzera italiano, IDSIA.

Friedman, M. and Savage, L.T. (1948). The utility analysis of choices involving risk. *Journal of Political Economy* 56 (4): 279–304.

Gazi, V. and Passino, K.M. (2011). *Swarm Stability and Optimization.* Berlin: Springer.

Giovanini, L. (2007). Cooperative-feedback control. *ISA Transactions* 46: 289–302.

Guckenheimer, J. and Homes, P.J. (2002). *Nonlinear Oscillations, Dynamical Systems, and Bifurcations of Vector Fields.* Berlin: Springer-Verlag.

Ha, S.-Y., Lee, K., and Levy, D. (2009). Emergence of time-asymptotic flocking in a stochastic Cucker-Smale system. *Communications in Mathematical Sciences* 7 (2): 453–469.

Hacohen, S., Shoval, S., and Shvalb, N. (2017). Applying probability navigation function in dynamic uncertain environments. *Robotics and Autonomous Systems* 87: 237–246.

Hansen, J. (2010, Oct 10). *Not eXactly C (NXC) Programmer's Guade*. Retrieved June 5, 2016, from http://bricxcc.sourceforge.net/nbc/nxcdoc/NXC_Guide.pdf

Hounslow, K. (2013). *OpenCV Object Tracking Library*. Retrieved from https://raw.githubusercontent.com/kylehounslow/opencv-tuts/master/object-tracking-tut/objecttrackingtut.cpp

Iocchi, L., Nardi, D., and Salerno, M. (2001). Reactivity and diliberation: a survey on multi-robot systems. In: *Balancing Reactivity and Social Diliberation in Multi-Agent Systems. From RoboCup to Real-World Applications* (ed. M. Hannebauer, J. Wendler and E. Pagello), 9–32. Berlin: Springer.

Kagan, E. and Ben-Gal, I. (2013). *Probabilistic Search for Tracking Targets*. Chichester: John Wiley & Sons.

Kagan, E. and Ben-Gal, I. (2015). *Search and Foraging. Individual Motion and Swarm Dynamics*. Boca Raton, FL: Chapman Hall/CRC/Taylor & Francis.

Kagan, E., Rybalov, A., Sela, A. et al. (2014). Probabilistic control and swarm dynamics of mobile robots and ants. In: *Biologically-Inspired Techniques for Knowledge Discovery and Data Mining* (ed. S. Alam), 11–47. Hershey, PA: IGI Global.

Kahneman, D. and Tversky, A. (1979). Prospect theory: an analysis of decision under risk. *Econometrica* 47 (2): 263–292.

Levy, O. (2016). *Model of Leadership in the Groups of Autonomous Mobile Agents. B.Sc. project*. Ariel: Ariel University.

Li, Z., & Ha, S.-Y. (2013, October 14). *Cucker-Smale flocking with alternating leaders*. Retrieved December 10, 2016, from arXiv:1310.3875v1.

Luce, R.D. and Raiffa, H. (1957). *Games and Decisions*. New York: John Wiley & Sons.

Meiss, J.D. (2007). *Differential Dynamical Systems*. Philadelphia, PA: SIAM.

Monahan, G.E. (1982). A survey of partially observable Markov decision processes: theory, models, and algorithms. *Management Science* 28 (1): 1–16.

Monderer, D. and Shapley, L.S. (1996). Potential games. *Games and Economic Behavior* 14: 124–143.

Monobrick. (2012). Retrieved June 5, 2016, from www.monobrick.dk

NASH, J. (1951). Non-cooperative games. *Annals of mathematics*, 286–295.

von Neumann, J. and Morgenstern, O. (1944). *Theory of Games and Economic Behavior*. Princeton: Princeton University Press.

Nikoy, L. and Buskila, M. (2017). *Potential Games of Stochastic Tsetlin Automata. B.Sc. project*. Ariel: Ariel University.

Nurmi, H. (1999). *Voting Paradoxes and How to Deal with Them*. New York: Springer Science & Business Media.

Nurmi, H. (2002). *Voting Procedures Under Uncertainty*. New York: Springer Science & Business Media.

OpenCV. (2016, May 19). Retrieved June 5, 2016, from http://www.opencv.org.

Osborne, M.J. and Rubinstein, A. (1994). *A Course in Game Theory*. Cambridge, MA: The MIT Press.

Petrosian, L.A., Zenkevich, N.A., and Shevkoplias, E.V. (2012). *The Games Theory*. Saint-Petersburg: BHV-Petersburg.

Qu, Z. (2009). *Cooperative Control of Dynamical Systems. Applications to Autonomous Vehicles*. London: Springer-Verlag.

Rabiner, L.R. (1989). A tutorial on hidden Markov models and selected applications in speech recognition. *Proceedings of the IEEE* 77 (2): 257–286.

Raybi, S. and Levy, R. (2017). *Social Choice and Decision Making in the Group of Flocking Agents. B.Sc. project.* Ariel: Ariel University.

Reynolds, C.W. (1987). Flocks, herds, and schools: a distributed behavioral model. In: *Computer Graphics (ACM SIGGRAPH'87 Conference Proceedings)*, vol. 21(4), 25–35. Association for Computing Machinery (ACM Digital Library).

Romanczuk, P., Bar, M., Ebeling, W. et al. (2012). Active Brownian particles. *The European Physical Journal Special Topics* 202: 1–162.

Ross, S.M. (2003). *Introduction to Probability Models.* San Diego, CA: Academic Press.

Rubin, E. and Shmilovitz, R. (2016). *Swarm Dynamics in Multi-Pursuer Pursuit-Evasion Game. B.Sc. Project.* Ariel: Ariel University.

Schweitzer, F. (2003). *Brownian Agents and Active Particles. Collective Dynamics in the Natural and Social Sciences.* Berlin: Springer.

Sery, N. and Feldman, E. (2016). *Collective Behavior with Decision-Making Driven by Irrational Judgments. B.Sc. Project.* Ariel: Ariel University.

Seuken, S. and Zilberstein, S. (2008). Formal models and algorithms for decentralized decision making under uncertainty. *Autonomous Agents and Multi-Agent Systems* 17 (2): 190–250.

Shen, J. (2007). Cucker-Smale flocking under hierarchical leadership. *SIAM Journal on Applied Mathematics* 68 (3): 694–719.

Shoham, Y. and Leyton-Brown, K. (2008). *Multiagent Systems. Algorithmic, Game-Theoretic, and Logical Foundations.* Cambridge, MA: Cambridge University Press.

Siegwart, R., Nourbakhsh, I.R., and Scaramuzza, D. (2011). *Introduction to Autonomous Mobile Robots*, 2e. Cambridge, MA: The MIT Press.

Smith, P. and Wickens, M. (2002). Asset pricing with observable stochastic discount factors. *Journal of Economic Surveys* 16 (3): 397–446.

Ton, T. V., Linh, N. T., and Yagi, A. (2015, August 27). *Flocking and non-flocking behavior in a stochastic Cucker-Smale system.* Retrieved December 10, 2016, from arXiv:1508.05649v2

Tsetlin, M.L. (1973). *Automaton Theory and Modeling of Biological Systems.* New York: Academic Press.

Weiss, G. (ed.) (1999). *Multiagent Systems. A Modern Approach to Distributed Artificial Intelligence.* Cambridge, MA: The MIT Press.

White, D.J. (1993). *Markov Decision Processes.* Chichester: John Wiley & Sons.

Wooldridge, M. (1999). Intelligent agents. In: *Multiagent Systems. A Modern Approach to Distributed Artificial Intelligence* (ed. G. Weiss), 27–78. Cambridge, MA: The MIT Press.

Wooldridge, M. (2009). *An Introduction to MultiAgent Systems*, 2e. Chichester, UK: John Wiley & Sons.

Zaslavsky, G.M. (2007). *The Physics of Chaos in Hamiltonian Systems.* London: Imperial College Press.

10

Collective Motion with Shared Environment Map

Eugene Kagan and Irad Ben-Gal

This chapter considers motion of the group of robots following common environment map, which is defined by the terrain or certain potential field, and stresses the differences between collective and swarm dynamics. In addition, the chapter considers the motion with and without leader. The discourse is illustrated by probabilistic search algorithms and the methods of obstacle and collision avoidance based on attraction/repulsion potentials.

10.1 Collective Motion with Shared Information

In the previous chapter, we considered general principles of swarm dynamics and indicated different types of communication and information transfer that are require for collective activity of the agents. In the simplest case, the agents communicate according to certain communication protocol and have complete information about the states of the environment and the actions of each agent in the group. In practice, such a situation is achieved by the use of central unit, which obtains information from each agent and shares it with the other agents. In the other case, the agents do not use central unit and transfer information about their local environment to other agents in the group. Then, motion planning is conducted using the environmental map and corresponding information, which are computed by each agent itself. These two situations are illustrated in Figure 10.1.

In the first case, the main communication and computation load lies on the central unit and the agents conduct decision-making tasks based on the received information, while in the second case, each agent has to be equipped with communication devices, which support information exchange with all the agents in the group, and onboard computer, which executes the computations required for creating a global map from separate local data obtained from the agents.

It is clear that for large groups of agents acting in complex environments, an implementation of such communication and computation is practically unrealistic; however,

Autonomous Mobile Robots and Multi-Robot Systems: Motion-Planning, Communication, and Swarming, First Edition. Edited by Eugene Kagan, Nir Shvalb and Irad Ben-Gal.

Companion website: www.wiley.com/go/kagan/robotsystems

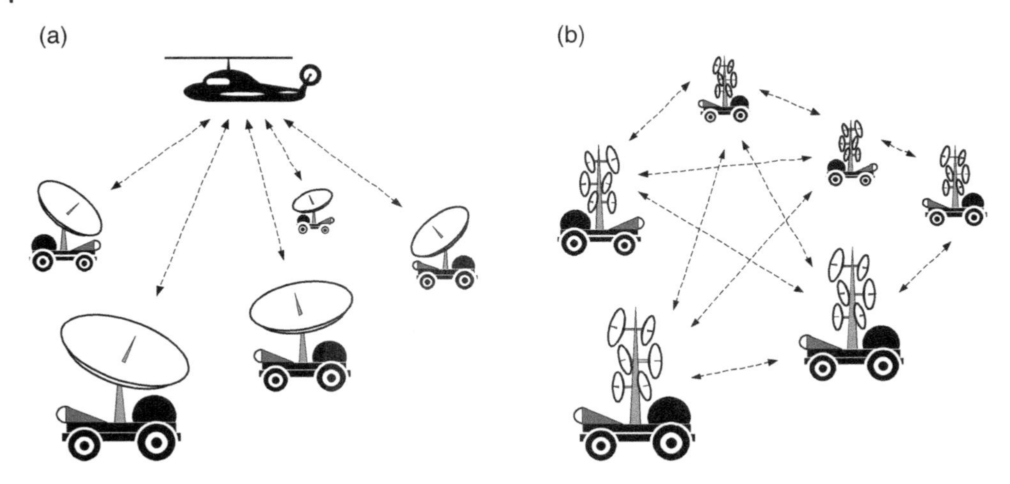

Figure 10.1 Sharing information: (a) using the central unit; and (b) by transferring local data between the agents.

for the small swarms there exist effective methods of modeling and control of swarm dynamics. In the next two sections, we consider such control using environmental and interaction potentials.

10.1.1 Motion in Common Potential Field

Let us start with the group of agents in which each agent has complete information about global environment. It is assumed that the states of the environment at each time are represented by an abstract potential field and that this field is used for motion control. The information transfer is conducted using a central unit, which is able to receive information from each agent and, after required computations, to transmit the results to all agents. The time of computation and information transfer is assumed to be infinitely smaller than the time of the agents' reactions and actions and, consequently, it is not taken into account.

Consider $m \geq 1$ agents acting in a two-dimensional domain $X \subset \mathbb{R}^2$ with the points $\vec{x} = (x, y)$. In the simulations and numerical examples, it is assumed that X is a grid $X = \left\{\vec{x}_1, \vec{x}_2, \ldots, \vec{x}_n\right\}$ consisting of $n \in \mathbb{N}$ points and each point $\vec{x}_i$, $i = 1, 2, \ldots, n$, is defined by a pair (i_x, i_y) of indices $i_x = 1, 2, \ldots, n_x$ and $i_y = 1, 2, \ldots, n_y$ such that $n = n_x \times n_y$ and $i = (i_x - 1)n_x + i_y$.

Following general methodology of motion-planning using potential field presented in Chapter 4 (see also original papers (Shahidi, Shayman, and Krishnaprasad 1991; Rimon and Koditschek 1992)), assume that each jth agent acts with respect to its potential function $U_j : X \times [0, T] \rightarrow \mathbb{R}$, $j = 1, 2, \ldots, m$. Consequently, potential in the point $\vec{x}$ at time t as it is observed by the jth agent is denoted by $U_j\left(\vec{x}, t\right)$. Similar to Chapter 4, it is assumed that the functions U_j are smooth enough over their arguments, so all required derivatives exist.

Assume that the agents are with differential drive; that is, in any position the agents are able to turn around their centers and to continue movement in the chosen direction. In addition, assume that the velocity of the agents can vary in certain range. In the other words, each jth agent, $j = 1, 2, \dots, m$, is considered as a point-like particle that moves in the potential field defined by the function U_j. There are several approaches of considering the dynamics and swarming of such particles. Here we follow the approach that is known as active Brownian motion (Schweitzer 2003; Romanczuk et al. 2012); the other slightly different approach is presented in the book by Gazi and Passino (2011). In Chapter 12 these approaches and their consequences are considered in detail.

In general, for each $j = 1, 2, \dots, m$ a random motion of jth active Brownian particle in potential field $U_j(\cdot,\cdot)$ is governed by the Langevin equation (Schweitzer 2003)

$$\frac{d}{dt}\vec{x}_j(t) = \vec{v}_j(t), \frac{d}{dt}\vec{v}_j(t) = -\gamma_j\left(\vec{x}_j, \vec{v}_j\right)\vec{v}_j(t) - \nabla U_j\left(\vec{x}_j, t\right) + \sqrt{2\sigma_j^2}G_{jt}, \tag{10.1}$$

where $\vec{x}_j(t) \in X$ is a position of the particle at time t, $\vec{v}_j(t)$ is its velocity at this time, $\gamma_j\left(\vec{x}_j, \vec{v}_j\right)$ is a friction coefficient, which in general is a nonlinear function of the coordinate $\vec{x}_j$ and velocity $\vec{v}_j$. Random factors of the motion are represented by Gaussian white noise G_{jt} and its strength is σ_j. For active particles, the friction function γ_j defines dissipation and pumping of the energy and its storage, while for simple Brownian motion the friction is constant $\gamma_j\left(\vec{x}_j, \vec{v}_j\right) = \gamma_{0j} > 0$, and the motion of the particle is defined only by the potential function U_j. For the potential constant over coordinates, the gradient is $\nabla U_j\left(\vec{x}_j, t\right) = 0$ and Eq. (10.1) with friction γ_{0j} defines simple Brownian motion, and if the potential varies in coordinates, then the gradient is $\nabla U_j\left(\vec{x}_j, t\right) \neq 0$, and the motion of the particle follows potential field $U_j(\cdot,\cdot)$. Trajectories of the particle with constant friction in the square domain X with constant and varying potential are shown in Figure 10.2.

It is seen that in the presence of the potential field, the original random walk (Figure 10.2a) follows the field and the particle is attracted by the areas with lower potential (Figure 10.2b). Such property allows an application of the functions $U_j, j = 1, 2, \dots, m$, both for specification of external potential fields that are used for motion planning and obstacle avoidance (see Chapter 4) and for implementation of the flocking rules (see Section 9.1.2) that result in aggregation of the particles and collision avoidance.

Assume that the agents act in homogeneous environment without obstacles and concentrate on swarming and the swarm mission. In terms of potential functions, the cohesion and separation rules of flocking are usually specified by the functions U_j^{atr} and U_j^{rep} that define the attraction and repulsion potentials, respectively. Often, the functions U_j^{atr} and U_j^{rep} are considered together as an aggregation or attraction/repulsion potential function U_j^{agr} such that (Gazi and Passino 2011; Romanczuk et al. 2012)

$$U_j^{agr}\left(\vec{x}_j, t\right) = U_j^{atr}\left(\vec{x}_j, t\right) + U_j^{rep}\left(\vec{x}_j, t\right).$$

Certainly, each of the indicated potentials can be defined in different forms with respect to the considered agents. Usually, these definitions follow an assumption of distance-based coupling (Schweitzer 2003), which prescribes that the aggregation depends

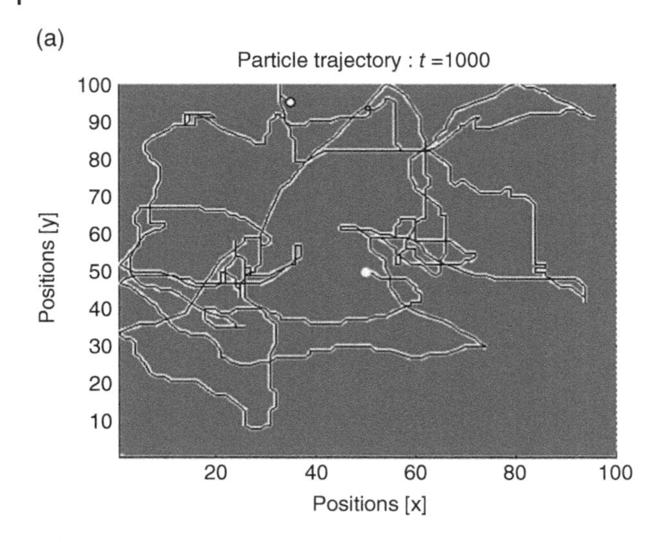

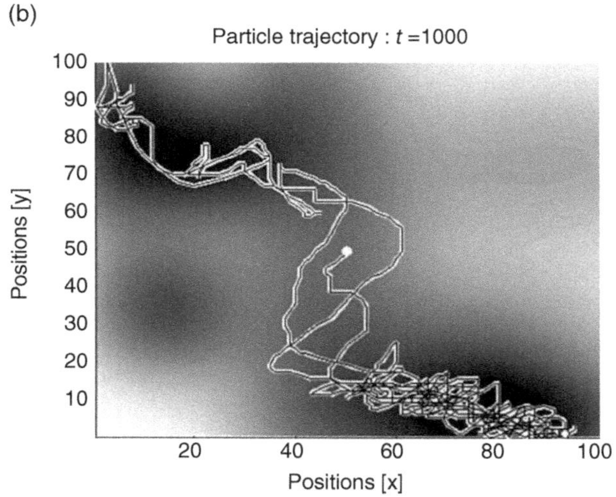

Figure 10.2 Trajectories of the particle: (a) simple Brownian motion; and (b) Brownian motion in potential field. White areas depict high potential and black areas low potential. In both cases, $\gamma\left(\vec{x}, \vec{v}\right) = \gamma_0 = 0.1$ and $\sigma = 0.5$. The size of the domain is 100×100, the starting point of the particle is $\vec{x}(0) = (50, 50)$ and motion time is $t = 0, 1, \ldots, 1000$.

on the distances between the agents, and is such that at long distances attraction is greater than repulsion, while at short distances repulsion increases and outperforms attraction (Gazi and Passino 2011).

Denote by $\mathrm{r}\left(\vec{x}_i, \vec{x}_j\right) = \left\|\vec{x}_i(t) - \vec{x}_j(t)\right\|$ the distance between ith and jth agents, $i, j = 1, 2, \ldots, m$. Then, following the indicated assumptions, one possible aggregation is defined by the function with quadratic attraction and exponential repulsion (Gazi and Passino 2011):

$$U_j^{agr}\left(\vec{x}_j,t\right)=\sum\nolimits_{i=1,i\neq j}^{m}\left\{\frac{\alpha_a}{2}\left[\mathrm{r}\left(\vec{x}_i(t),\vec{x}_j(t)\right)\right]^2+\frac{\alpha_r}{2}\exp\left[-\frac{1}{\beta_r}\left[\mathrm{r}\left(\vec{x}_i(t),\vec{x}_j(t)\right)\right]^2\right]\right\},\tag{10.2}$$

where $\alpha_a > 0$ is the amplitude of attraction and $\alpha_r > 0$ and $\beta_r > 0$ are the amplitude and the range of repulsion, respectively. The equilibrium distance for this aggregation is $\mathrm{r}^* = \sqrt{\beta_r \ln\frac{\alpha_r}{\alpha_a}}$ with $\alpha_r > \alpha_a$. Another useful aggregation function, which implements both exponential attraction and exponential repulsion is (Romanczuk et al. 2012)

$$U_j^{agr}\left(\vec{x}_j,t\right)=\sum\nolimits_{i=1,i\neq j}^{m}\left\{-\alpha_a\exp\left[-\frac{1}{\beta_a}\mathrm{r}\left(\vec{x}_i(t),\vec{x}_j(t)\right)\right]+\alpha_r\exp\left[-\frac{1}{\beta_r}\mathrm{r}\left(\vec{x}_i(t),\vec{x}_j(t)\right)\right]\right\},\tag{10.3}$$

where, respectively, $\alpha_a > 0$ and $\beta_a > 0$ are the amplitude and the range of attraction and $\alpha_r > 0$ and $\beta_r > 0$ are the amplitude and the range of repulsion. The equilibrium distance for this aggregation is $r^* = \frac{\beta_r\beta_a}{\beta_r-\beta_a}\ln\frac{\beta_r\alpha_a}{\beta_a\alpha_r}$ with $\beta_r > \beta_a$ and $\beta_r/\beta_a > \alpha_r/\alpha_a$. Certainly, in both equations the values of the parameters α_a, β_a and α_r, β_r can be different for different agents and can vary in space and time.

The mission of the agents with shared information is defined by the functions U_j^{act} that define the agents' activity and the influence of the potential fields of different agents to each other. In the case of common information, the functions $U_j^{act}, j = 1, 2, \ldots, m$, are the same for all agents in the swarm. This situation corresponds to the indicated above usage of central unit, which computes the function U^{act} and shares it with all m agents (Kagan and Ben-Gal 2015) that results in $U_j^{act} = U^{act}$ for all $j = 1, 2, \ldots, m$.

Finally, notice that in addition to aggregation potential, the Langevin equation (10.1) often includes a dissipative force F_j^{dis}, which directly specifies synchronization of the particles' velocities (Romanczuk et al. 2012). In the simplest case, this term is defined as follows (Kagan and Ben-Gal 2015):

$$\mathrm{F}_j^{dis}\left(\vec{x}_j,t\right)=-k_0^{diss}\sum\nolimits_{i=1,i\neq j}^{m}\exp\left[-\mathrm{r}\left(\vec{x}_i(t),\vec{x}_j(t)\right)\right]\left(\vec{v}_i(t)-\vec{v}_j(t)\right),\tag{10.4}$$

where the coefficient $k_0^{diss} \geq 0$ defines the influence of the force F_j^{dis}. This term is either combined with the nonlinear friction $\gamma_j\left(\vec{x}_j,\vec{v}_j\right)$ that gives

$$\gamma_j\left(\vec{x}_j,\vec{v}_j\right)\sim k_0^{diss}\sum\nolimits_{i=1,i\neq j}^{m}\exp\left[-\mathrm{r}\left(\vec{x}_i(t),\vec{x}_j(t)\right)\right]\left(\frac{\vec{v}_i(t)}{\vec{v}_j(t)}-1\right),$$

or is used as an additive term in the equation. For particular consideration of the influence of this term, see Section 6.1 in the overview by Romanczuk et al. (2012) and Section 5.3.2 in the book by Kagan and Ben-Gal (2015).

The next example illustrates an activity of the swarm of the particle-like agents governed by the Langevin equation (10.1) with the potential functions U_j defined by the aggregation and mission potential functions:

$$U_j\left(\vec{x}_j,t\right)=U_j^{agr}\left(\vec{x}_j,t\right)+U_j^{act}\left(\vec{x}_j,t\right), \tag{10.5}$$

for all $\vec{x}_j \in X$ and $t \in [0,\ T]$. The example implements two different activity functions U_j^{act}, which correspond to destructive and nondestructive probabilistic search (Viswanathan et al. 2011).

Example 10.1 Assume that $m = 25$ agents act in the square domain X of the size $n = 100 \times 100$ with the impermeable bounds. The motion of each agent j, $j = 1, 2, \ldots, m$, is governed by the Langevin equation (10.1); similar to the motion illustrated by Figure 10.2 it is assumed that for each agent $\gamma\left(\vec{x}_j,\vec{v}_j\right)=\gamma_0=0.1$ and $\sigma_j = 0.5$. Aggregation of the agents is governed by the attraction/repulsion potential function U_j^{agr} specified by the Eq. (10.3) with the parameters $\alpha_a = 30$, $\beta_a = 100$, $\alpha_r = 3$ and $\beta_r = 10$.

Let us consider the activity potential functions U^{act}, which correspond to the probabilistic search and foraging, and which are common for all agents that is $U_j^{act} = U^{act}, j = 1, 2, \ldots, m$. In such activity, the potential is defined as a probability of not finding the target in the area observed by the agent. In the other words, if $p\left(\vec{x},t\right)$ is the probability of the target's location in the point $\vec{x} \in X$ at time t, then the potential at this point in time t is $U^{act}\left(\vec{x},t\right)=1-p\left(\vec{x},t\right)$, and the agent is attracted to the point $\underset{\sim}{x}$ proportionally to the probability of finding the target in $\underset{\sim}{x}$.

With respect to the influence of the agents to the target location probabilities, usually it is considered a destructive search, in which it is assumed that the probabilities in the observed areas decrease according to the implemented detection function (with respective Bayesian renormalization of the probabilities in the other areas), and nondestructive search, where the probabilities in the observed areas remained unchanged (Viswanathan et al. 2011; Kagan and Ben-Gal 2015). In this example, it is assumed that in the destructive search each jth agent applies the Gaussian detection function centered on the agent's location $\vec{x}_j=\left(x_j,y_j\right)$. Thus, after detection the remained target location probabilities $\bar{p}\left(\vec{x},t\right)$ are specified as follows:

$$\bar{p}\left(\vec{x},t\right)=p\left(\vec{x},t\right)\frac{1}{2\pi s_x s_y}\exp\left[-\frac{\left(x-x_j\right)^2}{2\,s_x^2}-\frac{\left(y-y_j\right)^2}{2\,s_y^2}\right],$$

where deviations $s_x > 0$ and $s_y > 0$ define the form of the function over x and y axes. Finally, the probabilities $p\left(\vec{x},t+dt\right)$ at the next time $t + dt$ are obtained by the Bayesian rule as

$$p\left(\vec{x},t+dt\right)=\bar{p}\left(\vec{x},t\right)/\int_X \bar{p}\left(\vec{x},t\right)d\vec{x}.$$

For additional information about probabilistic search and foraging, see recent books by Kagan and Ben-Gal (2013, 2015).

Trajectories of the agents corresponding to nondestructive and destructive probabilistic search are illustrated in Figure 10.3. In the figures, $m = 25$ agents start from the points ordered by grid with the step of 10 units. Notice that the nondestructive search

(a)

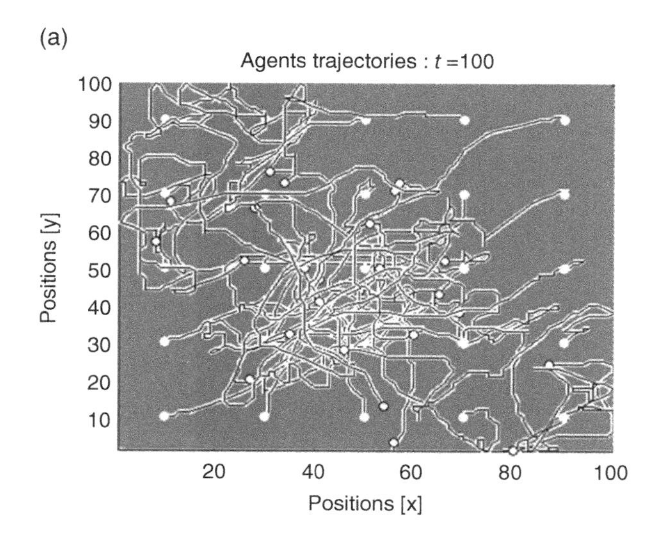

(b)

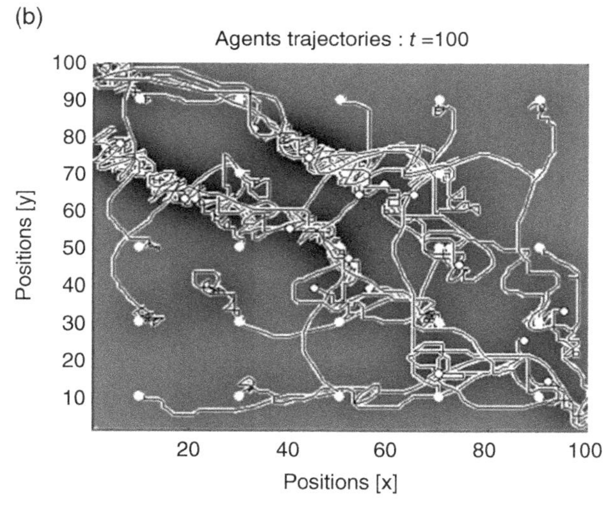

Figure 10.3 Trajectories of the swarm of $m = 25$ Brownian agents: (a) nondestructive search in which common activity potential is constant $U_j^{act}\left(\vec{x},t\right) = U^{act}\left(\vec{x},t\right) = const$ and the agents move only following the aggregation potential $U_j^{agr}\left(\vec{x},t\right) = U^{agr}\left(\vec{x},t\right), j = 1, \ldots, m$; (b) destructive search in which the agents move following, both common activity potential $U_j^{act}\left(\vec{x},t\right) = U^{act}\left(\vec{x},t\right)$ and common aggregation potential $U_j^{agr}\left(\vec{x},t\right) = U^{agr}\left(\vec{x},t\right), j = 1, \ldots, m$; white areas depict high potential and black areas – low potential. In both cases $\gamma\left(\vec{x}_j, \vec{v}_j\right) = \gamma_0 = 0.1$ and $\sigma_j = 0.5$. The size of the domain is 100×100 and motion time is $t = 0, 1, \ldots, 100$.

corresponds to the motion of the agents with respect only to the aggregation potential. For destructive search, the parameters of the detection function are $s_x = s_y = 5$.

It is seen that in the case of a nondestructive search with constant activity potential $U_j^{act}\left(\vec{x},t\right) = U^{act}\left(\vec{x},t\right) = const, j = 1, \ldots, m$, the agents follow their common aggregation potential $U_j^{agr}\left(\vec{x},t\right) = U^{agr}\left(\vec{x},t\right)$, which leads them to concentrate in the central areas of the domain where most of the agents are moving. In contrast, in the case of destructive search the agents change their common activity potential $U_j^{act}\left(\vec{x},t\right) = U^{act}\left(\vec{x},t\right)$ and, in addition to aggregation according to the potential $U_j^{agr}\left(\vec{x},t\right) = U^{agr}\left(\vec{x},t\right)$, are attracted by the regions with low values of the potential $U^{act}\left(\vec{x},t\right)$. ■

The presented models illustrate a general idea of motion-planning and swarm dynamics based on the potential field. The main difference between the motion of the agent governed by the potential function (see Chapter 4) and the motion of the swarm is in the use of the aggregation potential function, which defines the rules of collective motion. Notice that in the considered models, the agents implement common potential that implies that the information transfer and processing is conducted using a central unit (see Figure 10.1a). As a result, all agents apply the same potential and obtain an equal and complete knowledge about the environment. In contrast, if the agents interact directly without central unit, each agent obtains information about its local environment and shares it with its close neighbors. Consequently, different agents earn different information about the environment, and motion planning is conducted based on partially complete and partially predicted information about environmental states and locations of the agents. The next section considers this situation in the terms of motion planning using potential field techniques.

10.1.2 Motion in the Terrain with Sharing Information About Local Environment

In the previous section, we considered the motion of $m \geq 1$ particle-like agents governed by the Langevin equation (10.1) with the potential function U_j, $j = 1, 2, \dots, m$, which includes aggregation U_j^{agr} and activity U_j^{act} terms (see Eq. (10.5)). In the case of complete information provided by the central unit, the agents applied the equal potential functions considered as a common activity potential function $U^{act} = U_j^{act}$, $j = 1, 2, \dots, m$. Now, let us consider collective behavior of the agents, which communicate directly only with the close neighbors and share information about local environment.

Assume that the sensitivity of the agents' sensors decreases exponentially with the distance such that the perceived signals $z_j\left(\vec{x}\right)$ about the values $\mathrm{e}\left(\vec{x}\right)$ of the environmental states in the points $\vec{x} \in X$ are

$$z_j\left(\vec{x}\right) = \alpha_s \exp\left[-\frac{1}{\beta_s}\mathrm{r}\left(\vec{x}_j(t), \vec{x}\right)\right]\mathrm{e}\left(\vec{x}\right), \tag{10.6}$$

where $\vec{x}_j(t)$ is a position of the jth agent at time t and $\mathrm{r}\left(\vec{x}_j(t), \vec{x}\right)$ is a distance between the agent and the point $\vec{x}$. In addition, assume that the communication between the agents is limited by the communication or signaling radius $r_{\text{signal}}^{\max}$ that defines the neighboring agents, which are able to receive information from each other.

Then, for each agent j the activity potential is U_j^{act} combines the perceived information and the information obtained from the neighboring agents. In particular, if the potentials are additive, then in the points $\vec{x} \in X$ at time t the activity potential of the jth agent located in the point $\vec{x}_j(t)$ is

$$U_j^{act}\left(\vec{x},t\right) = \alpha_s \exp\left[-\frac{1}{\beta_s} r\left(\vec{x}_j(t), \vec{x}\right)\right] e\left(\vec{x}\right) + \sum\nolimits_{i=1,\; i \neq j,\; r\left(\vec{x}_i(t), \vec{x}_j(t)\right) \leq r_{\text{signal}}^{\max}}^{m} U_i^{act}\left(\vec{x},t\right), \tag{10.7}$$

where the first term corresponds to the information perceived by the agent's sensors and the second term specifies information about the environment obtained from the neighboring agents. The next example illustrates the motion of the agents using this activity potential function.

Example 10.2 Similar to the previous example, consider $m = 25$ agents acting in the square domain X of the size $n = 100 \times 100$ with the impermeable bounds. The motion of each agent j, $j = 1, 2, \ldots, m$, is governed by the Langevin equation (10.1) with the parameters $\gamma\left(\vec{x}_j, \vec{v}_j\right) = \gamma_0 = 0.1$, $\sigma_j = 0.5$ and aggregation function U_j^{agr} is given by the Eq. (10.3) with the parameters $\alpha_a = 30$, $\beta_a = 100$, $\alpha_r = 3$ and $\beta_r = 10$. In addition, it is assumed that the agents execute the same mission as in Example 10.1 of the probabilistic search and the activity potential is defined as a probability of nondetecting the targets in the point $\vec{x}$ at time t with the Gaussian detection function.

The agents implement activity potential U_j^{act} specified by the Eq. (10.7) with the parameters $\alpha_s = 10$ and $\beta_s = 100$; according to the mission, the environmental states are defined are $e\left(\vec{x}\right) = 1 - p\left(\vec{x},t\right)$, where $p\left(\vec{x},t\right)$ is the probability of finding the target in the point $\vec{x}$ at time t. Trajectories of the agents corresponding to the destructive search with different values of the signaling radius signaling radius $r_{\text{signal}}^{\max}$ are illustrated in Figure 10.4.

For a small signaling radius, $r_{\text{signal}}^{\max} = 10$ in Figure 10.4a and $r_{\text{signal}}^{\max} = 30$ in Figure 10.4b, the agents move in clusters around their initial locations, while for the larger signaling radius, $r_{\text{signal}}^{\max} = 50$ in Figure 10.4c and $r_{\text{signal}}^{\max} = 100$ in Figure 10.4d, each agent obtains more information about complete domain and the motion of the agents becomes similar to their motion with common activity potential function (see Figure 10.3b). ■

Example 10.2 illustrates the motion of the group of Brownian agents, which share information about their local environment; in the considered case it is a probability of nonfinding the target in the close surroundings of the agents. In addition, the example stresses the dependence of the agents' motion on the information obtained by the other agents; in Section 11.1.2 this dependence will be discussed again in the framework of sensor-based motion.

Notice that in the considered model each agent obtained information from the other agents and moved with respect to this information without using the knowledge about the mobility of its neighbors. However, following the assumption that the agents are identical, such knowledge is available: the neighbors of the agent react and move in exactly the same way as the agent itself, and the agent can estimate and predict their behavior and act correspondingly. For detailed consideration of the prediction methods, see Section 7.2.

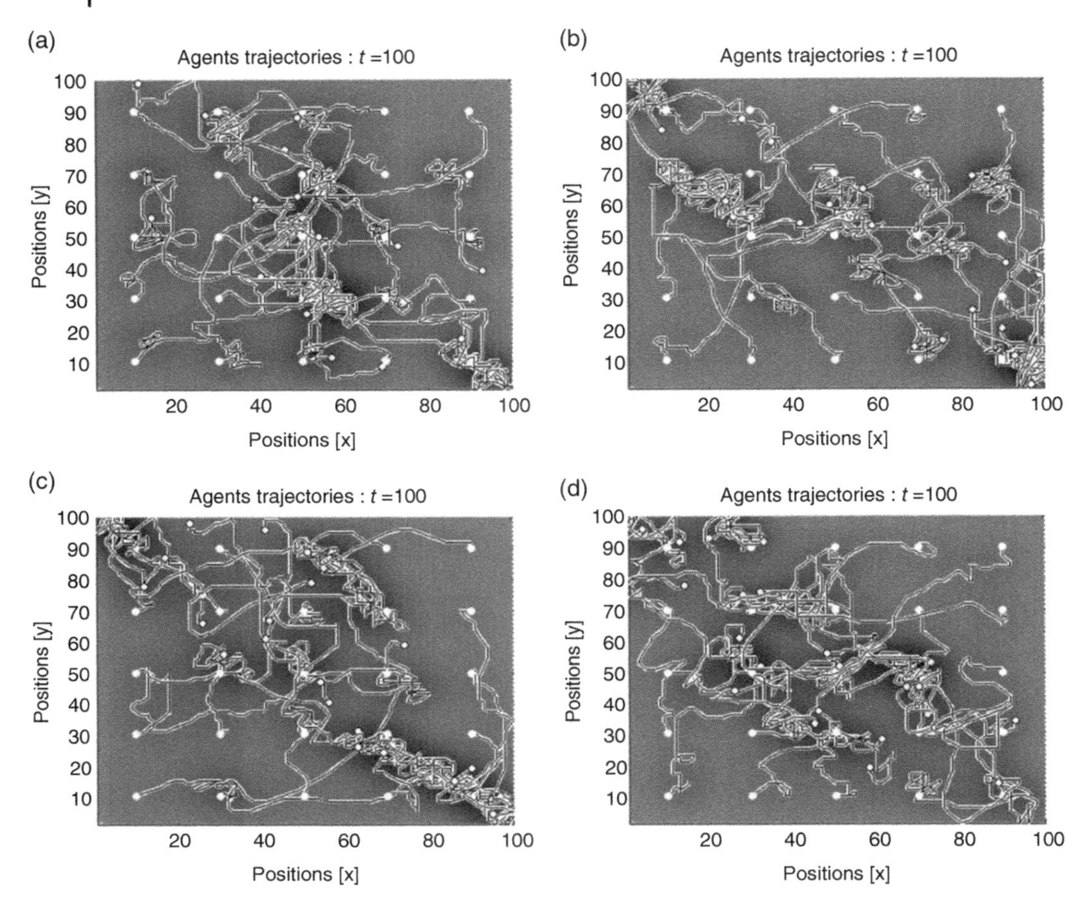

Figure 10.4 Trajectories of the swarm of $m = 25$ Brownian agents executing destructive search and sharing information about local environment among the neighbors with different signaling radius: (a) $r_{signal}^{max} = 10$; (b) $r_{signal}^{max} = 30$; (c) $r_{signal}^{max} = 50$; and (d) $r_{signal}^{max} = 100$. In the figures, white areas depict high potential and black areas – low potential. In both cases $\gamma\left(\vec{x}_j, \vec{v}_j\right) = \gamma_0 = 0.1$ and $\sigma_j = 0.5$, and the parameters of the activity potential, which defined the information sharing, are $\alpha_s = 10$ and $\beta_s = 100$. The size of the domain is 100×100 and motion time is $t = 0, 1, \ldots, 100$.

The considered models describe the activity of the swarm in which for each point $\vec{x} \in X$ and time $t \in [0, T]$ the activity potential $U^{act}\left(\vec{x}, t\right)$ is defined by the agents themselves and the aggregation potential $U_j^{agr}\left(\vec{x}, t\right), j = 1, 2, \ldots, m$, of the agents does not depend on the environmental states. Such description corresponds to the motion in homogeneous environment without obstacles, where motion of the agents is possible in any direction and both sensing and communication between them are not restricted by environmental conditions. In the next section, these models are extended to the motion in heterogeneous environment.

10.2 Swarm Dynamics in a Heterogeneous Environment

In contrast to the activity of the agents in the homogeneous environment, where the agents are not restricted in their motion, sensing, and communication abilities, swarm behavior in heterogeneous environments is restricted by the environmental states, which can accelerate or slow down the agents' movement, and by possible obstacles, which bound the agents' mobility. Following the model of swarm dynamics based on active Brownian motion, the first restriction is represented by nonlinear friction that can obtain both positive and negative values with respect to the environmental topography, and the second – by additional external potential.

10.2.1 Basic Flocking in Heterogeneous Environment and External Potential Field

Let us return to the Langevin equation (10.1) and consider the friction functions γ_j, which accelerate or slow down the motion of the agents. For simplicity, assumed that the agents are identical, so $\gamma_j = \gamma$ for all agents $j = 1, 2, \dots, m$. In contrast to the previously considered models, where the constants friction $\gamma\left(\vec{x}_j, \vec{v}_j\right) = \gamma_0 > 0$ was implemented, assume that the functions are varying with the location or velocity of the agents and, moreover, that the values $\gamma\left(\vec{x}_j, \vec{v}_j\right)$ can be both positive and negative. Here, these functions are applied for modeling the reaction of the agents to the topography of the environment.

There are several models underlying the nonconstant friction function γ (Romanczuk et al. 2012). In particular, the Schienbein-Gruler model implements the following linear friction function:

$$\gamma\left(\vec{x}, \vec{v}\right) = \gamma_0\left(\frac{\left|\vec{v}_0\right|}{\left|\vec{v}\right|} - 1\right), \tag{10.8}$$

where $\gamma_0 > 0$ is a friction constant and $\vec{v}_0$ is a critical velocity, and it is assumed that $\left|\vec{v}\right| > 0$. By contrast, the nonlinear Rayleigh-Helmholtz model applies the quadratic friction function

$$\gamma\left(\vec{x}, \vec{v}\right) = \gamma_0\left(\left|\vec{v}\right|^2 - \left|\vec{v}_0\right|^2\right), \tag{10.9}$$

where $\vec{v}_0$, as above, is a critical velocity. In both models, if velocity v of the agent is less than the critical velocity $\vec{v}_0$, then the friction is negative and the agent accelerates. In contrast, if velocity $\vec{v}$ is greater than the critical velocity $\vec{v}_0$, then the friction is positive and the agent slows down.

Finally, the Schweitzer model (Schweitzer 2003) implies that the agent is able to gain the energy from the environment, to accumulate it in the internal depot and then to use for self-propelling motion. In this model, the friction function is

$$\gamma\left(\vec{x},\vec{v}\right)=\gamma_0\left(1-\frac{r_0c_0}{\gamma_e+c_0\left|\vec{v}\right|^2}\right), \tag{10.10}$$

where $c_0>0$ is the energy conversion rate, $r_0>0$ is the energy intake rate, and $\gamma_e>0$ is the energy dissipation rate.

Notice that in all these models the friction depends on the velocity v of the agent, and does not depend on its location. On the other hand, given the topography $\mathcal{T}$ of the domain and assuming that the function $\mathcal{T}$ is smooth enough (cf. Chapter 5), friction can be defined directly with respect to the gradient $\nabla\mathcal{T}$ of the function $\mathcal{T}$. In the simplest case, it can be defined as follows:

$$\gamma\left(\vec{x},\vec{v}\right)=\gamma_0+\alpha_f\nabla\mathcal{T}, \tag{10.11}$$

where $\alpha_f>0$ is the normalization coefficient. If $\nabla\mathcal{T}<0$, then the agent accelerates, and if $\nabla\mathcal{T}>0$, then it slows down. Trajectories of the agents with friction governed by the Schweitzer model (Eq. (10.10)) and with friction defined directly by topography (Eq. (10.11)) are shown in Figure 10.5.

It is seen that in both cases, the agent's trajectory corresponds to random walk with varying length of the step. However, while in the movement with friction defined by the Schweitzer model (Figure 10.5a), the dependence of the step's length on the topography is rather light, in the movement with friction defined directly by the topography, this dependence is very strong: in the areas with large negative gradient $\nabla\mathcal{T}$ the agent moves by long steps and in the areas with large positive gradient $\nabla\mathcal{T}$ the steps of the agent are very short.

It is clear that for a group of agents, the appropriate definition of friction should be combined with the aggregation U_j^{agr} and activity U_j^{act} potentials, $j=1,2,\dots,m$, and, if it is needed, with additional external potential U. The motion of the group of $m=25$ agents with friction defined by topography and without and with aggregation is illustrated by the next example.

Example 10.3 Assume that $m=25$ agents acting in the square domain X of the size $n=100\times100$ with the impermeable bounds, and the motion of each agent j, $j=1,2,\dots,m$, is governed by the Langevin equation (10.1) with $\sigma_j=0.5$ and topography-dependent friction as defined by Eq. (10.11), in which $\gamma_0=0.1$ and $\alpha_f=0.1$. The aggregation function U_j^{agr} is given by Eq. (10.3) with the parameters $\alpha_a=30$, $\beta_a=100$, $\alpha_r=3$ and $\beta_r=10$. To stress dependence of the motion on the aggregation, assume that the agents does not execute any definite mission, so the activity potential is $U_j^{act}=0$ for each j. Trajectories of the agents without aggregation and with aggregation are shown in Figure 10.6a,b, correspondingly.

As it was expected, without aggregation (see Figure 10.6a) the agents move independently, and with aggregation (see Figure 10.6b), in contrast, the agents tend to cluster in the central part of the domain. The length of the steps is defined by the gradient $\nabla\mathcal{T}$ of the topography (cf. Figure 10.5b). Notice again that in both cases, topography does not influence the direction of the agents' motion. ■

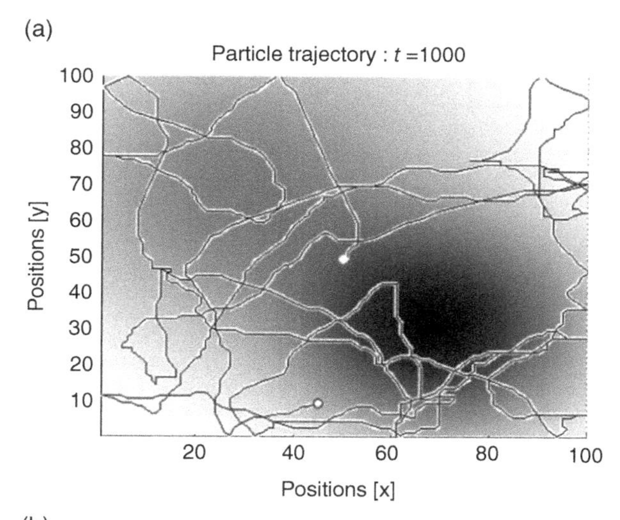

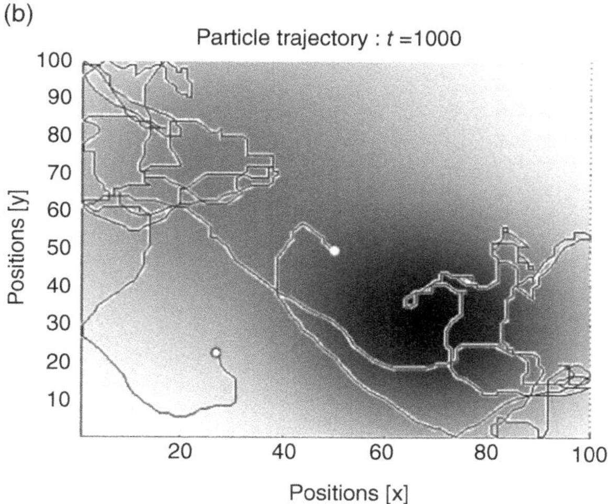

Figure 10.5 Trajectories of the agent with different friction functions: (a) Schweitzer model of friction with energy depot, in which $c_0 = 1$, $r_0 = 0.1$ and $\gamma_e = 0.1$; and (b) friction defined by topography with $\alpha_f = 0.1$. In both cases, $\gamma_0 = 0.1$ and $\sigma_j = 0.5$. The size of the domain is 100×100, the starting point of the particle is $\vec{x}(0) = (50,50)$ and motion time is $t = 0, 1, \ldots, 1000$.

The considered models provide a general description of motion of the group of mobile agents, and the presented examples demonstrate implementation of these models for specification of different types of motion and different dependences of the motion on the environment. In certain tasks, the motion can be defined using different aggregation, activity, and external potential functions and using different models of friction. In the next example, we demonstrate an implementation of one of the presented models for navigation of a group of mobile robots.

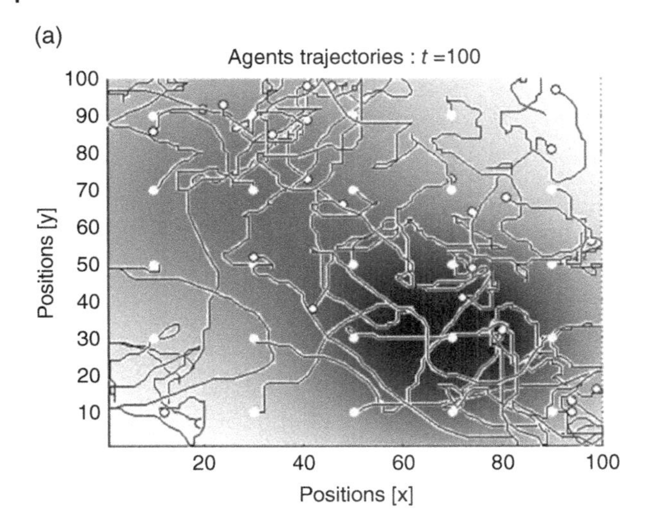

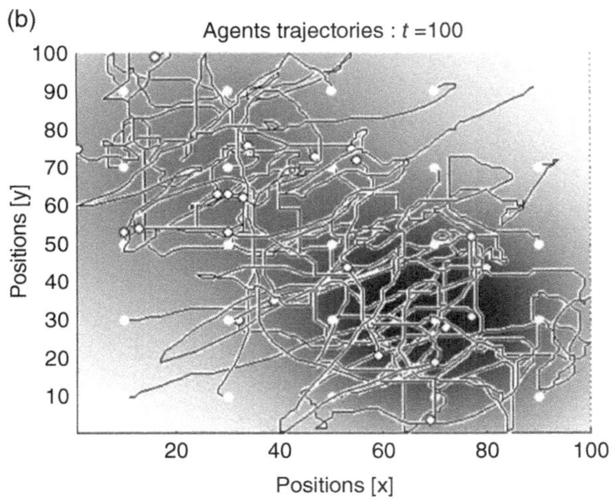

Figure 10.6 Trajectories of m = 25 Brownian agents with friction defined by topography: (a) agents without aggregation; and (b) swarm of the agents with aggregation with parameters $\alpha_a = 30$, $\beta_a = 100$, $\alpha_r = 3$ and $\beta_r = 10$. In both cases, $\alpha_f = 0.1$, $\gamma_0 = 0.1$ and $\sigma_j = 0.5$. The size of the domain is 100×100 and motion time is $t = 0, 1, \ldots, 100$.

Example 10.4 (prepared in collaboration with Eynat Rubin and Idan Hammer). Let us consider the simulated motion of the agents implementing aggregation potential function and following the environmental states. The motion of each agent is governed by the following routines inspired by the pseudocode supplementing the paper by Rubenstein, Cornejo, and Nagpal (2014).

follow_environment *(previous_brightness): current_brightness.*

1) Measure *current_brightness.*
2) If *current_brightness* ≥ *previous_brightness*
   then turn to random direction.

3) Step forward.
4) Measure *current_brightness.*

This routine models the motion of the agent in the external potential and governs the agent to go away from the brighter areas and to follow to the darker areas. Notice that in the homogeneous environment, this routine results in the simple random walk of the agent.

follow_neighbors *(previous_distance): current_distance.*

1) Measure distances to all neighbors.
2) Set $current\_distance = \min_{over\ all\ neighbors}\{measured\_distances\}$.
3) If $current\_distance > previous\_distance$ or $current\_current > MAX\_DISTANCE$ then turn to random direction.
4) If $current\_distance < MIN\_DISTANCE$ then turn to opposite direction.
5) Step forward.
6) Measure distances to all neighbors.
7) Set $current\_distance = \min_{over\ all\ neighbors}\{measured\_distances\}$.

According to this routine, the agent tends to move in group with its neighbors. However, if the agent is very close to the neighbor, it tries to avoid collision and moves away from the neighbor. Finally, if the agent is far from the group or moves away from its neighbors, it follows the simple random walk.

The simulated trajectories of the agents, which implement both following the environment and the neighbors, are shown in Figure 10.7a. In this example, the agents implement both conditions appearing in the presented routines such that

1) If $current\_distance < MIN\_DISTANCE$, then turn to opposite direction.
2) If $current\_brightness \geq previous\_brightness$ and ($current\_distance > previous\_distance$ or $current\_current > MAX\_DISTANCE$), then turn to random direction.

With respect to the parameters of the KBots and the used arena, in the simulations the size of the domain is defined as $n = 500 \times 500$ units and the size of the agents – as 3×3 units; the step length is $l = 10$. Minimal and maximal distances are $MIN\_DISTANCE = 5$ and $MAX\_DISTANCE = 10$ that corresponds to the size and communication abilities of the KBots. The number of the agents is $m = 9$ with random initial locations.

Similar to examples presented in Chapter 9, motion of the agents in heterogeneous environment was implemented using the group of KBots (K-Team Corporation 2012; Rubenstein, Ahler, and Nagpal 2012), where environmental state is represented by the brightness of the environment around the robot. The basic motion of the robots is defined by a simple random walk (see the Kilobot User Manual (Tharin 2013) and sample program KilobotSkeleton.c by Rubenstein and K-Team supplied by K-Team S.A.). Because of the limitations of the KBots' light sensors, the routine that governs the motion of the robot is based on the threshold, which distinguishes between light and darkness. The implemented routine was inspired by the Movetolight.c program appearing at the Kilobot Labs webpage (Kilobotics 2013) and written in usual C for KBots without usage

Figure 10.7 (a) Simulated trajectories of $m = 9$ agents with basic flocking, which follow the environmental states depicted by different levels of gray; (b) experimental setup for simulations with $m = 9$ KBots at their starting positions. Positions of the KBots at different times; (c) positions of the KBots at time $t = 3$ minutes and (d) their positions at time $t = 15$ minutes. Source: Photos by E. Kagan.

of the `kilolib` (see indicated page on Kilobot Labs). The routine is outlined as follows. Notice that at the turns the robot moves around one of its legs: at the turn left – around the left leg and at the turn right – around the right leg. In both directions the turns are to the angle of approximately 30°.

follow_darkness ( ).

1) Turn left and set *current_direction* = *LEFT*.
2) While(*true*)
3) Measure *current_brightness*.
4) If *current_direction* = *LEFT*, then
5) If *current_brightness* > *THRESHOLD*, then turn right and set *current_direction* = *RIGHT*.
6) Else turn left and set *current_direction* = *LEFT*.
7) End if.
8) Else turn left and set *current_direction* = *LEFT*.
9) End if.
10) End while.

Following the routine, if the robot is headed to the light source, it will move toward the light, then turn around and will move away from the light. If, by contrast, the robot is headed away from the light source, it will move directly to the darker regions. Notice that this routine is adjusted to the construction of the KBots and the placement of the light sensor. For the other robots, especially for the robots with two light sensors like the Braitenberg vehicles (see Figure 10.4), the routine should be slightly corrected with respect to the robot's motion and sensing abilities.

The simulations used $m = 9$ KBots (radius of each KBot is 33 mm) moving over the arena of the size 530×530 mm. The arena was illuminated by two lamps such that the brightness of the arena had the same form as the environmental states used in numerical simulations (see in Figure 10.7a); the experimental setup with the KBots in their initial positions is shown in Figure 10.7b. The robots executed the presented above routine and terminated after $t = 15$ minutes. The positions of the robots at time $t = 3$ minutes and at time $t = 15$ minutes before termination are shown in Figure 10.7c,d, respectively; both figures show the arena and the robots from the top view.

It is seen that similarly to the considered above examples of motion in potential field, the agents follow the environmental states (in the executed simulations – from brighter areas to the darker areas) and continue to move randomly in the darkest area. Certainly, the implemented functions of following environmental states and neighbors illustrate the simplest aggregation and dependence on the environmental states and in more complex tasks such functions can have exponential or any other appropriate form. The routine used in the KBots is the simplest one adjusted to the limitations of the robots; for the robots with more precise sensors and motion abilities, more sophisticated control can be used. ∎

Example 10.4 finalizes the consideration of the models of motion and flocking in heterogeneous environment. Detailed theoretical analysis of the models based on the Langevin equation and its correspondence to the probabilistic dynamics defined by the appropriate Fokker-Plank equation are considered in Chapter 12, and the next sections describe certain applications of the presented models.

10.2.2 Swarm Search with Common Probability Map

In the considered methods provided in this chapter, we have assumed that the agents move randomly with respect to the Langevin equation with appropriate potential functions, in particular – with the activity potential functions $U_j^{act}, j = 1, 2, \dots, m$, which specify the mission of agents. In Example 10.1, these functions were defined using certain detection function, which is used in the algorithms of probabilistic search (Stone 1975; Washburn 1989; Kagan and Ben-Gal 2013, 2015). Certainly, such definition of the activity potential is not unique.

Let us consider the probabilistic search, which follows a line of the algorithm of search presented in Section 7.3.1 and extends it to the search by several agents. This algorithm implies that the agents share information using a central unit (see Figure 10.1a) and apply a common map of the target's location probabilities; such a model corresponds to the motion in the common potential field considered in Section 10.1.1. This algorithm was developed in collaboration with Gal Goren (Kagan, Goren, and Ben-Gal 2010) and, as its predecessor considered in Section 7.3.1, follows general approach of Bayesian search or Bayesian tracking (Stone, Barlow, and Corwin 1999); detailed consideration of this algorithm appears in the recent book by Kagan and Ben-Gal (2015).

Assume that m agents are searching for a target in the domain $X \subset \mathbb{R}^2$ using the map of target's location probabilities $p\left(\vec{x}, t\right)$, $\vec{x} \in X$, $t \in [0, T]$. The observations are conducted using errorless sensors such that if at time t the agent checks certain observed area $a(t) \subset X$, then it obtains an exact result whether the target is a or not. Consequently, if the target is detected by one of the agents, then the search terminates, otherwise, following the approach of destructive search, the agent zeroes the target's location probabilities in the area $a(t)$ and transmits the map with zeroed probabilities to the central unit. The central unit obtains the maps of all agents, and using these maps builds a new map of the target's location probabilities, which represents the areas with zero probabilities, while the probabilities in the other areas are normalized following the Bayes rule. Then the new map is transmitted to the agents and they continue the search following this map. In addition, it is assumed that the agents are informed about the moving abilities of the target. More precisely, the algorithm of search is outlined as follows (detailed example of the algorithm's actions is presented in Section 10.3.1).

Algorithm 10.1 (collective probabilistic search; developed in collaboration with Gal Goren). (Kagan and Ben-Gal 2015). Given a domain X, initial map of the target's location probabilities $p\left(\vec{x},0\right)$, $\vec{x} \in X$, and the rule of the target's motion, do:

1) Start with $t = 0$ and initial target's location probabilities $p\left(\vec{x},t\right)$, $\vec{x} \in X$.
2) For each agent j, $j = 1, 2, \ldots, m$, do:
3) Define the probabilities map $p_j\left(\vec{x},t\right) = p\left(\vec{x},t\right)$ for agent j.
4) Choose observed area a_j such that $p_j\left(\vec{x},t\right) > 0$ in at least one point $\vec{x} \in a_j$ and such that the observation of a_j and decrease of the probabilities in the points $\vec{x} \in a_j$ down to zero will result in maximal difference between current probability map and the map, which will be obtained after the observation (including the expected target's motion).
5) If the choice in line 4 is impossible because of zero probabilities in all points of all available observed areas, then apply diffusion process based on the rule of the target's motion to the map of the target's location probabilities. The time of diffusion process is specified by the weighted distance between the agent and the center of distribution of the target's location probabilities, while the weight is defined as a lack of information about the target's location. After finishing diffusion process, return to choosing the observed area in line 4.
6) Move to the chosen area a_j and check whether the target is in a_j or not.
7) If the target is found, then return area a_j. Otherwise, decrease the probabilities $p_j\left(\vec{x},t\right)$ in all points $\vec{x} \in a_j$ down to zero.
8) End for.
9) If one of the agents detected the target, terminate search. Otherwise, update common map of the target's location probabilities $p\left(\vec{x},t\right)$ with respect to the probabilities $p_j\left(\vec{x},t\right)$, $j = 1, 2, \ldots, m$, specified by the agents, increase time t and continue acting starting from line 2.

As indicated, the presented algorithm extends Algorithm 7.7. The difference between these algorithms is in the use of different probabilities' maps for each agent and in combination of these maps into common map of the target's location probabilities. In addition, notice that in contrast to Algorithm 7.7, in this algorithm it is assumed that this target can also move over a domain according to a certain rule that, however, does not change the outline of the algorithm.

In the terms of motion in external potential field, this search implements common external potential function $U^{ext}\left(\vec{x},t\right)=1-p\left(\vec{x},t\right)$, $\vec{x}\in X$, and the activity potentials U_j^{act} such that $U_j^{act}\left(\vec{x},t\right)=1$ if target was not detected in the observed area $a(t)$, which includes $\vec{x}$, and $U_j^{act}\left(\vec{x},t\right)=U^{ext}\left(\vec{x},t\right)$ otherwise. If the observed areas are defined by infinitely small surroundings of the agents' positions, the choice of the area a_j is equivalent to choice of the direction by the gradient of the probabilities' map, or, equivalently, of the external potential U^{ext}, and activity potential U_j^{act} is specified by decreasing of the target's location probabilities. If, in addition, the estimation of the probabilities and corresponding choice of the observed area using the lack of information (see line 5 of the algorithm) is avoided, then the search process defined by the algorithm is equivalent to the search presented in Example 10.1 with obvious change of the detection function to the function of errorless detection.

Notice that Algorithm 10.1 of destructive search does not require specific repulsion potential for collision avoidance. Such property is based on the assumption that at each time the agent is located in the center of its currently observed area. Then, since the probabilities in the observed areas are decreased down to zero (see line 7 of the algorithm), these areas (as it follows from the rule of choice defined in lines 4 and 5) cannot be chosen at the time when the agents are in these areas that avoids collisions automatically. Certainly, without this assumption as well as for erroneous detections, collision avoidance should be provided by specific repulsion potential or by the use of additional probability maps, as it was suggested in the algorithms of search with erroneous detections (Chernikhovsky et al. 2012; Kagan, Goren, and Ben-Gal 2012; Kagan and Ben-Gal 2015). A numerical example illustrating the actions of Algorithm 10.1 is presented in Section 10.3.1.

10.3 Examples of Swarm Dynamics with Shared Environment Map

The considered above models of collective behavior form a basis for several algorithms that specify activity of the agents in different tasks. The next two sections present examples of such tasks: at first we consider a detailed implementation of Algorithm 10.1 of search with multiples searchers acting in discrete domain and then continue simulations presented in Example 10.4 and give an example of obstacle and collision avoidance and its implementation using KBots.

10.3.1 Probabilistic Search with Multiple Searchers

Let us consider an implementation of Algorithm 10.1 for the online search in discrete domain (Kagan and Ben-Gal 2015). The resulting routine is applicable for the search both for static and for moving targets, as well as for the target that changes its motion pattern during the search.

Let $X = \left\{\vec{x}_1, \vec{x}_2, \ldots, \vec{x}_n\right\}$ be the square grid and assume that $m \geq 1$ mobile agents are searching for a target in the grid X. The agents are not aware of the target's location up to detecting it in the chosen at time t = 0, 1, 2, ... observed area $a_j(t) \subset X$, j = 1, 2, ... , m, but are aware of the target's location probabilities $p\left(\vec{x}_i, t\right)$, $\vec{x}_i \in X$, which are known to all agents.

The motion of the target is governed by a Markov process with transition probabilities matrix $\rho = [\rho_{ik}]_{n \times n}$, where ρ_{ik} is the probability of the target's step to the point $\vec{x}_k$ given that the target is in the point $\vec{x}_i$, $i, k = 1, 2, \ldots, n$, and $\sum_{k=1}^{n} \rho_{ik} = 1$ for each $i = 1, 2, \ldots, n$. In the case of static target, transition probabilities matrix ρ is a unit matrix. It is assumed that the transition matrix ρ is also known to all agents.

The agents' motion in the grid X is defined as follows. Denote by $\mathcal{D}$ the set of all possible movements of the agent at a single step. In the grid, this set consists of at maximum five movements $\mathcal{D} = \{\delta_1, \delta_2, \delta_3, \delta_4, \delta_5\}$, where δ_1 stands for *move forward*, δ_2 – *move backward*, δ_3 – *move right*, δ_4 – *move left* and δ_5 – *stay in the current point*. Then, the agent's choice of the movement $\delta_l \in \mathcal{D}$ is based on the probability $\varphi\left(a_j(t)\right)$ of detecting the target in the observed area $a_j(t)$. In the considered case of errorless detection, it is assumed that $\varphi\left(a_j(t)\right) = 1$ if the target is in the area $a_j(t)$, and $\varphi\left(a_j(t)\right) = 0$ otherwise. The search is terminated if at least one of the agents j = 1, 2, ... , m detects the target.

Given the grid $X = \left\{\vec{x}_1, \vec{x}_2, \ldots, \vec{x}_n\right\}$, initial target location probabilities $p\left(\vec{x}_i, 0\right)$, $i = 1, 2, \ldots, n$, transition probabilities matrix ρ, the set $\mathcal{D}$ of possible movements and the detection function φ, the routine that implements Algorithm 10.1 is outlined as follows (developed in collaboration with Gal Goren).

collective_probabilistic_search$\left(\left\{\vec{x}_1, \vec{x}_2, \ldots, \vec{x}_n\right\}, \mathcal{D}, \varphi, \rho, \left\{p\left(\vec{x}_1, 0\right), \ldots, p\left(\vec{x}_n, 0\right)\right\}\right)$

Start search

1) Set t = 0.
2) Set initial locations of the agents and corresponding observed areas $a_j(0)$, j = 1, 2, ... , m.

   Continue search up to detecting the target.
3) While all detection probabilities are $\varphi\left(a_j(t)\right) = 0$, j = 1, 2, ... , m, do:

   Set current observed probabilities by each agent.
4) For each agent j = 1, 2, ... , m do:
5) Set $\left\{p\left(\vec{x}_1, t\right), \ldots, p\left(\vec{x}_n, t\right)\right\}$

   $= \text{current\_observed\_probabilities}\left(a_j(t), \left\{p\left(\vec{x}_1, t\right), \ldots, p\left(\vec{x}_n, t\right)\right\}\right).$
6) End for.

   Calculate common map of next location probabilities.
7) Set $\left\{p\left(\vec{x}_1, t+1\right), \ldots, p\left(\vec{x}_n, t+1\right)\right\}$

   $= \text{next\_location\_probabilities}\left(\rho, \left\{p\left(\vec{x}_1, t\right), \ldots, p\left(\vec{x}_n, t\right)\right\}\right).$

Choose next observed area by each agent.

8) For each agent j = 1, 2, … , m do:
9) Set $a_j(t+1) = \text{next\_observed\_area}\left(\vec{x}_j(t), \mathcal{D}, \rho, \left\{p\left(\vec{x}_1, t+1\right), \ldots, p\left(\vec{x}_n, t+1\right)\right\}\right)$.
10) End for.

Increment time.

11) Set $t = t + 1$.

Move to the next location and observe the chosen area.

12) For each agent j = 1, 2, … , m do:
13) Set location $\vec{x}_j(t)$ with respect to the area $a_j(t)$.
14) Set detection probability $\varphi(a_j(t))$ with respect to observation of the area $a_j(t)$.
15) End for.
16) End while.

Return the target's location.

17) Return the area $a_j(t)$, for which detection probability is $\varphi(a_j(t)) = 1$, j = 1, 2, … , m.

It is clear that the presented routine follows the same steps as the routine, which implements Algorithm 7.7, and, consequently, uses the same functions *current_observed_probabilities*(…), *next_location_probabilities*(…) and *next_observed_area*(…). Notice that in this routine the agents always deal with the common map of the target's location probabilities that is possible only for the case of errorless detection. For erroneous detections, the collective motion is defined by the procedures, which specify sequential choices and updates (Kagan and Ben-Gal 2015).

The trajectories of the agents acting according to Algorithm 10.1 and the routine *collective_probabilistic_search*(…) while searching for a static target are illustrated in Figure 10.8. For comparison, Figure 10.8a,b depict the examples of the trajectories (depicted by solid lines) created by the routine for a single agent (cf. trajectory shown in Figure 7.6b) and Figure 10.8c,d depicts the trajectories (depicted by dotted lines) of these agents acting in collaboration with two additional agents. The size of the domain X is $n = 100 \times 100$. In Figure 10.8a,b the starting point of the agent is $\vec{x}_1(0) = (15, 1)$, and in Figure 10.8c,d the starting points of the agents are $\vec{x}_1(0) = (15, 1)$, $\vec{x}_2(0) = (85, 1)$ and $\vec{x}_3(0) = (50, 99)$. The target's location is specified by random; in the figures it is depicted by a white star. Higher target's location probabilities over the domain X are specified by white color and lower – by black color. Similar to Algorithm 7.7, it is assumed that at all times t = 0, 1, … the observed areas $a_j(t)$ for all agents j = 1, 2, 3 are the squares 3×3 around the agents.

The first example (Figure 10.8a,b) stresses the difference between the trajectory of the agent acting individually and in group. It is seen that in the individual search (Figure 10.8a) the agent spends a short time for searching in the region, which is close to its starting point, and then moves to the farther regions. In contrast, in the collective search (Figure 10.8b) this agent concentrates on searching in its region, while the search in the other regions is conducted by the other agents.

The second example (Figure 10.8c,d) illustrates the collision avoidance. It is seen that in the individual search (Figure 10.8c) in which the problem of collision avoidance does

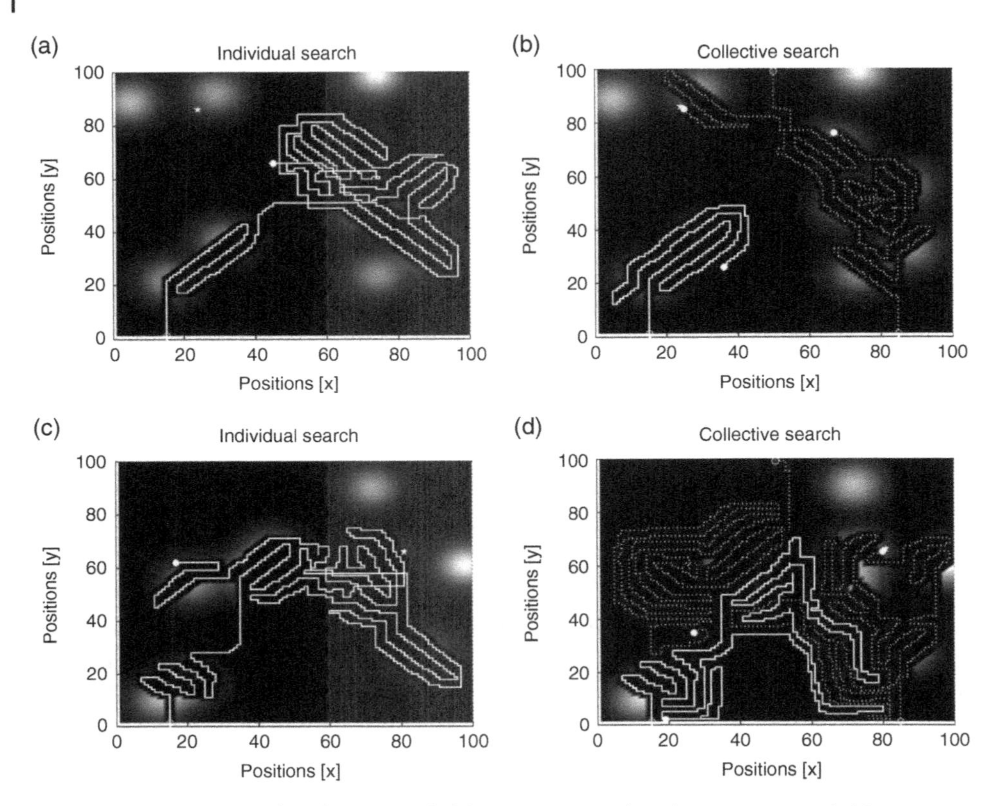

Figure 10.8 Trajectories of single agent and of three agents searching for a static target: (a, b) – trajectories of a single agent and (c, d) – trajectories of the same agents acting in collaboration with two additional agents. The search is conducted up to finding the target or terminated after $t = 1000$ steps.

not appear, trajectory of the agent has several intersections with itself. In contrast, in the collective search (Figure 10.8d), intersections of the trajectories of different agents are extremely rare (in the considered case, the trajectory of the first agent has two intersections with the trajectories of the other agents) and hold only when the agents are far from each other, resulting in effective collision avoidance.

Notice again that such properties essentially use an assumption that the detection is errorless and so the agents zero the probabilities of finding the target in the observed areas. In the case of erroneous detections (Chernikhovsky et al. 2012; Kagan, Goren, and Ben-Gal 2012; Kagan and Ben-Gal 2015), collision avoidance requires additional methods. Possible examples of such methods are considered in the next section.

10.3.2 Obstacle and Collision Avoidance Using Attraction/Repulsion Potentials

Finally, let's consider the modification of Algorithm 10.1, which provides obstacle and collision avoidance. The idea of such modification follows directly from the potential function approach (Rimon and Koditschek 1992); in slightly different manner was used in the paper by Sosniki et al. (2013). For simplicity, it is assumed that the agents follow from their initial positions to the known target position and the goal is to locate the swarm around this position.

Recall that in Algorithm 10.1 $m \geq 1$ agents act in the domain $X \subset \mathbb{R}^2$ and chose their movements according to the map of the target location probabilities $p\left(\vec{x}, t\right)$, $\vec{x} \in X$, $t \in [0, T]$. If the target's location $\vec{x}_0 = (x_0, y_0)$ is known, the probabilities' map can be immediately defined by the binormal distribution with the center in the point $\vec{x}_0$ and the deviations σ_{0x} and σ_{0y} over the axes x and y such that at each point $\vec{x} \neq \vec{x}_0$ the gradient of the target's location probability $p\left(\vec{x}, 0\right)$ is nonzero. It is clear that such definition corresponds to the application of diffusion process to the probabilities' map with permeable bounds such that $p\left(\vec{x}_0, 0\right) = 1$ and $p\left(\vec{x}, 0\right) = 0$, $\vec{x} \neq \vec{x}_0$, up to the moment, at which in all boundary points of $\mathcal{X}$, except may be a single point, the target location probabilities are greater than zero.

Attraction between the agents is defined similarly as the attraction by the target that is by definition of the binormal distributions with the centers at the agents' locations $p\left(\vec{x}_j, t\right)$, $j = 1, 2, \ldots, m$. Finally, since following Algorithm 10.1 the agents avoid from following to the areas with zero probabilities, both obstacle and collision avoidance are provided by specifying at each time t zero target location probabilities to the points in which the obstacles and the agents are located and, if needed, around the obstacles and the agents (cf. aggregation potentials defined by Eqs. (10.2) and (10.3)). Notice that in contrast to Algorithm 10.1 of destructive search, in Algorithm 10.2 the agents conduct nondestructive search and, consequently, are not required to decrease the probabilities in the observed areas down to zero.

Algorithm 10.2 (follow to the known target location with obstacle and collision avoidance). Given a domain X, initial the target position $\vec{x}_0 \in X$, initial locations of the agents $\vec{x}_j \in X$, $j = 1, 2, \ldots, m$, and the areas $o_k \subset X$, $k \geq 0$, occupied by the obstacles, do:

1) Start with $t = 0$ and specify initial target's location probabilities such that $p\left(\vec{x}_0, t\right) = 1$ and $p\left(\vec{x}, t\right) = 0$, $\vec{x} \neq \vec{x}_0$, $\vec{x} \in X$.
2) Apply diffusion process to the map of the target's location probabilities $p\left(\vec{x}, t\right)$, $\vec{x} \in X$, up to the moment in which the resulting probabilities in all points $\vec{x} \in X$ except maybe a single point are greater than zero, and set probabilities $p\left(\vec{x}, t\right)$, $\vec{x} \in X$, as the probabilities resulted by the diffusion.
3) For each obstacle area o_k, $k = 1, 2, \ldots,$ do:
4) Set the probabilities $p\left(\vec{x}, t\right)$ in the points $\vec{x} \in o_k$ as zero and normalize the map of the target's location probabilities.
5) End for.
6) Set probabilities map $p'\left(\vec{x}\right) = p\left(\vec{x}, t\right)$, $\vec{x} \in X$.
7) Set the probabilities $p\left(\vec{x}, t\right)$ in the points $\vec{x}$ occupied by agents as zero and normalize the map of the target's location probabilities.

8) For each agent j, $j = 1, 2, \ldots, m$, do:
9) Choose the new location of the agent j such that $p\left(\vec{x}, t\right) > 0$ in at least one point $\vec{x}$ in the area occupied by the agent and such that the decrease of the probabilities in its points down to zero will result in maximal difference between current probability map and the map, which will be obtained after the decrease of the probabilities.
10) Move to the chosen location and check whether the target location is reached – that is, whether the target location $\vec{x}_0$ is in the area occupied by the agent.
11) If the target location $\vec{x}_0$ is reached, then terminate motion of the agent j.
12) End for.
13) Set probabilities map $p\left(\vec{x}, t\right) = p'\left(\vec{x}\right)$, $\vec{x} \in X$.
14) If all m agents terminated their motion, then terminate. Otherwise, increase t and continue acting starting from line 7.

In general, the presented algorithm follows the approach used in Algorithm 10.1, where the obstacles areas and the areas occupied by the agents are indicated by zero target location probabilities. However, since Algorithm 10.2 implements nondestructive search, it, in contrast to Algorithm 10.1, does not imply that the agents can fail in choosing their new locations and, consequently, does not require diffusion of target's location probabilities during decision-making.

An implementation of Algorithm 10.2 is rather obvious and in discrete time and gridded space it uses the same lines as presented in the routine *collective_search*(...), which implements Algorithm 10.1. The next example illustrates the actions of Algorithm 10.2 in the similar setup as in Section 10.3.1.

Example 10.5 Assume that $m = 4$ agents act in the square-gridded domain X of the size $n = 100 \times 100$ and follow to the known target position by conducting the nondestructive search according to Algorithm 10.2. The agents start in the points $\vec{x}_1(0) = (15,1)$, $\vec{x}_2(0) = (85,1)$, $\vec{x}_3(0) = (15,99)$, and $\vec{x}_4(0) = (85,99)$ and follow the target's location probabilities. It is assumed that the domain X includes 20 square and circle obstacles located in random points of X; in the Figure 10.9 the obstacles are depicted by black color that corresponds to zero target location probabilities, while the higher location probabilities are represented by white color.

The trajectories of the agents obtained in two setups are shown in Figure 10.9a,b. In the first setup (Figure 10.9a), the target is located in the central point of the domain $\vec{x}_0 = (50,50)$ and the target's location probabilities are defined by the bi-normal distribution with the center in the point $\vec{x}_0$ and the deviations $\sigma_{0x} = 50$ and $\sigma_{0y} = 50$. In second setup (Figure 10.9b), the target is located in the point $\vec{x}_0 = (70,30)$ and the target location probabilities are defined by two binormal distributions with the centers in the points $\vec{x} = (30,70)$ and $\vec{x} = (70,30)$ and the variances $\sigma_{0x} = 30$, $\sigma_{0y} = 30$ and $\sigma_{0x} = 20$, $\sigma_{0y} = 20$, respectively (cf. potential functions used in Section 10.2.1).

It is seen that in both cases the agents avoid obstacles following local maximum of the gradient of the target location probabilities. In the first setup with convex function of the

(a)

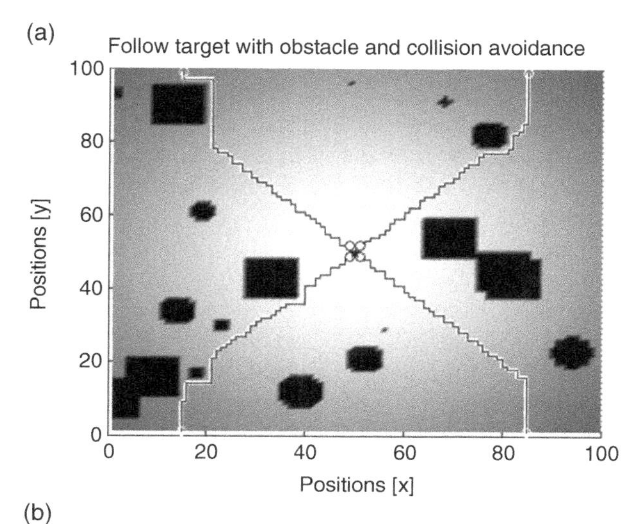

(b)

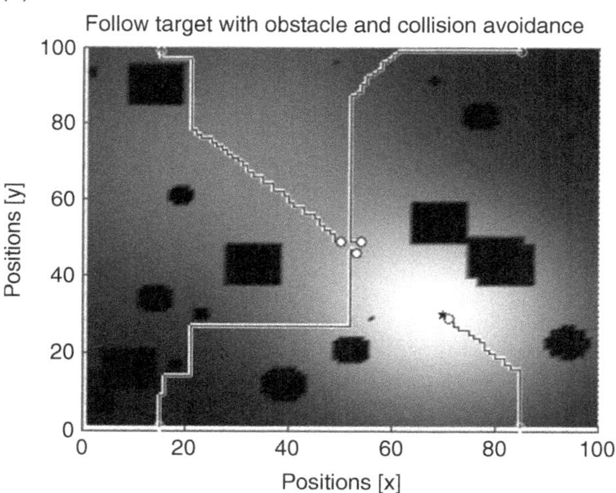

Figure 10.9 Trajectories of m = 4 agents following the known target position with obstacle and collision avoidance: (a) the target's location probabilities function is convex and the target is located in maximal point; (b) the target's location probabilities function have local maximum in the point $x = (30, 70)$ and global maximum in the point $x = (70, 30)$. In both cases, the target is located in the point of global maximum, the size of the domain is 100×100, and the agents start in the points $x_1(0) = (15, 1)$, $x_2(0) = (85, 1)$, $x_3(0) = (15, 99)$, and $x_4(0) = (85, 99)$. Square and circle obstacles are depicted by black color.

target location probabilities (see Figure 10.9a), the agents arrive to the target location, but because of the collision avoidance stay around the target position. In the second setup, in which the target's location probabilities function has local and global maxima, the first, third, and fourth agents initially follow to the minimum of the target's location probabilities function with avoiding collision, and the second agent, which is closer to the target position, follows directly to this position that is the global maximum of the target's location probabilities function. ■

The considered example of obstacle and collision avoidance follows a potential function approach (see Chapter 4) with the potential defined by the target's location probabilities. Together with the previous examples, it, on one hand, demonstrates the close relation between the search and foraging problems (Viswanathan et al. 2011; Kagan and Ben-Gal 2015). On the other hand, the methods of swarm navigation and control (Hamann 2010; Gazi and Passino 2011) are based on the active Brownian motion approach (Schweitzer 2003; Romanczuk et al. 2012), which allows us to consider these methods and tasks following general dynamical systems framework. Chapter 12 describes this framework with additional details.

Finally, let us consider an implementation of obstacle and collision avoidance methods using the group of KBots (K-Team Corporation 2012; Rubenstein, Ahler, and Nagpal 2012). Because of the limited abilities of the KBots, instead of the exponential attraction and repulsion potentials defined by Eq. (10.2) or (10.3), the obstacle and collision avoidance algorithm applies the threshold techniques used in Example 10.4. Similar to Algorithm 10.2, the KBots are required to follow the known target location represented by the light source. The KBots' version of Algorithm 10.2 is outlined as follows.

Algorithm 10.3 (follow to the known target location with obstacle and collision avoidance: KBots' version). Given an arena with light source and definite KBots representing the obstacles, do:

1) Start from initial locations of the robots.
2) While(*true*)
3) For each robot j, $j = 1, 2, \ldots, m$, do:
4) Step toward the light.
5) If distance to the obstacle is greater than threshold, then avoid collision by turning left.
6) If distance to the other robot is greater than threshold, then avoid collision by random turn.
7) End for.
8) End while.

Certainly, the presented algorithm is an obvious version of Algorithm 10.2, which coincides with the abilities of the KBots. For navigation of more complex robots with advanced sensing abilities it can be modified and, instead of the turns with respect to the thresholds, it can include appropriate lines that govern the robots according to the exponential or the other potentials. The next example illustrates the actions of Algorithm 10.3 (cf. Example 10.4).

Example 10.6 Let us implement Algorithm 10.3 in the form of the pseudocode that can be directly used for the KBots. Here it is assumed that the arena is illuminated by a single lamp, which defines the target position of the robots. In addition, one immobile KBot with the definite ID represents an obstacle that should be avoided by the other robots. The ID of the KBot is transmitted to the other robots through KBots messaging techniques.

The running routine includes the lines 10.16, which mimic the lines of the Movetolight.c program appearing at the Kilobot Labs webpage (Kilobotics 2013). Lines 3–9 that provide obstacle and collision avoidance are similar to the corresponding lines appearing in the routines used in the simulations in Example 10.4. The routine is outlined as follows.

follow_light_avoid_collisions( ).

1) Turn left and set *current_direction* = *LEFT*.
2) While(*true*)
3) Measure *obstacle_distance* (by receiving message from the KBot representing obstacle).
4) If *obstacle_distance* < *THRESHOLD_OBSTACLE*,
   then turn left and set *current_direction* = *LEFT*.
5) End if.
6) Measure distances to all neighbors.
7) Set *neighbor_distance* = $\min_{over\ all\ neighbors}\{measured\_distances\}$.
8) If *neighbor_distance* < *THRESHOLD_NEIGHBOR*,
   then turn to random direction (left or right)
   and set *current_direction* to *LEFT* or *RIGHT*, respectively.
9) End if.
10) Measure *current_brightness*.
11) If *current_direction* = *LEFT* and *current_brightness* < *THRESHOLD_LOW*,
    then turn right and set *current_direction* = *RIGHT*.
12) Else turn left and set *current_direction* = *LEFT*.
13) End if.
14) If *current_direction* = *RIGHT* and *current_brightness* > *THRESHOLD_HIGH*,
    then turn left and set *current_direction* = *LEFT*.
15) Else turn left and set *current_direction* = *LEFT*.
16) End if.
17) End while.

Following the routine, the robot avoids collision with the obstacle by going around its left side and collision with the neighbors by turning at random direction. Notice that such techniques do not guarantee the collision avoidance. Rather, they decrease the number of such collisions.

The motion toward the light is conducted as follows. If the measured brightness is less than the *THRESHOLD_LOW*, it turns right approximately 30° around the right leg. If after turning right, the measured brightness is greater than *THRESHOLD_HIGH*, the robot turns left approximately 30° around left leg. Such turns result in the robot moving toward the light with certainty, based on the location of the light sensor. The correction of this error is provided by the difference *THRESHOLD_HIGH* − *THRESHOLD_LOW* > 0 between the thresholds, which is defined with respect to the strength of the light.

The presented algorithms and examples illustrate some of the methods considered in the previous sections and can form a basis for further implementations of these methods using appropriate mobile robots. As already indicated, the implemented routines are written with respect to the structure of the KBot robots and their sensing and motion abilities; for more complicated robots, the routines can be modified. Chapter 11 provides additional examples of swarm dynamics based on the Lego NXT robots with stronger sensing and communication abilities. ■

10.4 Summary

In this chapter we considered the motion of the group of mobile robots over the plane with the shared map of the environmental states. Such motion assumes that the robots communicate either with central unit, which obtains information regarding the agents' movements and observed environment and after processing transmits the results to all robots in the group, or with each robot in the group such that the information processing is conducted by each robot by itself. Following the assumption that the information exchange and processing is errorless, the chapter looked at three general cases of swarm dynamics with shared information:

1) Motion with common potential field, which implies that the robots have complete information about the environment obtained either from the central unit or from each of the robots in the group (Section 10.1.1). In this case, motion of the swarm is governed by external potential field that, if it is needed, is combined with common activity field, which defined the mission of the swarm.
2) Motion with shared information about local environment, which is obtained from the neighboring robots via central unit or directly from the neighbors (Section 10.1.2). Such motion implies that the robots transmit at least the distances between them and correct their motion with respect to these distances; such situation is modeled using attraction and repulsion potentials.
3) Motion in heterogeneous environment where the robots both follow certain potential fields and change their motion characteristics with respect to the parameters of the environment using either nonlinear friction or topography function (Section 10.2.1). Such motion combines two previous cases with additional dependence of motion on the terrain, in which the swarm acts.

The considered models of swarm dynamics are based on the general Langevin equation with corresponding potentials and were illustrated by numerical examples (Sections 10.1 and 10.2.1). The principles of these models are implemented using the swarms of simplest mobile robots (KBots) (see Examples 10.4 and 10.6). The algorithmic solutions are illustrated by the procedure of probabilistic search by a group of robots using common probability map (Sections 10.2.2 and 10.3.1) and by the obstacle and collision avoidance methods implemented numerically and using the swarm of KBot robots (Section 10.3.2).

References

Chernikhovsky, G., Kagan, E., Goren, G., and Ben-Gal, I. (2012). Path planning for sea vessel search using wideband sonar. In: *Proc. 27-th IEEE Conv. EEEI*. https://doi.org/10.1109/EEEI.2012.6377122. Institute of Electrical and Electronics Engineers (IEEE).

Gazi, V. and Passino, K.M. (2011). *Swarm Stability and Optimization*. Berlin: Springer.

Hamann, H. (2010). *Space-Time Continuous Models of Swarm Robotic Systems: Supporting Global-to-Local Programming*. Berlin: Springer.

Kagan, E. and Ben-Gal, I. (2013). *Probabilistic Search for Tracking Targets*. Chichester: Wiley.

Kagan, E. and Ben-Gal, I. (2015). *Search and Foraging. Individual Motion and Swarm Dynamics*. Boca Raton, FL: Chapman Hall/CRC/Taylor & Francis.

Kagan, E., Goren, G., and Ben-Gal, I. (2010). Probabilistic double-distance algorithm of search after static or moving target by autonomous mobile agent. In: *Proc. 26-th IEEE Conv. EEEI*, 160–164.

Kagan, E., Goren, G., and Ben-Gal, I. (2012). Algorithm of search for static or moving target by autonomous mobile agent with erroneous sensor. In: *Proc. 27-th IEEE Conv. EEEI*. https://doi.org/10.1109/EEEI.2012.6377124.

K-Team Corporation. (2012). Kilobot. http://www.k-team.com/mobile-robotics-products/kilobot.

Kilobotics. (2013). Kilobot labs. http://www.kilobotics.com/labs.

Rimon, R. and Koditschek, D.E. (1992). Exact robot navigation using artificial potential functions. *IEEE Transactions of Robotics and Automation* 8: 501–518.

Romanczuk, P., Bar, M., Ebeling, W. et al. (2012). Active Brownian particles. *The European Physical Journal Special Topics* 202: 1–162.

Rubenstein, M., Ahler, C., and Nagpal, R. (2012). Kilobot: a low cost scalable robot system for collective behaviors. In: *IEEE International Conference on Robotics and Automation*, 3293–3298. Saint Paul, MN. Institute of Electrical and Electronics Engineers (IEEE).

Rubenstein, M., Cornejo, A., and Nagpal, R. (2014). Programmable self-assembly in a thousand-robot swarm. *Science* 345 (6198): 795–799.

Schweitzer, F. (2003). *Brownian Agents and Active Particles. Collective Dynamics in the Natural and Social Sciences*. Berlin: Springer.

Shahidi, R., Shayman, M., and Krishnaprasad, P.S. (1991). Mobile robot navigation using potential functions. In: *IEEE International Conference on Robotics and Automation*, 2047–2053. Sacramento, California. Institute of Electrical and Electronics Engineers (IEEE).

Sosnicki, T., Turek, W., Cetnarowicz, K., and Zabinska, M. (2013). Dynamic assignment of tasks to mobile robots in presence of obstacles. In: *18th International Conference on Methods and Models in Automation and Robotics (MMAR'13)*, 538–543. Miedzyzdroje, Institute of Electrical and Electronics Engineers (IEEE).

Stone, L.D. (1975). *Theory of Optimal Search*. New York: Academic Press.

Stone, L.D., Barlow, C.A., and Corwin, T.L. (1999). *Bayesian Multiple Target Tracking*. Boston: Artech House Inc.

Tharin, J. (2013). *Kilobot User Manual*. Vallorbe, Switzerland: K-Team S.A.

Viswanathan, G.M., da Luz, M.G., Raposo, E.P., and Stanley, H.E. (2011). *The Physics of Foraging*. Cambridge: Cambridge University Press.

Washburn, A.R. (1989). *Search and Detection*. Arlington, VA: ORSA Books.

11

Collective Motion with Direct and Indirect Communication

Eugene Kagan and Irad Ben-Gal

This chapter considers two types of communication between the robots: direct communication based on certain communication protocols and indirect communication using environmental changes, which leads to stigmergy in multi-robot systems. Dynamics of swarms with respect to these types of communication are illustrated by intermittent strategies motion of the robots and by the pheromone robotics phenomena.

11.1 Communication Between Mobile Robots in Groups

As indicated in Chapters 9 and 10, there are several types of communication between the mobile robots acting in groups, and according to the robots' abilities and mission, information transfer can be implemented in different manners and using different means. In particular, in Chapter 10, it was assumed that the robots either transfer information by the use of direct communication or by the use of changes of the environmental states (e.g., by decreasing the target's location probabilities as it is done in the destructive search considered in Sections 10.2.2 and 10.3.1). The second type of communication is known as indirect communication. The schemes of direct and indirect communications are shown in Figure 11.1.

Usually, direct communication between mobile robots follows general principles of information transfer used in computer networks and implements certain per-to-per or broadcasting protocols (Weiss 1999). Often, especially for simple mobile robots, such communication mimics the signaling observed in the groups of animals. Notice that in this case, the transmitted information is not stored in the environment and is available only in the time of communication.

Indirect communication, by contrast, assumes that the agents can interact via the environment and the environment can be considered as a common memory of the group, in which the information is stored and is available during a certain period. The resulting behavior of the group with indirect communication is usually called stigmergy (Hamann 2010; Sumpter 2010). Finally, in the framework of communication abilities

Autonomous Mobile Robots and Multi-Robot Systems: Motion-Planning, Communication, and Swarming, First Edition. Edited by Eugene Kagan, Nir Shvalb and Irad Ben-Gal.

Companion website: www.wiley.com/go/kagan/robotsystems

Figure 11.1 Types of communication between the mobile agents: (a) direct communication using certain communication protocol; and (b) indirect communication using changes of the environmental states.

of the agent, direct receiving of information about the relative positions of the other agents using certain sensors (e.g., cameras or ultrasonic distance sensors) is often considered as sensor-based communication. Certainly, in real multi-robot systems, the agents apply different types of communication and combine the obtained information for individual motion planning and collective decision-making in the swarm. In the next sections, we consider general principles of the indicated types of communication and their inferences to swarm dynamics.

Let us consider indirect communication between the agents in some details. As indicated above, indirect communication is conducted using the changes of the environmental states and assumes that the agents are equipped with the appropriate activators and sensors (Balch and Arkin 1994). The environment is this case is considered as a common shared memory of the agents that stores and provides information regarding the work-in-progress and influences the decision-making by each agent in the group. Usually, the consideration of the environment as a shared memory of the swarm is known as embodiment (Clark 1997, 2003) or of percolation (De Gennes, 1976), and the corresponding phenomena observed in the swarm dynamics are called stigmergy (see e.g. (Trianni 2008; Hamann 2010)).

Following Beckers, Holland, and Deneubourg (1994, p. 181),

> the principle is that of stigmergy, recognized and named by the French biologist P. P. Grassé (Grasse 1959) during his studies of the nest building in termites. Stigmergy is derived from the roots 'stigma' (goad) and 'ergon' (work), thus giving the sense of 'incitement to work by the products of work'. It is essentially the production of a certain behavior in agents as a consequence of the effects produces in the local environment by previous behavior.

The most known example of indirect communication and corresponding swarm activity is a deploying of pheromone trails by foraging ants that mark the paths to the food that are used by the other ants of the colony (Gordon 2010; Sumpter 2010). This mechanism forms a basis for the recently developed approach widely known as pheromone robotics, in which communication between the mobile robots in the group mimics the pheromone-based communication between the ants.

Practically, such approach is implemented using direct per-to-per communication and a certain scheduling scheme that allows distribution of the messages from one robot to another via the other robots used as intermediate transmitters (Payton et al. 2001;

Rubenstein, Cornejo, and Nagpal 2014). Then, in the Payton's et al. implementation, when one of the robots detects the target or arrives to the desired position, it starts broadcasting the message, including the "pheromone concentration," to the closest neighbors; these neighboring robots transmit the received message with slightly decreased "concentration" to the next neighbors and so far. The decreasing of the "pheromone concentration" is defined with respect to the distance between the robots and can also depend on time for modeling the drying of real pheromone. In the Rubenstein's et al. swarm, the continuous broadcasting was conducted by some of the robots, while the other robots only transmitted the received messages. The direction, from which the message is received, and the value of "pheromone concentration" define the pheromone gradient (Payton et al.) or the gradient of potential field (Rubenstein et al.), and given this gradient the robot makes a decision regarding its further steps.

Additional property of the pheromone robotics approach is a reaction of the swarm to the environmental changes that are resulted by the dynamics of the environment itself. Such reaction is usually considered by the same methods and following the same framework as a reaction to the sensed pheromone trails and changes of these trails in time. Together with general principles of swarming considered in Section 9, these two properties – pheromone deployment and sensing and reactivity to the environmental changes – form a basis for the approach widely known as swarm intelligence (SI) (Bonabeau, Dorigo, and Theraulaz 1999).

The term *swarm intelligence* was introduced in 1990 by Beni and Wang (1990) in the context of swarm robotics (see Section 4 and mentioned there the Beni work (Beni 2005)). Informally, it is defined as follows (White 1997):

> Swarm Intelligence (SI) is the property of a system whereby the collective behaviors of (unsophisticated) agents interacting locally with their environment cause coherent functional global patterns to emerge. SI provides a basis with which it is possible to explore collective (or distributed) problem solving without centralized control or the provision of a global model.

For more precise specification of the swarm intelligence often there are applied five principles formulated Millonas (1994, pp. 418–419):

1) *Proximity principle.* The group should be able to do elementary space and time computations... Computation is understood as a direct behavioral response to the environmental stimuli which is in some sense maximizes the utility to the group as a whole of some type of activity.
2) *Quality principle.* The group should be able to respond not only to time and space considerations, but to quality factors, for instance, to the quality of foodstuffs or of location.
3) *Principle of diverse response.* The group... should seek to distribute its resources along many modes as insurance against the sudden change in any one of them due to environmental fluctuations.
4) *Principle of stability.* The group should not shift its behavior from one mode to another upon every fluctuation of the environment, since such changes take energy, and may not produce a worthwhile return for the investment.

5) *Principle of adaptability*. When the rewards for changing a behavioral mode are likely to be worth the investment in energy, the group should be able to switch. The best response is likely to be a balance between complete order, and total chaos, and therefore, the level of randomness in the group is an important factor.

It is clear that since, in contrast to the artificial intelligence based on computer multi-agent systems (Weiss 1999), the necessary condition for existence of the swarm intelligence is the changes of the environment it can be implemented only using multi-robot systems and cannot be built on the basis of non-robotic agents (for discussion of necessary properties of robots see Section 1). In addition, notice that indirect communication and stigmergy themselves do not imply self-organization that is spatial or temporal ordering of the agents in swarms (Bonabeau et al. 1997; Bonabeau, Dorigo, and Theraulaz 1999); for such phenomena also certain flocking abilities (see Section 9.1.2) or long distance communication (Theraulaz et al. 2003) are required. The detailed review and analysis of stigmergic phenomena and related inferences to robotic systems is presented by Trianni (2008); additional considerations and simulations are presented in the recent books by Hamann (2010), Gazi and Passino (2011), and Kagan and Ben-Gal (2015).

The stigmergic phenomena are illustrated in Figure 11.2. In the figure it is assumed that $m = 50$ agents (particles of zero dimension) move over the squire domain X of the size 100×100. The basic motion of the agents is a random walk with the length of the step $l = 3$. During the motion, each agent deploys pheromone trails and senses pheromone deployed by all other agents except the pheromone deployed by the agent itself. The width of the pheromone trail is $w = 1$. In addition, the agent senses the environment states, which are common for all agents; the sensing radius of each agent is equal to the length of the step that is $r_{\text{sense}} = 3$. The direction of the next step is chosen with respect to the maximum amount of pheromone and maximal value of the environment state around the agent. Figure 11.2a,b illustrate the trajectories of the agents in the homogeneous environment at times $t = 50$ and $t = 500$, and Figure 11.2c,d show the trajectories of the agents in heterogeneous environment at the same times; in the figure the whiter areas are preferable.

It is seen that the agents are extremely sensitive to the pheromone trails and the motion pattern obtained in relatively small number of steps does not change essentially and stay for a long time up to drying the pheromone. In the case of heterogeneous environment, the agents follow both the environmental states and deployed pheromone. In particular, the agents move toward the preferable whiter areas, but if a certain agent meets the pheromone trail, it follows the pheromone even if it leads the agent to the inferior area.

The presented example illustrates the effect of local communication and stigmergy in the simplest case of randomly moving particles. However, in addition to basic phenomena, it shows that in practice the motion based on the communication using pheromone trails depends on the size of the agent and the sensing radius. In particular, if the steps length l and the sensing radius r_{sense} are equal and if the agents are able to sense the pheromone deployed by themselves, then they move one step forward and one step backward around their initial locations, following the trail deployed by each agent itself. As indicated above, in the considered example, as indicated above, each agent senses only the pheromone deployed by all other agents that makes the motion possible. Below, in Section 11.3.1, we will return to the pheromone robotics approach and consider the motion of the ant-like agents with certain geometrical size and definite sensing abilities.

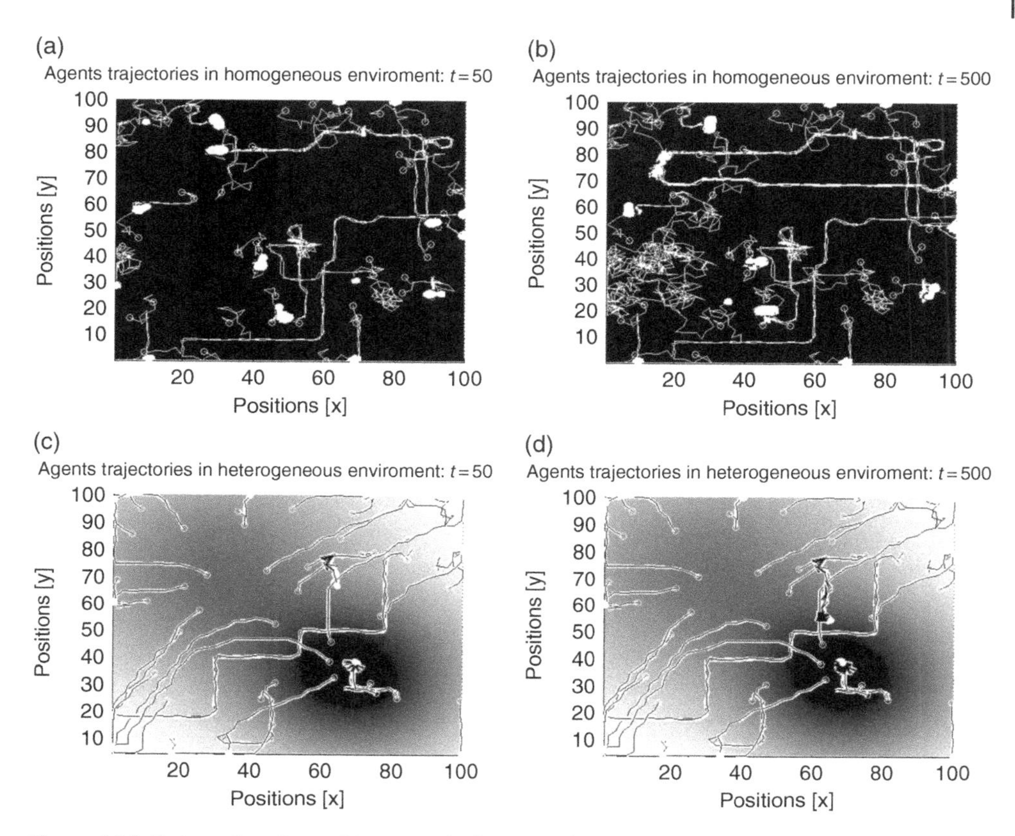

Figure 11.2 Trajectories of $m = 50$ agents deploying and sensing pheromone trails of the width $w = 1$: (a, b) trajectories of the agents in homogeneous environment after $t = 50$ and $t = 500$ time units; (c, d) trajectories of the agents in heterogeneous environment after $t = 50$ and $t = 500$ time units. In both cases, the size of the domain is 100×100, the basic motion of the agents is random walk with the step $l = 3$ and the sensing radius is $r_{sense} = 3$.

11.2 Simple Communication Protocols and Examples of Collective Behavior

In the previous section, we considered general principles of direct and indirect communication between the mobile robots acting in group and the phenomena led by different types of communication. This section presents an example of simple communication protocol, which implements the Mobile Ad Hoc Networking (MANET's) principles for communication between toy Lego NXT robots, and slight modification of this protocol for asynchronous communication in PAN, which supports flocking and preserving the robot's group. The basic protocol and software design were developed and implemented in collaboration with Jennie Steshenko; modifications of the protocol and corresponding software design were conducted in collaboration with Mor Kaizer and Sharon Makmal and then implemented in collaboration with Amir Ron and Elad Mizrahi.

11.2.1 Examples of Communication Protocols for the Group of Mobile Robots

Following general framework of MANET (Frodigh, Johansson, and Larsson 2000; Conti and Giordano 2014), assume that the mobile robots communicate using close-range radio technology like Bluetooth. In addition, assume that the communication abilities of the robots are restricted such that each robot is able to communicate with certain small number of the neighbors and this number is essentially smaller than the number of the robots in group. In particular, in the implemented example with the toy Lego NXT robots, the master agent can communicate with at maximum three predefined slave agents (one agent per connection), while the slave agents can communicate with a single agent, which is a definite master agent (Demin 2008). The other scenarios and implementations of the MANET principles for navigation of the groups of mobile robots are presented, for example, in the works of Aguero et al. (2007), Das et al. (2002, 2005), Jung et al. (2009), Niculescu and Nath (2004), Wang et al. (2003, 2005), and Yoshida et al. (1998).

11.2.1.1 Simple Protocol for Emulating One-to-One Communication in the Lego NXT Robots

In the models of swarms and corresponding mobile networks, the agents have to be able to change the connectivity dynamically, and allow arbitrary connections between the agents. The network available with the Lego NXT robots and the desired network are shown in Figure 11.3a,b, respectively.

The presented below simple protocol and corresponding scheduling scheme illustrate a possible solution of the indicated problem following the methodology presented by Huhns and Stephens (1999). The description of the protocol follows the paper by Steshenko, Kagan, and Ben-Gal (2011); the figures are adapted from the report by Steshenko (2011).

The suggested protocol follows a line of the ARPA (address and routing parameter area) text messages standard (Crocker 1982), which was adapted for the considered communication scheme and the requirements. The protocol assumes the star-type wireless network such that the master agent is implemented by a regular PC, and the slave agents are implemented by Lego NXT robots. The communication between the master and the slaves is conducted by the use of standard Bluetooth technology (Martone 2010), while

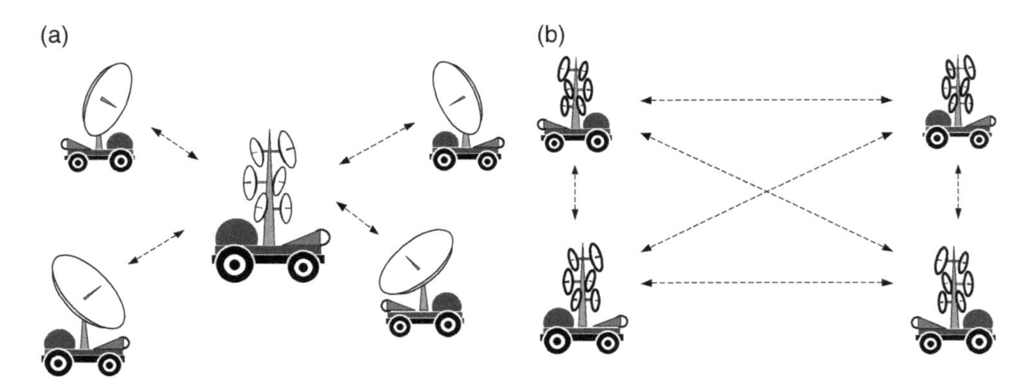

Figure 11.3 (a) The network available with the Lego NXT robots; and (b) the desired network.

the direct communication between the slaves is not allowed. Following the suggested protocol, the master agent sequentially sends queries to the slave agents, and the slave agents reply either with specific messages, which have to be relayed to the other agents, or with informational messages for the master.

In the protocol, the messages are defined as ASCII character strings, which includes a header and a body. Since the Bluetooth technology supported by the basic Lego NXT robots uses the size-limited messages (up to 58 bytes) and the controllers of the robots have rather weak computational abilities, during the development of the protocol it was required to meet the following general restrictions:

- The size of the message header, which is required for technical purposes, has to be as minimal as possible.
- The message has to contain all required information for the definition of the next communication stage, so that the slave agent is not required to store the communication history.
- The messages are required to have a unified structure in order to specify unified message processing procedures both for the header and for the body.

In general, the structure of the message is defined as follows. The first nine characters of the message are preserved for the header, which consists of three mandatory fields that specify the type of the message, the sender, and recipient. The remaining characters are used for a body of the message. The fields are divided by the delimiter ":" (colon). The structure of the message is:

`<msgtype>:<src>:<dst>:<body>`

The fields of the message are defined as follows:

`<msgtype>` is a two-chars identifier of the message type that is:
`IR` – information request
`SV` – sensor value
`EC` – execute command
`MT` – message transmit

where both characters in the values were selected in such a manner, so they are different from one another both in a single message type and in between all message types, causing as little confusion as possible for both an agent and a user.

`<src>` is a two-chars identifier of the sender; in particular:
`00` – identifier of the master agent
`XY` – identifier of the slave agent, where `X` and `Y` are one-digit literals `0`, ..., `9` defined as chars

`<dst>` is a two-chars identifier of the recipient that can obtain the following values:
`00` – identifier of the master agent
`XY` – identifier of the slave agent, where `X` and `Y` are, as above, one-digit literals
`AL` – identifier of all agents, including master agent and slave agents
`GR` – identifier of all agents in the predefined group

`<body>` is a message body, which can include any number of characters up to the available 49 characters. The first part of the body can include certain information regarding communication or specific commands and data. In particular, in the reply messages the body starts with the definite two-char identifier `<replytype>`, which specifies the type of the reply message. The `<replytype>` identifier obtains the following values:

`NU` – identifier of the absence of the message or non-availability of the request
`AK` – identifier of acknowledgement for the received message
`NK` – identifier of noncompleted processing of the message

The message `IR` (information request) is sent by the master agent to a certain slave agent as an initial request for information and initializes a session of negotiations. The structure of the message is the following:

```
IR:<src>:<dst>:<infotype>
```

where `<infotype>` is a two-char identifier of the requested information.

The message `EC` (execute command) is sent by the master agent to a certain slave agent or a group of slave agents in order to command it to conduct the action that is defined by the message. The structure of the message is the following:

```
EC:<src>:<dst>:<commandtype><:args>
```

where `<commandtype>` is a two-chars identifier of the command to be executed and `<:args>` is a nonmandatory string of the arguments of the specific command. The values of the identifier `<commandtype>` without arguments `<:args>` are the following:

`RM` – remove message. Upon message reception, the master agent notifies the slave agent to remove the message from the stack.
`MW` – message write. After the master agent receives the `"MT"` message as a reply from the slave agent, the master agent notifies the slave agent to write the message that has to be forwarded to the other slave agent.

The message `MT` (message transmit) is sent by one agent (master or slave agent) to another agent (slave or master agent) in order to transmit specified information. The structure of the message is as follows:

```
MT:<src>:<dst>:<messageinfo>
```

where `<messageinfo>` is a string that contains an actual message up to 49 chars long. Upon the reception of the `MT` message, the slave agent replies to the master agent by the acknowledgment:

```
MT:<src>:<dst>:AK
```

where `<dst>` is the identifier `"00"` of the master agent.

The message `SV` (sensor value) is sent by the master agent to a certain slave agent in order to obtain current data from the specified sensor of the slave agent. The structure of the message is as follows:

```
SV:<src>:<dst>:<sensortype>
```

where `<sensortype>` is a two-char identifier of the sensor, from which the value has to be obtained.

According to the suggested protocol, each communication session is initiated by the master agent as it is induced by the master-slave scheme. Below we present the appropriate sequences of the messages. The master agent has an identifier "00" and the slave agents have identifies "XX" and "YY."

The session is initiated by the IR (information request) message with the suffix SQ (send query), which is sent by the master agent "00" to a certain slave agent XX. The slave agent replies with an IR message with either the suffix NU, which specifies the absence of information, or with the suffix MT, which indicated that the slave agent has a message ready for transmission. The negotiations are as follows:

- master agent 00 to slave agent XX: IR:00:XX:SQ
- slave agent XX to master agent 00: IR:XX:00:SQ:NU, or IR:XX:00:SQ:MT

The first reply of the slave agent leads to finishing the session, and the second reply continues the session with master processing the MT message (see below). Data flow for the IR message is shown in Figure 11.4.

The EC message is processed in two manners. While the master agent sends this message with the suffixes TN or GO, the session is initiated, the slave agent executes the corresponding command, replies with the acknowledgement AK, and then the session is finished. The following examples illustrate such negotiations.

Request to turn 45° clockwise:

- master agent 00 to slave agent XX: EC:00:XX:TN:+045
- slave agent XX to master agent 00: EC:XX:00:TN:AK

Request to move 10 cm forward and turn 45° counterclockwise:

- master agent 00 to slave agent XX: EC:00:XX:G0:+010:-045
- slave agent XX to master agent 00: EC:XX:00:GO:AK

Data flow for the EC message with the suffixes TN and GO is shown in Figure 11.5.

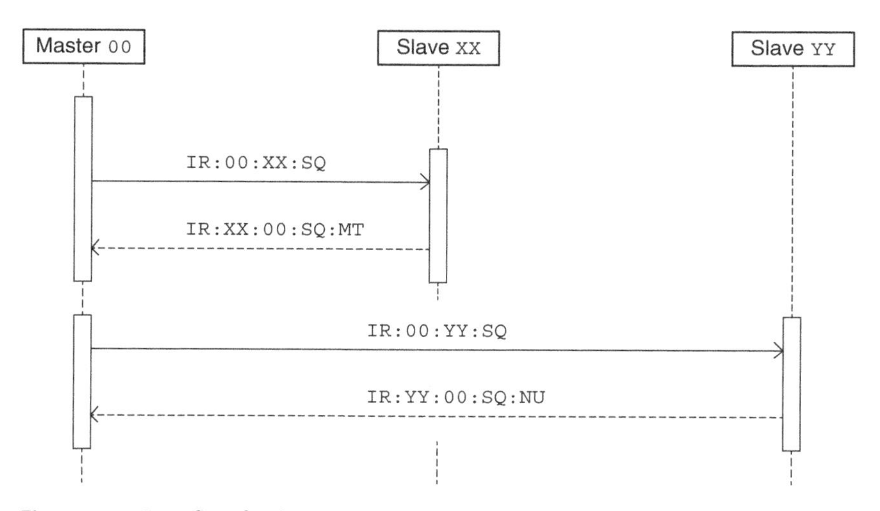

Figure 11.4 Data flow for the master-slave negotiations with the IR (information request) message.

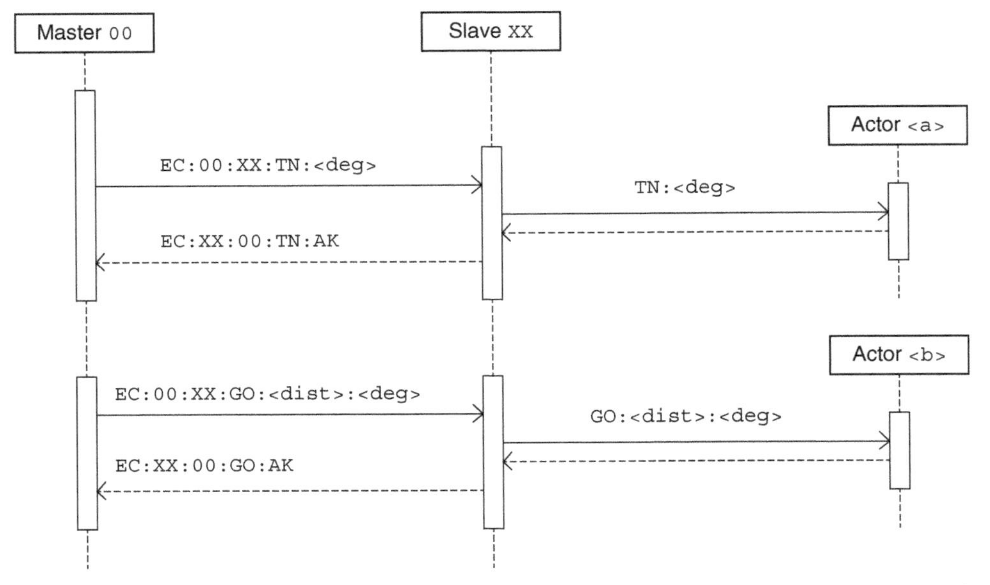

Figure 11.5 Data flow for the master-slave negotiations with the EC (execute command) message with the suffixes TN and GO.

In the figure, actors <a> and <b> stand for the devices or equipment that execute the required command. Notice that the EC message with the suffixes RM or MW is a part of communication session using the MT message, and is processed during the negotiations between the slave agents via the master agent.

The SV message is processed similarly to the EC message with the suffixes TN or GO. While the master agent sends this message to the slave agent, the session is initiated and the slave agent replies with the value of the requested sensor or with the suffix NU when the required sensor does not exist. After the slave agent's reply, the session is finished. An example of the negotiations is as follows:

Request for distance value (usually, it is provided by ultrasonic sensor) and reply with the distance value (e.g. 30 cm) or with NU if the sensor not installed:

- master agent 00 to slave agent XX: SV:00:XX:VD
- slave agent XX to master agent 00: SV:XX:00:VD:+030 or SV:XX:00:VD:NU

In both cases, the session is finished after the reply. Data flow for the SV message with the distance value and the suffix NU is shown in Figure 11.6.It is assumed that the robots support different types of sensors. In particular, for the Lego NXT robots, the following sensors and corresponding suffixes are used:

VD – distance sensor
VL – light sensor
VC – color sensor
VA – acceleration sensor

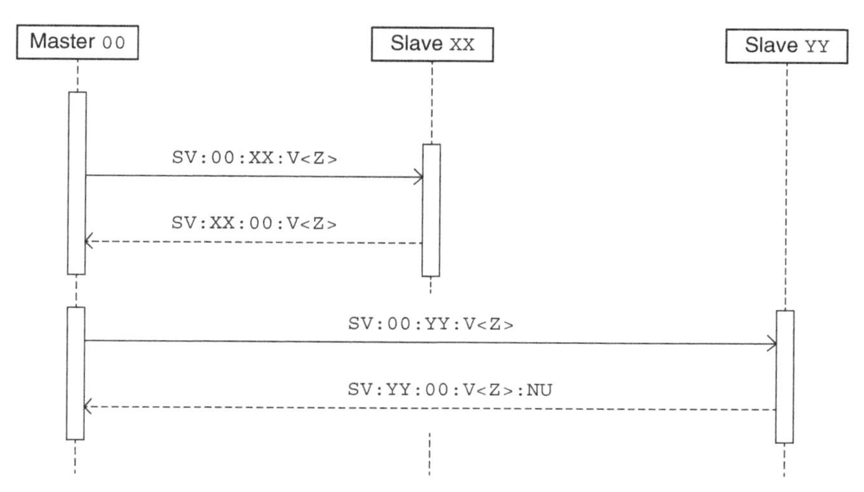

Figure 11.6 Data flow for the master-slave negotiations with the SV (sensor value) message with the distance value and the suffix NU.

If the robots are equipped with the other sensors, the corresponding suffixes in the format V<Z>, where Z stands for a single character indicating the sensor, can be added, respectively.

Finally, let us consider to the session of transmitting the message from one slave agent to another via the master agent. Such session is started, when the reply of the slave agent to the IR message is as follows (see Figure 11.4):

- slave agent XX to master agent 00: IR:XX:00:SQ:MT
  By receiving this reply from the slave agent XX the master agent 00 replies by the message EC with the suffix MW:
- master agent 00 to slave agent XX: EC:00:XX:MW
  The next negotiations are the following:
- slave agent XX to master agent 00: MT:XX:YY:<message>
  that means that the message <message> has to be forwarded from the slave agent XX to the slave agent YY.
- master agent 00 to slave agent XX: EC:00:XX:RM
  that acknowledges the reception of the message and requires the slave to remove this message from the message queue.
- master agent 00 to slave agent YY: MT:XX:YY:<message>
- slave agent YY to master agent 00: MT:XX:YY:AK
  which acknowledges the reception of the message and finishes the session.

  For each message, an error of the message's execution or uncompleted processing results in the same message with the suffix NK.

  Data flow for the session of message transmitting is shown in Figure 11.7.

  The suggested protocol completely defines the negotiations between the agents in the network and includes the messages that can be used for the control of the slave agents by the master agent. In the next section, we present an adaptation of this protocol for the flocking tasks.

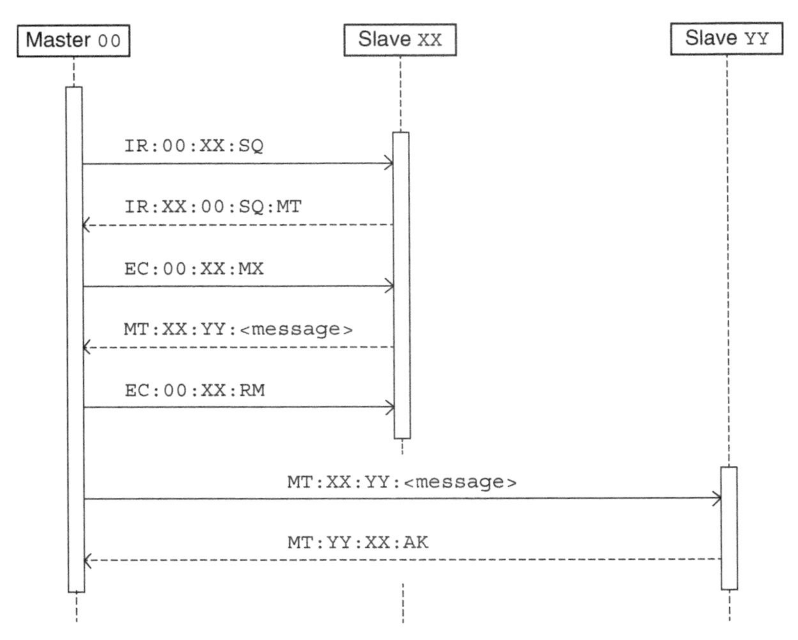

Figure 11.7 Data flow for the session of message transmitting.

11.2.1.2 Flocking and Preserving Collective Motion of the Robot's Group

The presented above protocol formed a basis for development of the next protocol and corresponding software, which supports flocking and preserving of the group of mobile robots. The description of the protocol follows the report by Kaizer and Makmal (2015).

Assume that the group of mobile robots forms a MANET with definite central unit such that the communication between the agents – between the robots and between robots and central unit – is provided by the Bluetooth technology and maximal distance between the robots is less than the maximal distance supported by the Bluetooth. In the other words, it is assumed that radio communication between the robots is always possible. In addition, assume that the robots are equipped with certain distance sensors (e.g., ultrasonic sensors), which are able to measure distances between the robots and between the robots and the obstacles, but are not able to distinguish between the robots and the obstacles.

Following the mission, the robots move simultaneously in unknown environment with obstacles, and a group is preserved if the distance between each robot and at least one another robot is smaller than maximal distance supported by the distance sensors. Otherwise, the group is considered as broken, where the robot or robots, which are not sensed by the distance sensors of any other robot, are lost, and the group should inform the central unit about this situation. Notice, that both robots and central unit have no access to global coordination system, so absolute positions of the robots are unknown. For the group of three robots, its possible structures in the sense of the distances between the robots are shown in Figure 11.8a,b. In contrast, Figure 11.8c depicts situation, in which the robots measure distance to the obstacle and since the sensors do not

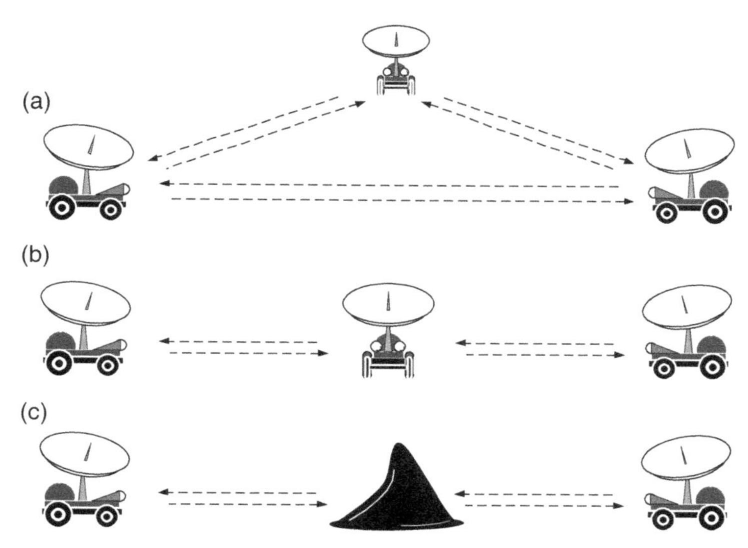

Figure 11.8 (a) "Triangle" group, in which the distances between all the robots are smaller than maximal distance supported by the distance sensors; (b) linear group, in which distances of the pairs of robots are smaller that maximal supported distance; (c) erroneously recognized group, in which both robots consider an obstacle as the third robot.

distinguish between the robots and the obstacles the robots erroneously recognize an existence of the group.

The protocol, which supports the indicated mission, follows the approach of the protocol presented in Section 11.2.1.1 and is based on the messages of three types – `IR` message, `SV` message, and `EC` message. Format of the messages is the following.

Information request message:

`IR:<src>:<dst>:<arg>`

where `<arg>` is a parameter, which controls the negotiations and is defined as follows:

- if `arg` is empty, then the robot replies by the `SV` message;
- if `arg` is equal to the two-chars string appearing in the `dst` field, then the agent replies by the acknowledge message:

`IR:<dst>:<src>:AK`

- if `arg` is `AK`, then the agent closes the session.

`SV` message:

`SV:<src>:<dst>:<distance`$_1$`>:<distance`$_2$`>:...:<distance`$_n$`>`

where each `<distance`$_i$`>`, $i = 1, 2, \ldots, n$, is a two-char value of the distance between the robot and the observed object – robot or obstacle. The number of these values determines the number of the observed objects. However, notice that since the in the Lego

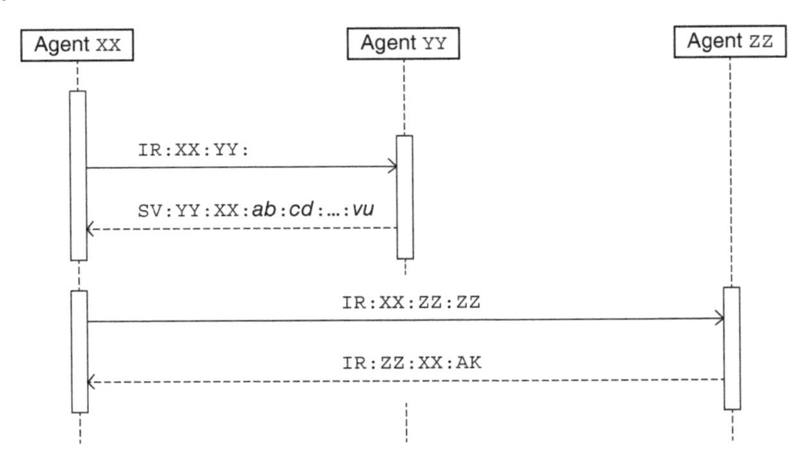

Figure 11.9 Data flow for the agents' negotiations with the IR and SV messages. The variables *<ab>*, *<cd>*, ..., *<vu>* in the SV message stand for the two-char values of the distances.

NXT robots the total length of the message body is 49 characters, the maximal number of the observed objects is limited by 16. In addition, notice that the value of the distance defined by two characters follows the abilities of the Lego NXT ultrasonic sensor, in which maximal sensed distance is 40 cm. Notice that since this message supports only distance sensor, the SV message does not include the field <sensortype>, and, certainly, this field can be added according to the needs. Data flow for the IR and SV messages is shown in Figure 11.9.

Notice that since the considered protocol is applied only for the mission of preserving the group, the data flow shown in Figure 11.9 combines the corresponding data flows of the basic protocol shown in Figures 11.4 and 11.6.

Execute command message:

```
EC:<src>:<dst>:<commandtype>:<arg>
```

where the possible values of the <commandtype> and the <arg> fields are the following:

GO – start the regular activity that is movement, sending messages, screening environment, etc.
GO:<ZZ> – send message to the central unit that the robot with the two-chars identifier <ZZ> got lost;
ST – stop the current activity of the agent;
AK – acknowledgment of finalizing the mission required by the EC message;

If central unit is an agent that can start seeking the lost robot, e.g. helicopter or any other robot with definite mission of search, then the message GO:<ZZ> sent to the central unit is interpreted as a command to start seeking the lost robot <ZZ>. Consequently, the <commandtype> equal to ST requires to stop seeking and to return to the base; this value of the <commandtype> sent to any other robot in the group requires stopping the current movement. Data flow for the EC message is shown in Figure 11.10.

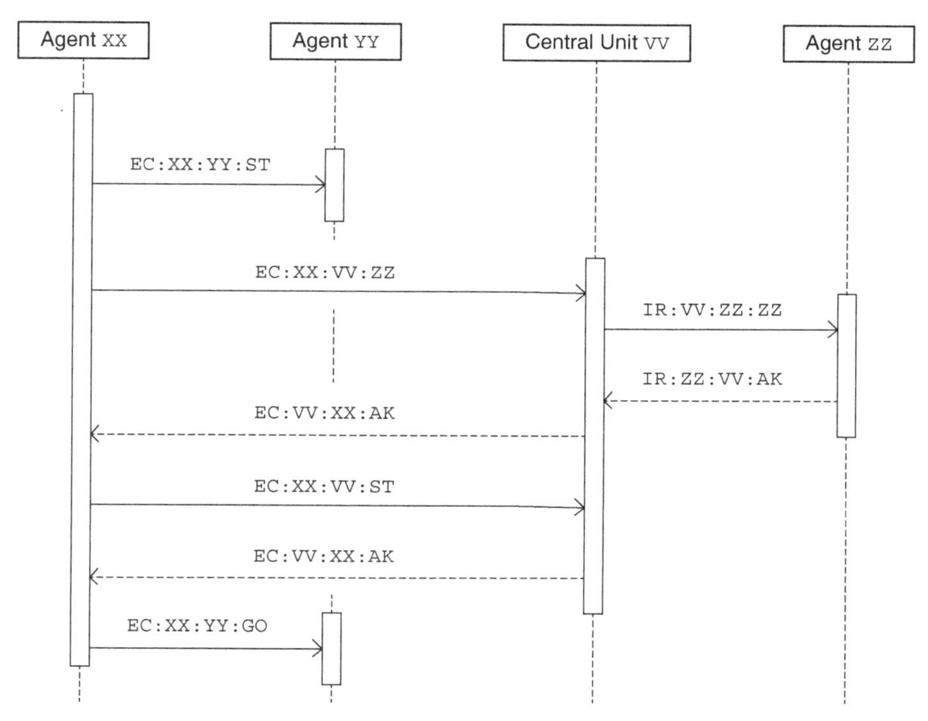

Figure 11.10 Data flow for the agents' negotiations with the EC message.

As indicated above, the presented protocol is aimed to support a single mission that is the preserving of the motion of the robots in the group. However, following the direction of the basic protocol presented in Section 11.2.1.1, it can be extended for supporting additional tasks that imply mobility control of the robots and require information transmission between the agents.

11.2.2 Implementation of the Protocols and Examples of Collective Behavior of Mobile Robots

The considered protocols can be implemented using any type mobile robots that support Bluetooth or even infrared communication. In the next sections, we describe the implementation of the protocols in the group of Lego NXT robots that communicate using Bluetooth technology.

11.2.2.1 One-to-One Communication and Centralized Control in the Lego NXT Robots

As already indicated, the protocol presented in Section 11.2.1.1, assumes that the master PC and mobile robots form a star-type wireless network (with the master PC in the center of the network). The master agent is used for transmitting information and scheduling the messages, and the slave agents that are the Lego NXT robots communicate with the master agent, move over the environment with respect to the received messages and sensed environmental states and conduct additional actions using available equipment. General architecture of the system and with some of the functions is shown in Figure 11.11.

Notice that the diagram shown in the figure mostly includes the functions that implement the protocol; and the functions that define the motion of the robots are contained in the objects, which created using the class Actor and its successors. Since the Lego NXC does not support C++ data structures, the code implemented in the robots is written using the C structures; however, the architecture of the system is still the same.

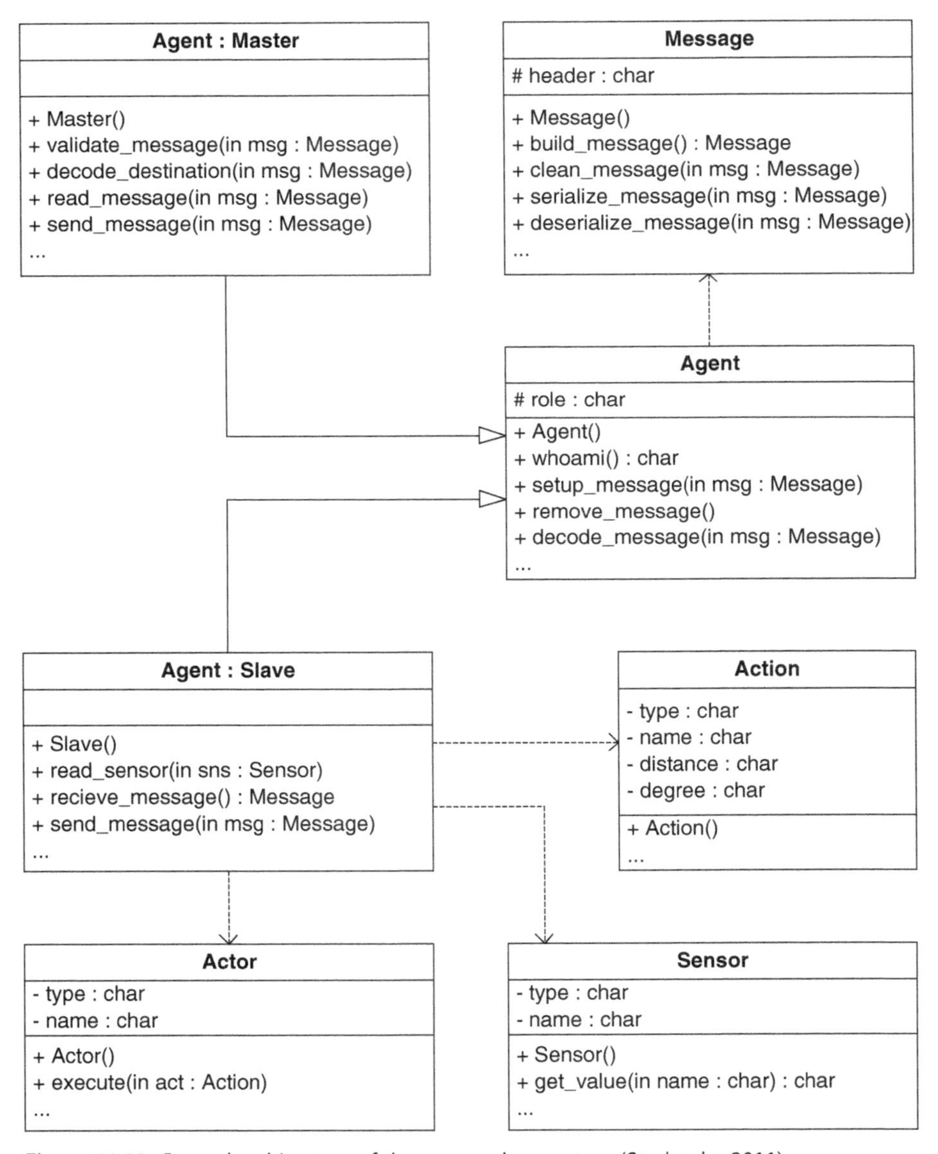

Figure 11.11 General architecture of the master-slave system (Steshenko 2011).

In the implementation of the protocol, the software modules running on the ordinary personal computer (PC) with MS Windows operating system (MS Windows NXT or higher) were written using the C++ communication library provided by Borg (2009), and the modules running on the Lego NXT bricks were programmed using the NXC communication library BTLib provided by Benedettelli (2007).

In order to clarify the activity of the robots' group, let us consider the following simple mission (Steshenko 2011). Assume that there is a group of three Lego NXT robots such that the first robot (GRP) is equipped with gripper and with an ultrasonic distance sensor, the second robot (COL) – with a color sensor and the third robot (DST) – with an ultrasonic distance sensor. In addition, all robots are equipped with odometers, which measure the length of the passed paths, and with compass sensors, which provide executing exact turns. In order to obtain certain direct indication of the robots' states, the robots are also equipped with three color LEDs (red, green, and blue), which are the part of the Lego NXT color sensors. The mission of the robots is defined as follows:

1) The GRP robot opens the gripper and sends initial message to the other robots.
2) After receiving the message and the acknowledgment, all robots turn on the LEDs, indicating the readiness to execute the mission.
3) The GRP robot moves up to the obstacle, catches it with the gripper, moves back to the initial position removing the obstacle, and after returning, sends message to the other robots (COL and DST) with the distance between the initial position and the obstacle.
4) After receiving the message, the robots COL and DST start moving forward up to the obstacles. The COL robot senses the obstacle by the color sensor, and the DST robot does so with the distance sensor.
5) When the obstacles were reached, the COL robot distinguishes the color of the obstacle and sends the corresponding message to the other robots (GRP and DST).
6) After receiving the message with the color of the obstacle, all robots switch on the LED with corresponding color and move backward such that the distance of the movement is equal to the initial distance between the GRP robot and the obstacle measured at step 3.
7) When the robots finalize their motion, the mission is finished.

To simplify the implementation, in the presented messages `EC` (execute command) and `MT` (message transmit) the following extensions were used. The fourth two literal token of the message, which consists of two literals, is the first token of the message body was defined as `GM` (game) in order to indicate that the messages should be parsed following the indicated mission. If the remainder of the message's body is parsed successfully, a positive response is transferred to the agent (slave or master), and an execution is proceeded according to the contents of the message. Otherwise, the message is marked as an invalid, and the agent responses with the corresponding message with the suffix `NK`.

For example, the message, which is sent by the master agent (master PC) to all slave agents (Lego NXT robots) for initialization of the mission, is:

```
EC:00:AL:GM:GAMEINIT
```

The expected reply to this message by each slave agent, e.g. agent `XX`, is:

```
MT:XX:00:GM:INITDONE
```

Similarly, the commands CLEAROBSTACLE, GOTOBARRIER, and IDENTIFYCOLOR with corresponding replies OBSTACLECLEARED, BARRIERREACHED, and COLOR<*u*>, where the value <*u*> is R (Red), G (Green), or B (Blue), are defined.

The states of the mobile robots at different stages of the mission (see stages 3–6 above) and the screen of the master PC are shown in Figure 11.12.

The considered example of the mobile robots' activity demonstrates the effectiveness of the suggested simple communication protocol as a tool for definition of the robots' negotiations in swarms and possibility of implementation of the mobile robot networks using regular equipment and even toy Lego NXT robots. Certainly, in order to enrich the abilities of the robots' negotiations and to solve additional tasks, the protocol can be extended with additional messages and additional parameters. Notice also that for specific tasks, the system can implement only the part of the messages and corresponding encoding procedures as it was done for the protocol and mission considered in Section 11.2.1.2; the next section presents some remarks on the robots' system, which implements this protocol and mission.

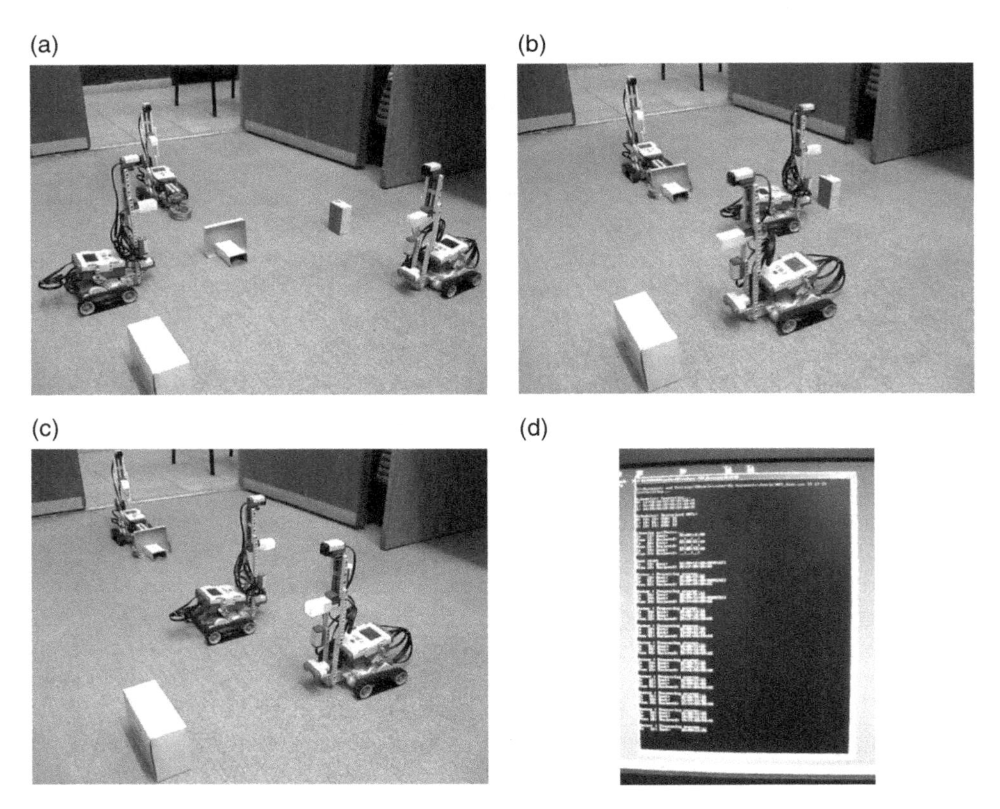

Figure 11.12 Example of collective behavior of the group of mobile robots: (a) initial positions of the robots and obstacles; (b) the robot with gripper removed obstacle and the other robots started motion; (c) the robot distinguished the color of the obstacle and informed the other robots, which turn on the corresponding light; (d) the screen of the master agent (PC) with the messages at the beginning of the session. Source: Photos by E. Kagan and J. Steshenko.

11.2.2.2 Collective Motion of Lego NXT Robots Preserving the Group Activity

Finally, let us consider the group of Lego NXT robots in which the mission is to move together and preserve the group. General mission and the variant of the protocol, which is used for the robots' negotiations, are presented in Section 11.2.1.2. In the system, the star-type network with master-slave communication scheme is used; in the network one of the robots is defined as master agent and the other robots are slave agents. It is assumed, that all robots are equipped with ultrasonic distance sensors and odometers and are able to turn 90° left and right and to turn around 360°. The programs running on the Lego NXT bricks were written using the NXC communication library BTLib provided by Benedettelli (2007) and are based on the procedures used in the system presented in Section 11.2.2.1.

According to the mission, while the robots are in group, they execute the following actions:

1) The robots turn around and check whether they are in group or not.
2) If the robots are in group, then
   2.1 Each robot moves forward 10 cm.
   2.2 If during the motion the robot distinguishes an obstacle,
       then the robot turns 90° left or right (direction is chosen randomly) and sends the message to the other robots regarding its turn.
   2.3 If robot receives the message, which informs it about the turn of the other robot,
       then the robot turns 90° left or right according to the message.
3) Otherwise, the robots terminate motion and inform central unit that the group is broken and the robot with certain ID is lost.

Following the mission (see Section 11.2.1.2), the check whether the robots are in group or not, which is used in line 1 of the activity definition, is conducted by the next routine:

check_group ( ): *check_resul*

1) Turn around 360° and create array of distances to the obstacles; the distance is included into array when it is smaller than a certain threshold (as indicated in Section 11.2.1.2, for the Lego NTX ultrasonic sensor it is 40 cm).
2) Send by the `SV` message the created array of distances to the other robots in the group according to their `IR` requests (see Figure 11.9).
3) Send the `IR` requests and by `SV` messages receive arrays of distances from the other robots in the group (see Figure 11.9). The received arrays are indicated by the IDs of the robots from which the array was received.
4) Compare the arrays of distances, including the own array, as follows (see Figure 11.13):
   If the arrays of different robots include equal distances (because of the sensing errors, the equivalence means that the distances are in the same small interval of values), then the robots are in the group; otherwise the group is broken.
5) Return the result of comparison.

The comparison of the arrays of distances defined in line 4 is illustrated in Figure 11.13.

In the first configuration (Figure 11.13a,c), the distances' array of each robot includes a distance equivalent to the distance appearing in one of the other distances' arrays, and such equivalence is interpreted as a preserved sensual contact between the robots. In

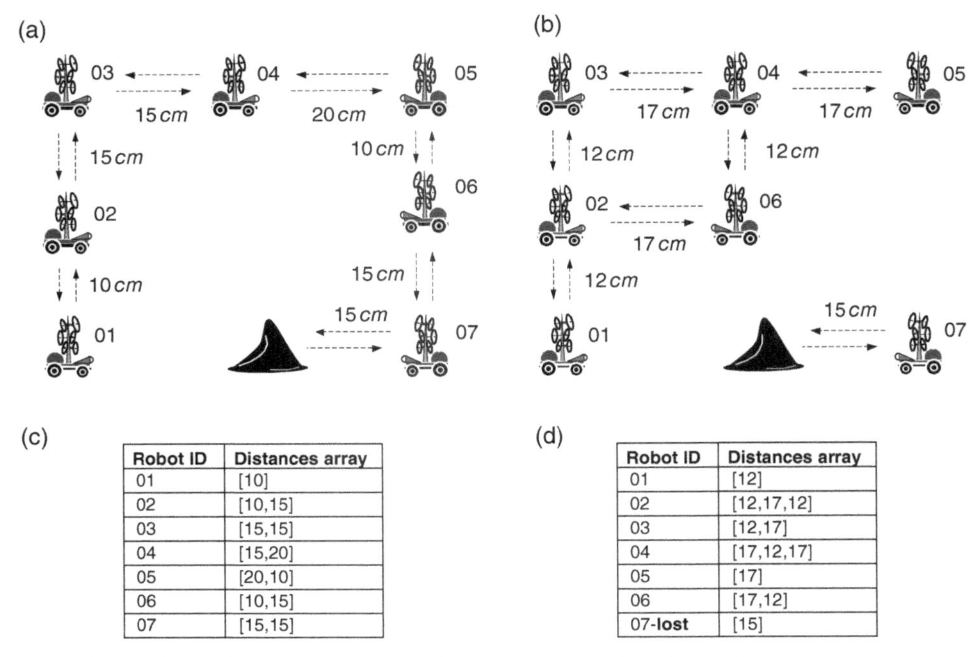

Robot ID	Distances array
01	[10]
02	[10,15]
03	[15,15]
04	[15,20]
05	[20,10]
06	[10,15]
07	[15,15]

Robot ID	Distances array
01	[12]
02	[12,17,12]
03	[12,17]
04	[17,12,17]
05	[17]
06	[17,12]
07-**lost**	[15]

Figure 11.13 Comparison of the arrays of distances in the routine of checking the group: the configuration of the robots' group with IDs of the robots and distances are depicted in figures (a, b); figures (c, d) show the corresponding arrays of distances of each robot with stressed array of the lost robot (ID 07).

contrast, in the second configuration (Figure 11.13b,d), the distances' array of the robot 07 includes a distance that does not appear in any other distances' array that indicates that this robot is not sensed by the other robots and so it is lost.

It is clear that since it is assumed that the robots' distance sensors are not able to distinguish the IDs of each other and, in general, to distinguish between the robot and the obstacle, the obtained distances' arrays can be interpreted as a description of several different configurations, as well as can represent an erroneous situation, in which the obstacle is recognized as a neighboring robot (see Figure 11.8c). However, because of extremely small probability that the distances between the robots are equal eventually, the probability of the ambiguous recognition of the group configuration and of the erroneous group is also extremely small.

The presented example finalizes the consideration of the collective behavior of the group of robots with direct communication. As already indicated, such communication follows a general approach of MANET with certain limitations led by the abilities of the robots, their motion, and requirements on the interactions and sensing. However, since mobile robots act in the environment and environmental states and their changes play an important role in navigation and path planning of single robots and the robots' groups, they are often considered as an additional or even a single channel of communication between the robots (see Section 11.1.2). The next section presents some examples of the behavior of the robots in groups with indirect communication via environmental changes and with interactions, which include both direct and indirect communication.

11.3 Examples of Indirect and Combined Communication

Indirect communication between mobile robots in the group implies that the robots change the environmental states and they act with respect to these changes. In the case of combined communication, the robots communicate both directly using certain protocol and indirectly via environmental changes. This section presents two examples of the collective behavior of the mobile robots with indirect and combined communication. The first example addresses the pheromone robotic system by simulating collective motion of ants (developed in collaboration with Alex Rybalov, Jennie Steshenko, Hava Siegelmann, and Alon Sela; simulation by Lego NXT robots was programmed in collaboration with Rakia Cohen and Harel Mashiah), and the second example considers the model of biosignaling (developed in collaboration with Hava Siegelmann).

11.3.1 Models of Ant Motion and Simulations of Pheromone Robotic System

According to the general approach of the pheromone robotics (see Section 11.1.2), the robots during their motion change the environmental states such that the other robots can interpret these changes as signs, which lead to corresponding decisions regarding the further motion. Since such approach is inspired by the behavior of foraging ants and their abilities to sign the trails by deploying pheromone, the presented examples deal with the behavior of the group of modeling ants and corresponding simulations.

At first, let us consider the simulated motion of the group of modeling ants (Kagan et al. 2014; Kagan and Ben-Gal 2015). Following the odometry and stepping scheme of the ants (Wohlgemuth, Ronacher, and Wehner 2001; Wittlinger, Wehner, and Wolf 2006, 2007), in the model it is assumed that the motion of the ant results from consequent steps of the left and right side legs. In addition, it is assumed that the sensors are located at constant left and right positions at the front of the ant.

Control of the modeling ant is organized as a control of the simple Braitenberg vehicle (Braitenberg 1984) with cross connectivity between the sensors and the activators (see Figure 11.4b) such that the ant follows toward the target. In the considered example, the control functions are defined using the tangent-based uninorm $\oplus_\theta$ and absorbing norm $\otimes_\vartheta$ aggregators that for any $a, b \in (0, 1)$ are defined as (Rybalov, Kagan, and Yager 2012)

$$a \oplus_\theta b = g^{-1}(g_\theta(a) + g_\theta(b)) \text{ and } a \otimes_\vartheta b = g^{-1}(g_\vartheta(a) \times g_\vartheta(b))$$

with the generating function $g(a) = \tan\left(\pi\left(a^\beta - \frac{1}{2}\right)\right), a \in (0, 1), \beta > 0$. Parameters $\theta, \vartheta \in [0, 1]$ are neutral and absorbing elements, respectively. Then:

$$a \oplus_\theta b = \frac{1}{2} + \frac{1}{\pi}\arctan\left(\tan\left[\pi\left(a^\beta - \frac{1}{2}\right)\right] + \tan\left[\pi\left(b^\beta - \frac{1}{2}\right)\right]\right), \tag{11.1}$$

$$a \otimes_\vartheta b = \frac{1}{2} + \frac{1}{\pi}\arctan\left(\tan\left[\pi\left(a^\beta - \frac{1}{2}\right)\right] \times \tan\left[\pi\left(b^\beta - \frac{1}{2}\right)\right]\right), \tag{11.2}$$

where $\theta = \vartheta = (1/2)^{1/\beta}$. Notice that if $\theta = \vartheta = 1/2$, the function g^{-1} has the same form as the Cauchy distribution $\mathcal{F}_{m,\alpha}(\xi) = \frac{1}{2} + \frac{1}{\pi}\arctan\left(\frac{\xi - m}{k}\right)$ with the median $m = 0$ and the parameter $k = 1$. For general consideration of the probability-based uninorm and absorbing norm aggregators, see Kagan et al. (2013). In the following simulations the neutral and absorbing elements were defined as $\theta = 1/2 + \varepsilon_\theta$ and $\vartheta = 1/2 + \varepsilon_\vartheta$, where ε_θ and ε_ϑ are uniformly distributed random variables drawn from the interval $[-0.1, 0.1]$.

Using the uninorm $\oplus_\theta$ and absorbing norm $\otimes_\vartheta$ aggregators, the input, transition, and output functions of the left and right sensors and activators are defined as follows (Kagan et al. 2014; Kagan and Ben-Gal 2015):

1) Input functions, which define the fusion of the sensed environmental states $z_{left}^{env}(t)$, $z_{right}^{env}(t)$ and pheromone signs $z_{left}^{phm}(t)$, $z_{right}^{phm}(t)$ at time t by the left and right side sensors of the ant:

$$z_{left}(t) = z_{left}^{env}(t) \oplus_\theta z_{left}^{phm}(t),\ z_{right}(t) = z_{right}^{env}(t) \oplus_\theta z_{right}^{phm}(t). \tag{11.3}$$

2) State transition functions, which define the states $s_{left}(t)$ and $s_{right}(t)$ of the left and right side controllers at time t given the inputs $z_{right}(t)$ and $z_{left}(t)$ at that this and the previous states $s_{left}(t-1)$ and $s_{right}(t-1)$:

$$s_{left}(t) = z_{right}(t) \otimes_\vartheta s_{left}(t-1),\ s_{right}(t) = z_{left}(t) \otimes_\vartheta s_{right}(t-1). \tag{11.4}$$

3) Output functions, which define the outputs $y_{left}(t)$ and $y_{right}(t)$ of the left and right side controllers at time t given the states of the controllers at this and previous times:

$$y_{left}(t) = s_{left}(t) \otimes_\vartheta s_{left}(t-1),\ y_{right}(t) = s_{right}(t) \otimes_\vartheta s_{right}(t-1). \tag{11.5}$$

Consequently, the steps l_{left}^{step} and l_{right}^{step} of the left and right legs of the ant are defined proportionally to the output values:

$$l_{left}^{step}(t) \sim y_{left}(t), l_{right}^{step}(t) \sim y_{right}(t), \tag{11.6}$$

with the exact proportionality depending on the size of the ant.

The movement of five modeling ants in discrete time $t = 1, 2, \ldots, 1000$ is illustrated in Figure 11.14a. In the figure it is assumed that the ants start in the center $\vec{x} = (250{,}250)$ of the square domain X of the size 500×500. The favorable environmental states are depicted in white color, unfavorable – in black color and neutral states – in gray color; the pheromone trails are depicted by dotted white curves. For comparison, Figure 11.14b shows the trajectories of the ants acting in the same as in Figure 11.14a setup, but without sensing pheromone.

It is seen that the ants that sense the deployed pheromone start moving in random directions and following the environmental states. However, while the ants sense the pheromone trail, they tend to follow this trail. In contrast, without sensing pheromone trails the ants start moving in random directions and then continue motion with respect to the environmental states. This example illustrates the simplest motion of artificial agents with indirect communication following the pheromone robotics approach (Payton et al. 2001) (see Section 11.1.2). Additional examples of motion, in which the agents are able to sense the neighboring and to follow the group as it is required by general flocking rules (see Section 9), are presented in the work by Kagan et al. (2014) and the book by Kagan and Ben-Gal (2015).

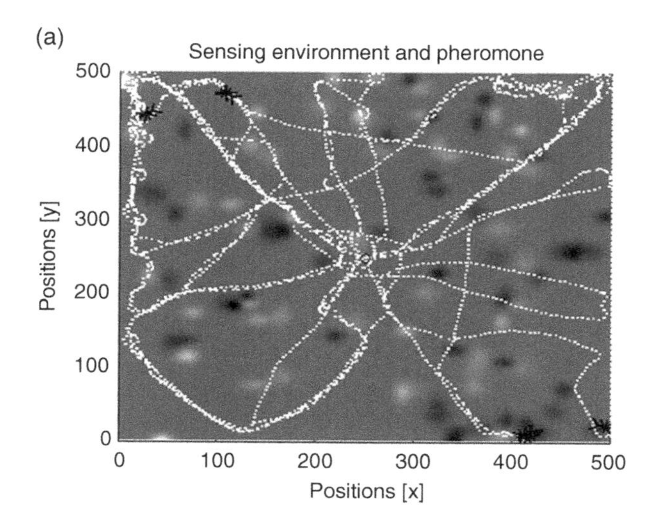

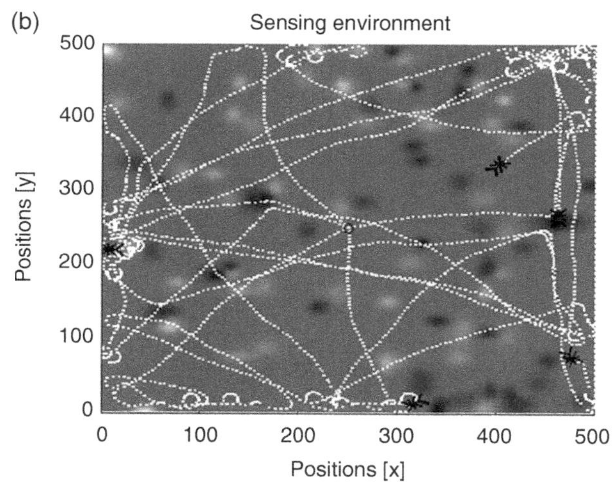

Figure 11.14 The pheromone trails deployed by five modeling ants starting in the center of the domain and moving in random environment: (a) trails of the ants, which follow the environmental states and deployed pheromone trails; (b) trails of the ants, which follow only environmental states without sensing the pheromone trails. Favorable environmental states are depicted by white color, unfavorable states – by black color and neutral states – by gray color. The trails of the ants are depicted by white dotted curves.

The next example demonstrates the implementation of the pheromone-governed motion using the toy Lego NXT robots. Each robot in the group is equipped with the ultrasonic distance sensor, which is located in front of the robot and measures distance to the obstacle up to 40 cm, with two light sensors located in front of the robot and headed down such that the robot is able to follow the black curve drawn on the white arena, and with the pen, which is located at the tail of the robot and can be moved up and down such that the robot is able to draw the black curve on the arena according to the needs. Drawing the curve simulates the pheromone deploying by the ant and the resulted curve emulates the pheromone trail.

In the mission the robots is to seek the arena for the "food" and after finding – to sign the path to the "food" by the "pheromone" trail. In the simulations, the robots act in the closed arena bounded by the walls, and the food is simulated by the relatively small object located in the center of the arena. The mission is outlined as follows (suggested by R. Cohen and H. Mashiah):

1) The robots move randomly over the arena, checking the distance between the robots and the obstacles in front of them.
2) If the robot distinguishes the obstacle, it checks whether it is the wall, another robot, or a "food." The check is conducted as follows:
   2.1 If at the sides of the obstacle are other obstacles, then the obstacle is the wall.
   2.2 If at the sides of the obstacle there are no other obstacles but the obstacle disappeared at the end of the check, then it is another robot.
   2.3 If at the sides of the obstacle there are no other obstacles and the obstacle is still at the same position at the end of the check, then it is a "food."
3) If the "food" is distinguished, then the robot turns backward, puts the pen down, and moves straightforward, drawing the curve up to the wall.
4) Otherwise, the robot turns backward and continues random motion (see line 1).

The snapshots of two simulations of the robot motion are shown in Figure 11.15 (because of weak sensitivity of the light sensors, the curves depicting the "pheromone" trails were stressed additionally). In Figure 11.15a one of the robots moves from the "food" drawing the "pheromone" trail (the previous trail is stressed) and the second robot moves randomly (directed toward the "food"). In Figure 11.15b, the robot follows the

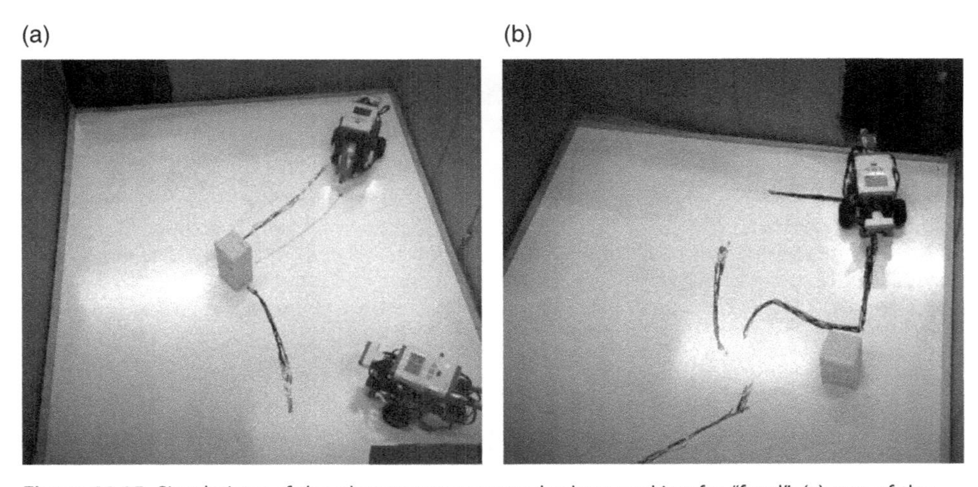

Figure 11.15 Simulations of the pheromone-governed robots seeking for "food": (a) one of the robots moves from the "food" drawing the "pheromone" trail and the second robot moves randomly; (b) the robot follows the "pheromone" trail. The white arena is bounded by the walls and the food is simulated by the object located in the arena. Pheromone trails are simulated by the black curves drawn by the robots. Source: Photo by R. Cohen and H. Mashiah; represented by their kind permission.

"pheromone" trail, which leads it toward the "food." In the considered example, the control of the robots was programed in the simplest manner using standard algorithm of following the line; threshold distance, at which the obstacle is distinguished is 10 cm. The size of the arena is 117 × 86.5 cm.

It is clear that the presented toy example of pheromone robotics is rather far from real practical applications. However, the studies in general pheromone robotics approach can bring to fruitful insights in developing mathematical models of animal behavior and to understanding stigmergy and self-organization phenomena in the swarms of living organisms. At the same time, the communication using infrared and the close-range radio technology like Bluetooth and mimicking the real pheromones by message propagation over the swarm, as it is suggested by Payton et al. (2001) (see Section 11.1.2), can lead and has already led (Rubenstein, Cornejo, and Nagpal 2014) to effective techniques of swarm navigation and control.

In contrast to the considered communication using pheromone trails, real robotic systems apply both indirect and direct communication. In particular, in the search tasks indirect communication is represented by changes of the probabilities of finding the target in certain observed areas (see Sections 10.2 and 10.3), while direct communication is used for informing the other agents in the group about the current decisions and observed environmental states. The next section presents an example of such combined communication implementing an approach widely known as biosignaling.

11.3.2 Biosignaling and Destructive Search by the Group of Mobile Agents

The last example illustrates collective motion of the mobile agents with direct communication, which implements a protocol widely known as biosignaling (Smith 1991; Johnstone and Grafen (1992); Bergstrom and Lachmann 1997, 1998a,b), and indirect communication provided by the destructive search (see Chapters 7 and 10), in which the searcher acts as a predator and "eats" the target or the prey when it is detected (Viswanathan et al. 2011). The presented discourse follows the paper by Siegelmann, Kagan, and Ben-Gal (2014) and the book by Kagan and Ben-Gal (2015).

Formally, biosignaling is defined as an extensive one-stage game between two players; in the framework of biosignaling such a game is known as the Sir Philips Sydney game. In the game, the first player – the signaler – decides whether to send a costly request for a resource that will increase the signaler's fitness, and the second player – the donor – decides whether to provide the resource to the signaler according to the received request, thus decreasing its own fitness. In the probabilistic setup, it is assumed that at the beginning of the game, the signaler is needy for the resource with probability p and not needy for the resource with probability $1 - p$.

Parameters of the game are defined as follows. If the signaler is needy for the resource, then without the resource it has its fitness $1 - a$; if the signaler is not needy for the resource, then without of the resource its fitness is $1 - b$. If the signaler sends the signal to the donor, then it pays the signaling cost c. If the donor transfers the resource, then the donor's fitness decreases by d and becomes $1 - d$. In addition, the relativeness between the players is defined by the value k. It is assumed that all these parameters are taken from the interval $[0, 1]$ and their values are known both for the signaler and the donor.

With respect to the parameters, the Sir Philips Sydney game allows four equilibrium states (Bergstrom and Lachmann 1997; Hutteger and Zollman 2009):

1) "Newer signal"/"Newer donate," for which $d > k(pa + (1 - p)b)$
2) "Always signal"/"Newer donate," for which $d < k(pa + (1 - p)b)$
3) "Signal only if needy"/"Donate only if no signal," for which $a \geq d/k \geq b$ and $a \geq kd - c \geq b$
4) "Signal only if needy"/"Donate only if signal," for which $a \geq d/k \geq b$ and $a \geq kd + c \geq b$

The considered example addresses the last equilibrium, also called the *signaling equilibrium,* that defines the most natural scenario of communication between animals (Smith 1991; Bergstrom and Lachmann 1997; Hutteger and Zollman 2009).

In the terms of search, these strategies are defined as follows (Siegelmann, Kagan, and Ben-Gal 2014; Kagan and Ben-Gal 2015). Assume that during the search, the agent (signaler) finds the region that contains a certain amount of prey. Then the agent has two possibilities: either to continue searching in this region individually (that will require relatively long time), or to signal to the other agents about the found prey and to request their help in the search (so the search will require less search time). If the request was sent, then each of the agents (donors) that received the request also have two possibilities: either to move to the region found by the signaler or to ignore the request and to continue searching in its current region.

Similar to Chapters 7 and 10, assume that the agents act in the bounded two-dimensional square discrete domain $X = \left\{\vec{x}_1, \vec{x}_2, \ldots, \vec{x}_n\right\} \subset \mathbb{R}^2$ over which the map of the target location probabilities $p\left(\vec{x}, t\right)$, $\vec{x} \in X$, $t \in [0, T]$, is defined. The observations are conducted using errorless sensors such that at time t the agent checks certain observed area $a(t) \subset X$ and, following the destructive search scenario, zeroes the target's location probabilities in the area $a(t)$. In addition, denote by $A(t) \subset X$ a patch, which can be sensed by the agent but cannot be checked by immediate observation. In terms of sensing abilities of the agent located in the point $\vec{x}(t) \in X$ at time t, the observed area is usually defined as $a(t) = \left\{\vec{x}: \left\|\vec{x}(t) - \vec{x}\right\| \leq r_{\text{vis}}\right\}$, where $r_{\text{vis}} > 0$ is the visibility radius, and the sensed patch – as $A(t) = \left\{\vec{x}: \left\|\vec{x}(t) - \vec{x}\right\| \leq r_{\text{sense}}\right\}$, where $r_{\text{sense}} \geq r_{\text{vis}}$ is the sensing radius; so $a(t) \subseteq A(t)$.

Assume that at time t the signaler is located in the point $\vec{x}_{\text{s}}(t)$ and the agent, which can decide to be a donor, is located in the point $\vec{x}_{\text{d}}(t)$. Then, parameters of the game at time t are defined with respect to the target location probabilities $p\left(\vec{x}, t\right)$ and the distance between the signaler and the potential donor:

$$a(t) \sim \sum\nolimits_{\vec{x} \in A_{\text{s}}(t)} p\left(\vec{x}, t\right),\ d(t) \sim \sum\nolimits_{\vec{x} \in A_{\text{d}}(t)} p\left(\vec{x}, t\right),\ c \sim \mathrm{r}_{\text{sense}},\ k(t) \sim \left\|\vec{x}_{\text{s}}(t) - \vec{x}_{\text{d}}(t)\right\|. \tag{11.7}$$

Regarding the neediness b of the signaler it is assumed that if the signaler is not needy for the resource, its neediness is $b \ll 1$.

Finally, denote by $r_{\text{signal}}^{\min} > 0$ and by $r_{\text{signal}}^{\max} > r_{\text{signal}}^{\min}$ the minimal and maximal distances, to which the searcher can send request for help, respectively. These values limit the signaling abilities of the agents and specify that the signaler can rend a request only to the neighbors that are not too close to and not too far from the signaler's current location. It is clear that such limitations plat the same role as the attraction and repulsion potentials considered in Chapter 10. Using the defined parameters, the signaling and donating scenarios according to the Sir Philips Sydney game applied to the probabilistic search are outlined as follows (Kagan and Ben-Gal 2015):

signaling $\left(\vec{x}_s(t)\right)$

1) Given location $\vec{x}_s(t)$, consider the patch $A_s(t)$ defined by the sensing distance r_{sense} and calculate the neediness $a(t)$ and the cost c by the Eq. (11.7).
2) Send the request, which includes location $\vec{x}_s(t)$, neediness $a(t)$ and the cost c, to all the agents located in the ring defined by the signaling distances r_{signal}^{min} and r_{signal}^{max}.

donating $\left(\vec{x}_d(t)\right)$: *decision*

1) Given location $\vec{x}_d(t)$ and the received location $\vec{x}_s(t)$, neediness $a(t)$, and cost c, consider the patch $A_d(t)$ defined by the sensing distance r_{sense} and calculate the fitness decrement $d(t)$ and the relativeness $k(t)$ by the Eq. (11.10).
2) If the signaling equilibrium $a(t) \geq d(t)/k(t) \geq b$ and $a(t) \geq k(t)d(t) + c \geq b$ (see equilibrium 4 above) holds, then decide donate to the signaler, i.e., set *decision* = *donate*; else decide not to donate; i.e., set *decision* = *not donate*.

Notice that since the signaler has no information about the regions in which the other agents are acting, in the procedure *signaling*(...) it is assumed that the signaler always sends requests to the neighbors. In contrast, in the procedure *donating*(...) the agents have all required information for the decision-making whether to donate or not.

Using the defined procedures, the mission of the agents conducting destructive probabilistic search with direct communication using biosignaling is defined as follows (Kagan and Ben-Gal 2015):

1) Each agent in the group seeks over the domain according to certain search algorithm and zeroes the target's location probabilities in the areas that have been observed.
2) During the search, the agent locating in the point $\vec{x}\,(t)$ conducts the following actions:
   2.1 As a signaler, the agent signals for help according to the procedure *signaling*$\left(\vec{x}\,(t)\right)$.
   2.2 As a potential donor, the agent listens to the requests from the other agents, and if the request was received, then it decides whether to donate using the procedure *donating*$\left(\vec{x}\,(t)\right)$.
3) If *decision* is *donate*, then the agent stops searching in its current region, and as a donor moves toward the signaler's location with maximal velocity.
4) By arrival to the signaler's patch, the agent continues searching as a signaler as it is defined in line 1.

The presented mission follows general procedures of destructive search with biosignaling (Siegelmann, Kagan, and Ben-Gal 2014; Kagan and Ben-Gal 2015) and illustrates the main principles of the combined communication. The search indicated in line 1 can be conducted following different methods that certainly influence to the agents motion and effectiveness of the execution of the mission in general.

Figure 11.16a illustrates the activity of $m = 10$ agents with signaling (an example was developed in collaboration with Hava Siegelmann). In the figure, it is assumed that the agents act on the squire gridded domain X of the size $n = 100 \times 100$ with impermeable

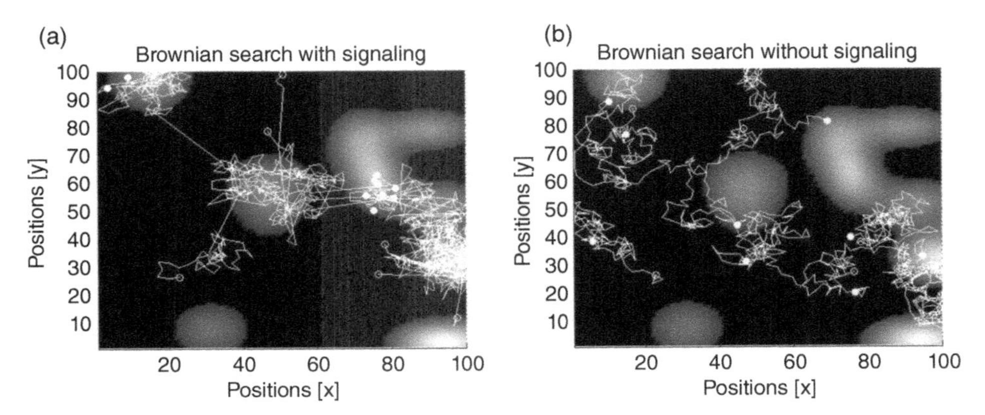

Figure 11.16 Trajectories of $m = 10$ agents conducting Brownian search (a) with signaling and (b) without signaling. Parameters of the signaling in figure (a) are $r_{vis} = 1$, $r_{sense} = 5$, $r_{signal}^{min} = 10$ and $r_{signal}^{max} = 25$. In both cases, the size of the domain is 100×100. The whiter regions correspond to greater target's location probabilities and the black regions – to zero location probabilities. The basic motion of the agents is Brownian walk with the step $l = 3$ and the time of motion is $t = 1, \dots, 100$.

bounds and conduct Brownian destructive search for the prey randomly distributed over the domain. The whiter regions include greater target's location probabilities and black regions represent zero location probabilities. The step of the Brownian motion of the agents is $l = 3$, visibility and sensing radiuses are $r_{vis} = 1$ and $r_{sense} = 5$, and minimal and maximal signaling distances are $r_{signal}^{min} = 10$ and by $r_{signal}^{max} = 25$. The motion time is $t = 1, \dots, 100$. For comparison, Figure 11.16b shows the trajectories of the agents starting from the same random locations and acting without signaling over the domain with same map of target location probabilities.

It is seen that in contrast to the Brownian search without signaling, which is represented by usual random walk shown in Figure 11.16b, the search with signaling includes long jumps of the donor agents to the regions where the signalers act. As a result, the agents concentrate in the regions with relatively large amount of prey that certainly leads to more effective search. Detailed information about probabilistic search with signaling and its correspondence to different mobility models see the indicated paper by Siegelmann, Kagan, and Ben-Gal (2014) and the book by Kagan and Ben-Gal (2015).

The considered example finalizes consideration of the collective behavior of mobile agents with different types of communication. Certainly, the presented discourse does not exhaust the wide range possible definitions of such communication and includes only general techniques of its implementation, which can be used as a basis for different applications. Summarizing the chapter, we will discuss some of further directions in this field.

11.4 Summary

In the chapter we considered the motion of the group of mobile robots with direct and indirect communication and with communication, which includes both direct and indirect techniques. In the first case, it is assumed that the robots can transmit information directly from one to another following a definite communication protocol and may be via some central unit. In the second case, in contrast, the information is transmitted using the changes of the environmental states such that the environment is considered as a kind of shared memory of the robots. In the combined communication, the robots implement both direct and indirect communication. Certainly, the methods of communication between mobile robots follow general approach of the MANET with reasonable limitations led by the robots' computation and communication abilities, while mobility of the robots and their swarm dynamics are specified by corresponding navigation algorithms and flocking techniques.

With respect to such correspondence between robots' communication and networking, the chapter started with a brief review of communication techniques and of the stigmergic phenomena implied by the indirect communication (Section 11.1), and considers communication protocols and collective motion of mobile agents with direct, indirect, and combined communication as follows:

1) Collective motion of the mobile robots with direct communication based on simple communication protocols based on the Bluetooth close-range radio technology. The protocols (Section 11.2.1) allow both distributed and centralized control of the execution of the robots' mission and support the basic flocking techniques, which provide preserving of the swarm. Examples of collective motion of the mobile robots that implement the indicated protocols were presented in Section 11.2.2.
2) Motion of the group of agents with indirect communication, which follows the approach of pheromone robotics and mimics the collective behavior of ants (Section 11.3.1). In such motion, the agents sign the environment (in the presented implementation – draw the trajectory of the robot) such that the other agents make decisions regarding their further steps according to these signs.
3) Motion of the agents with combined communication, in which the agents both transmit certain information from one to another and change the environmental states that influence the agents' decisions. In Section 11.3.2 such motion was illustrated by destructive probabilistic search with biosignaling, which mimics foraging behavior of animals.

The considered models are illustrated by numerical examples (ants' motion in Section 11.3.1 and search with biosignaling in Section 11.3.2) and by the collective motion of the Lego NXT robots (direct communication and preserving the group in Section 11.2.2 and pheromone robotics in Section 11.3.1).

References

Aguero, C., Canas, J.M., and Ortuno, M. (2007). Design and implementation of an ad-hoc routing protocol for mobile robots. *Turkish Journal of Electrical Engineering & Computer Sciences* 15 (2): 307–320.

Balch, T. and Arkin, R.C. (1994). Communication in reactive multiagent robotic systems. *Autonomous Robots* 1 (1): 27–52.

Beckers, R., Holland, O.E., and Deneubourg, J.L. (1994). From local actions to global tasks: stigmergy and collective robotics. In: *Artificial Life IV* (ed. R.A. Brooks and P. Maes), 181–189. Cambridge, Masachussetts: MIT Press.

Benedettelli, D. (2007, Jan). *NXC Bluetooth Library*. Retrieved from http://robotics/benedettelli.com/bt_nxc.htm (currently it is a part of the Lego firmware and supported by the bricxcc studio: http://bricxcc.sourceforge.net)

Beni, G. (2005). From swarm intelligence to swarm robotics. In: *Swarm Robotics. Lecture Notes in Computer Science*, vol. 3342 (ed. E. Sahin and W.M. Spears), 1–9. Berlin: Springer.

Beni, G. and Wang, J. (1990). Self-organizing sensory systems. In: *Proc. NATO Advanced Workshop on Highly Redundant Sensing in Robotic Systems, Il Cioco, Italy (June 1988)*, 251–262. Berlin: Springer.

Bergstrom, C.T. and Lachmann, M. (1997). Signaling among relatives I. Is signaling too costly? *Philosophical Transactions of the Royal Society of London, Series B: Biological Sciences* 352: 609–617.

Bergstrom, C.T. and Lachmann, M. (1998a). Signaling among relatives II. Beyond the tower of Babel. *Theoretical Population Biology* 54: 146–160.

Bergstrom, C.T. and Lachmann, M. (1998b). Signaling among relatives III. Talk is cheap. *Proceedings of the National Academy of Sciences of the United States of America* 95: 5100–5105.

Bonabeau, E., Theraulaz, G., Deneubourg, J.-L. et al. (1997). Self-organization in social insects. *Trends in Ecology and Evolution* 12 (5): 188–193.

Bonabeau, E., Dorigo, M., and Theraulaz, G. (1999). *Swarm Intelligence: From Natural to Artificial Systems*. New York: Oxford University Press.

Borg, A. (2009, Apr 6). *The C++ Communication Library*. Retrieved from http://www.norgesgade14.dk/bluetoothlibrary.php (currently the C# version is available at http://www.monobrick.dk/software/monobrick; accessed at Aug 23 2015)

Braitenberg, V. (1984). *Vehicles: Experiments in Synthetic Psychology*. Cambridge, Massachusetts: The MIT Press.

Clark, A. (1997). *Being There: Putting Brain, Body and World Together Again*. Cambridge, Massachusetts: The MIT Press.

Clark, A. (2003). *Natural-Born Cyborgs – Minds, Technologies, and the Future of Human Intelligence*. Oxford: Oxford University Press.

Conti, M. and Giordano, S. (2014). Mobile ad hoc networking: milestones, challenges, and new research directions. *IEEE Communications Magazine* 52 (1): 85–96.

Crocker, D. H. (1982, Aug 13). *RFC 822 - standard for the format of ARPA internet text messages*. http://www.w3.org/protocols/rfc822

Das, A., Spletzer, J., Kumar, V., and Taylor, C. (2002). Ad hoc networks for localization and control. In: *Proc. 41-th IEEE Conf. Decision and Control*, 2978–2983. Las Vegas, CA.

Das, S.M., Hu, Y.C., Lee, G.C., and Lu, Y.-H. (2005). An efficient group communication protocol for mobile robots. In: *Proc. IEEE Conf. Robotics and Automation*, 88–93. Barcelona, Spain.

De Gennes, P.G. (1976). La percolation: un concept unificateur. *La Recherche* 7: 919–927.

Demin, A. (2008, Feb 1). *NXT Brick remote control over Bluetooth*. Retrieved Aug 18, 2015, from http://demin.ws/nxt/bluetooth

Frodigh, M., Johansson, P., and Larsson, P. (2000). Wireless ad hoc networking - the art of networking without a network. *Ericsson Review* 4: 248–263.

Gazi, V. and Passino, K.M. (2011). *Swarm Stability and Optimization*. Berlin: Springer.

Gordon, D.M. (2010). *Ant Encounters: Interaction Networks and Colony Behavior*. Princeton: Princeton University Press.

Grasse, P.-P. (1959). La reconstruction du nid et less coordinations interindividuelles chez bellicositermes natalensis et cubitermes sp. la theorie de la stigmergie: essai d'interpretation du comportement des termites constructeurs. *Insectes Sociaux* 6: 41–83.

Hamann, H. (2010). *Space-Time Continuous Models of Swarm Robotic Systems: Supporting Global-to-Local Programming*. Berlin: Springer.

Huhns, M.N. and Stephens, L.M. (1999). Multiagent systems and societies of agents. In: *Multiagent Systems. A Modern Approach to Distributed Artificial Intelligence* (ed. G. Weiss), 79–120. Cambridge, Massachusetts: The MIT Press.

Hutteger, S.M. and Zollman, K.J. (2009). Dynamic stability and basins of attraction in the Sir Philip Sydney game. *Proceedings of the Royal Society of London Series B* 277 (1689): 1915–1922.

Johnstone, R.A. and Grafen, A. (1992). The continuous Sir Philips Sydney game: a simple model of biological signaling. *Journal of Theoretical Biology* 156: 215–234.

Jung, T., Ahmadi, M., and Stone, P. (2009). Connectivity-based localization in robot networks. In: *Proc. Int. Workshop Robotic Sensor Networks (IEEE DCOSS' 09)*. Marina Del Rey, CA.

Kagan, E. and Ben-Gal, I. (2015). *Search and Foraging. Individual Motion and Swarm Dynamics*. Boca Raton, FL: Chapman Hall/CRC/Taylor & Francis.

Kagan, E., Rybalov, A., Siegelmann, H., and Yager, R. (2013). Probability-generated aggregators. *International Journal of Intelligent Systems* 28 (7): 709–727.

Kagan, E., Rybalov, A., Sela, A. et al. (2014). Probabilistic control and swarm dynamics of mobile robots and ants. In: *Biologically-Inspired Techniques for Knowledge Discovery and Data Mining* (ed. S. Alam), 11–47. Hershey, PA: IGI Global.

Kaizer, M. and Makmal, S. (2015). *Information Transfer in the Network of Mobile Agents for Preserving the Swarm Activity. B.Sc. Project*. Ariel: Ariel University.

Martone, A. (2010). *NXT to NXT communication using PC programs*. http://www.alfonsomartone.itb.it/yepuji.html

Millonas, M.M. (1994). Swarms, phase transitions, and collective intelligence. In: *Artificial Life III* (ed. C. Langton), 417–445. Santa Fe: Perseus Books.

Niculescu, D. and Nath, B. (2004). Position and orientation in ad hoc networks. *Ad Hoc Networks* 2 (2): 133–151.

Payton, G.H., Daily, M., Estowski, R. et al. (2001). Pheromone robotics. *Autonomous Robots* 11: 319–324.

Rubenstein, M., Cornejo, A., and Nagpal, R. (2014). Programmable self-assembly in a thousand-robot swarm. *Science* 345 (6198): 795–799.

Rybalov, A., Kagan, E., and Yager, R. (2012). Parameterized uninorm and absorbing norm and their application for logic design. In: *Proc. 27-th IEEE Conv. EEEI*. https://doi.org/10.1109/EEEI.2012.6377125.

Siegelmann, H., Kagan, E., and Ben-Gal, I. (2014). Honest signaling in the cooperative foraging in heterogeneous environment by a group of Levy fliers.

Smith, J.M. (1991). Honest signaling: the Philip Sydney game. *Animal Behavior* 42: 1034–1035.

Steshenko, J. (2011). *Centralized Control of Cooperative Behavior of NXT Robots via Bluetooth. B.Sc. Thesis*. Tel-Aviv: Tel-Aviv University.

Steshenko, J., Kagan, E., and Ben-Gal, I. (2011). A simple protocol for a society of NXT robots communicating via Bluetooth. In: *Proc. IEEE 53-rd Symp. ELMAR*, 381–384.

Sumpter, D.J. (2010). *Collective Animal Behaviour*. Princeton: Princeton University Press.

Theraulaz, G., Gautrais, J., Camazine, S., and Deneubourg, J.L. (2003). The formation of spatial patterns in social insects: from simple behaviours to complex structures. *Philosophical Transactions of the Royal Society of London, Series A: Mathematical, Physical and Engineering Sciences* 361: 1263–1282.

Trianni, V. (2008). *Evolutionary Swarm Robotics: Evolving Self-Organising Behaviors in Groups of Autonomous Robots*. Berlin: Springer-Verlag.

Viswanathan, G.M., da Luz, M.G., Raposo, E.P., and Stanley, H.E. (2011). *The Physics of Foraging*. Cambridge: Cambridge University Press.

Wang, Z., Zhou, M., and Ansari, N. (2003). Ad-hoc robot wireless communication. In: *Proc. IEEE Conference on Systems, Man and Cybernetics*, vol. 4, 4045–4050.

Wang, Z., Liu, L., and Zhou, M. (2005). Protocols and applications of ad-hoc robot wireless communication networks: an overview. *International Journal of Intelligent Control and Systems* 10 (4): 296–303.

Weiss, G. (ed.) (1999). *Multiagent Systems. A Modern Approach to Distributed Artificial Intelligence*. Cambridge, Massachusetts: The MIT Press.

White, T. (1997). *Swarm Intelligence*. Retrieved Oct 4, 2015, from Network Management and Artificial Intelligence Laboratory: www.sce.carleton.ca/netmanage/tony/swarm.html

Wittlinger, M., Wehner, R., and Wolf, H. (2006). The ant odometer: stepping on stilts and stumps. *Science* 312: 1965–1967.

Wittlinger, M., Wehner, R., and Wolf, H. (2007). The desert ant odometer: a stride integrator that accounts for stride length and walking speed. *Journal of Experimental Biology* 210: 198–207.

Wohlgemuth, S., Ronacher, B., and Wehner, R. (2001). Ant odometry in the third dimension. *Nature* 411: 795–798.

Yoshida, E., Arai, T., and Yamamoto, M.O. (1998). Local communication of multiple mobile robots: design of optimal communication area for cooperative tasks. *Journal of Robotic Systems* 15 (7): 407–419.

12

Brownian Motion and Swarm Dynamics

Eugene Khmelnitsky

12.1 Langevin and Fokker-Plank Formalism

The interactions between the environment and the Brownian particle results in three forces that act on the Brownian particle: a damping force, a force associated with the exogenous fields, and a fluctuating force. The forces affect the dynamics of the location and speed of the Brownian particle, as given in the one-dimensional Langevin equation:

$$\dot{x} = v \text{ and } \dot{v} = -\gamma(x,v)v + F(x,t) + \xi(t), \tag{12.1}$$

where $\xi(t)$ is a Gaussian noise, i.e. the values of $\xi(t)$ at any pair of times are identically distributed, $\xi(t) \sim N(0,\sigma^2)$, and uncorrelated $F(x,t)$ represents additional forces (Sjögren 2015, Schimansky-Geier et al. 2005).

Since the dynamics are random, we are interested in the probability density $p(x, v, t)$ to find the Brownian particle in the interval $(x, x + dx)$ and $(v, v + dv)$ at time t. The derivation of the dynamic equation for $p(x, v, t)$ – the Fokker-Planck equation – is carried out in two steps. The first one derives the equation of motion for the probability density $\rho(x, v, t)$ for one realization of the random force $\xi(t)$. The second step averages the value of $\rho(x, v, t)$ over many realizations:

$$p(x,v,t) = E[\rho(x,v,t)].$$

For simplicity, we derive the Fokker-Planck equation, assuming no external fields, i.e. $F(x, t) = 0$.

The Brownian particle is located in the area $dxdv$ of the state space (x, v) with probability $\rho(x,v,t)dxdv$. Since the particle must lie somewhere in the state space, we have that for any t,

$$\int_{-\infty}^{\infty}\int_{-\infty}^{\infty} \rho(x,v,t)dvdx = 1.$$

Autonomous Mobile Robots and Multi-Robot Systems: Motion-Planning, Communication, and Swarming, First Edition. Edited by Eugene Kagan, Nir Shvalb and Irad Ben-Gal.

Companion website: www.wiley.com/go/kagan/robotsystems

Consider a finite volume V in the state space. The change in the probability contained in V must be due to a flow of probability through the surface S surrounding V. That is,

$$\frac{d}{dt}\iint_V \rho(x,v,t)dxdv = -\int_V \rho(x,v,t)(\dot{x},\dot{v})d\vec{S}.$$

We can now use the Gauss theorem to change the surface integral into a volume integral for the vector field $\rho(x,v,t)(\dot{x},\dot{v})$,

$$\iint_V (\nabla\cdot(\rho(x,v,t)(\dot{x},\dot{v})))dV = \int_V \left(\rho(x,v,t)(\dot{x},\dot{v})d\vec{S}\right),$$

and obtain

$$\iint_V \frac{\partial}{\partial t}\rho(x,v,t)dxdv = -\iint_V (\nabla\cdot(\rho(x,v,t)(\dot{x},\dot{v})))dV.$$

Since V is arbitrary, the last equation results in the partial differential equation for $\rho(x, v, t)$:

$$\frac{\partial}{\partial t}\rho(x,v,t) = -\nabla\cdot(\rho(x,v,t)(\dot{x},\dot{v})) = -\frac{\partial}{\partial x}(\rho\dot{x}) - \frac{\partial}{\partial v}(\rho\dot{v}) \tag{12.2}$$

This is a general equation just stating that probability is conserved. To write it explicitly for a Brownian particle, we substitute its dynamics determined by the Langevin equation (12.1) into (12.2):

$$\frac{\partial}{\partial t}\rho(x,v,t) = -\frac{\partial}{\partial x}(\rho v) + \frac{\partial}{\partial v}(\rho\gamma(x,v)v) - \frac{\partial}{\partial v}(\rho\xi(t)) \tag{12.3}$$

By introducing the operators,

$$A = v\frac{\partial}{\partial x} - \frac{\partial}{\partial v}(\gamma(x,v)v) - \gamma(x,v)v\frac{\partial}{\partial v} \text{ and } B = \xi(t)\frac{\partial}{\partial v},$$

(12.2) reduces to

$$\frac{\partial}{\partial t}\rho(x,v,t) = -A\rho - B\rho \tag{12.4}$$

By changing variable,

$$\varphi(x,v,t) = exp(At)\rho(x,v,t)$$

the differential equation (12.4) reduces further:

$$\frac{\partial}{\partial t}\varphi(x,v,t) = -D(t)\varphi(t), \tag{12.5}$$

where

$$D(t) = exp(At)Bexp(-At).$$

The solution of (12.5) is

$$\varphi(x,v,t) = C\cdot exp\left(-\int_0^t D(\tau)d\tau\right), \tag{12.6}$$

where C is a constant determined from the boundary conditions. This determines the dynamics of $\rho(x, v, t)$.

However, since we monitor an actual Brownian particle, we are observing the average effect of the random force. The averaging of $\varphi(x, v, t)$ requires the expectation and the variance of $\int_0^t D(\tau)d\tau$. Since $D(t)$ is proportional to $\xi(t)$ at each t, we note that $\int_0^t D(\tau)d\tau$ must also be a Gaussian variable with $E\left[\int_0^t D(\tau)d\tau\right] = 0$ and

$$Var\left[\int_0^t D(\tau)d\tau\right] = E\left[\left(\int_0^t D(\tau)d\tau\right)^2\right] = \int_0^t\int_0^t E[D(\tau_1)D(\tau_2)]d\tau_1 d\tau_2$$

$$= \sigma^2 \int_0^t exp(A\tau)\frac{\partial}{\partial v} exp(-A\tau)\, exp(A\tau)\frac{\partial}{\partial v} exp(-A\tau)d\tau = \sigma^2 \int_0^t exp(A\tau)\frac{\partial^2}{\partial v^2} exp(-A\tau)d\tau$$

Now, by applying expectation to both sides of (12.6), we obtain the following:

$$E[\varphi(t)] = C\cdot E\left[exp\left(-\int_0^t D(\tau)d\tau\right)\right] = C\cdot exp\left(\frac{1}{2}Var\left[\int_0^t D(\tau)d\tau\right]\right)$$

$$= C\cdot exp\left(\frac{\sigma^2}{2}\int_0^t exp(A\tau)\frac{\partial^2}{\partial v^2} exp(-A\tau)d\tau\right)$$

By changing back to the original variable $p(x, v, t)$, we obtain

$$\frac{\partial}{\partial t}p(x,v,t) = -Ap(x,v,t) + \frac{\sigma^2}{2}\frac{\partial^2}{\partial v^2}p(x,v,t). \tag{12.7}$$

12.2 Examples

Example 12.1 Consider a free, noninteracting particle, i.e. a particle for which $(x, v) = 0$ $\forall x, v$. The Fokker-Plank equation (12.7) simplifies in this case to

$$\frac{\partial}{\partial t}p(x,v,t) = -v\frac{\partial}{\partial x}p(x,v,t) + \frac{\sigma^2}{2}\frac{\partial^2}{\partial v^2}p(x,v,t). \tag{12.8}$$

Suppose that the initial location and initial speed of the particle are x_0 and v_0, respectively. Then, the solution of (12.8) is (see, e.g., Tanski 2004)

$$p(x,v,t) = \frac{\sqrt{3}}{\pi\sigma^2 t^2} exp\left(-\frac{6\hat{x}^2 - 6\hat{x}\hat{v}t + 2\hat{v}^2 t^2}{\sigma^2 t^3}\right), \tag{12.9}$$

where $\hat{x} = x - x_0 - v_0 t$ and $\hat{v} = v - v_0$.

By integrating (12.9) with regard to x, we obtain the distribution of speed, i.e. the probability density of particle's speed regardless of its location:

$$p(v,t) = \frac{1}{\sqrt{2\pi t\sigma}} exp\left(-\frac{\hat{v}^2}{2\sigma^2 t}\right), \tag{12.10}$$

and, similarly, by integrating (12.9) with regard to v, we obtain the probability density of particle's location regardless of its speed,

$$p(x,t) = \frac{\sqrt{3}}{\sqrt{2\pi t^3\sigma}} exp\left(-\frac{3\hat{x}^2}{2\sigma^2 t^3}\right). \tag{12.11}$$

We observe that both speed and location are distributed normally with the mean and variance being equal to

$$E(v) = v_0, \quad Var(v) = \sigma^2 t, \; E(x) = x_0 + v_0 t, \quad Var(x) = \sigma^2 t^3/3.$$

The dynamics of speed follow the Brownian motion, growing linearly with time. The variance of the location grows much faster, as the third power of time. Figure 12.1 shows the distributions $p(v, t)$ and $p(x, t)$ for a given set of parameters. ■

Example 12.2 This example adds a damping force proportional to the speed with a constant coefficient, $(x, v) = \gamma \;\; \forall x, v$. The Fokker-Plank equation (12.7) becomes

$$\frac{\partial}{\partial t}p(x,v,t) = \left(-v\frac{\partial}{\partial x} + \gamma + \gamma v\frac{\partial}{\partial v} + \frac{\sigma^2}{2}\frac{\partial^2}{\partial v^2}\right)p(x,v,t). \tag{12.12}$$

Suppose again that the initial location and initial speed of the particle are x_0 and v_0, respectively. Then, the solution of (12.12) is

$$p(x,v,t) = \frac{1}{2\pi\sigma^2\sqrt{c_1(t)}} exp\left(-\frac{c_2(t)\hat{x}^2 + 2c_3(t)\hat{x}\hat{v} + c_4(t)\hat{v}^2}{2\sigma^2 c_1(t)}\right), \tag{12.13}$$

where

$$\hat{x} = x - x_0 - \frac{v_0}{\gamma}(1 - e^{-\gamma t}), \qquad \hat{v} = v - v_0 e^{-\gamma t},$$

$$c_1(t) = \frac{1}{2\gamma^4}\left(\gamma t(1 - e^{-2\gamma t}) - 2(1 - e^{-\gamma t})^2\right), \qquad c_2(t) = \frac{1}{2\gamma}(1 - e^{-2\gamma t}),$$

$$c_3(t) = \frac{1}{2\gamma^2}(1 - e^{-2\gamma t}) - \frac{1}{\gamma^2}(1 - e^{-\gamma t}), \qquad c_4(t) = \frac{t}{\gamma^2} + \frac{1}{2\gamma^3}(1 - e^{-2\gamma t}) - \frac{2}{\gamma^3}(1 - e^{-\gamma t}).$$

By integrating (12.13) w.r.t x, we obtain that the probability of particle's speed regardless of its location is normally distributed with

$$E(v) = v_0 e^{-\gamma t}, \quad Var(v) = \frac{\sigma^2}{2\gamma}(1 - e^{-2\gamma t}).$$

Similarly, by integrating (12.13) w.r.t v, we obtain that the probability of location regardless of speed is normally distributed with

$$E(x) = x_0 + \frac{v_0}{\gamma}(1 - e^{-\gamma t}), \quad Var(x) = \frac{\sigma^2}{2\gamma^3}(2\gamma t - 3 + 4e^{-\gamma t} - e^{-2\gamma t}).$$

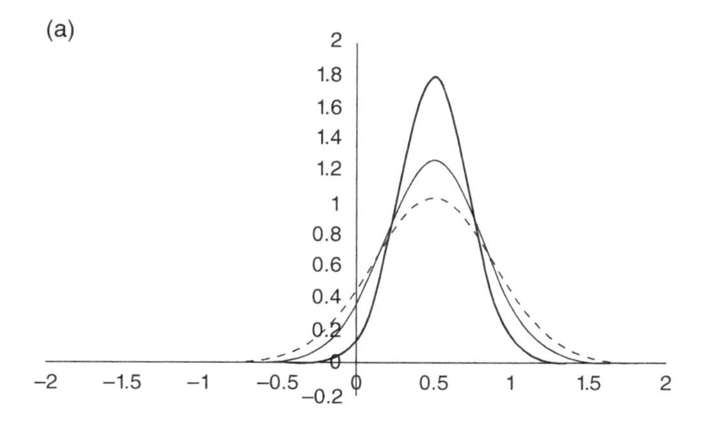

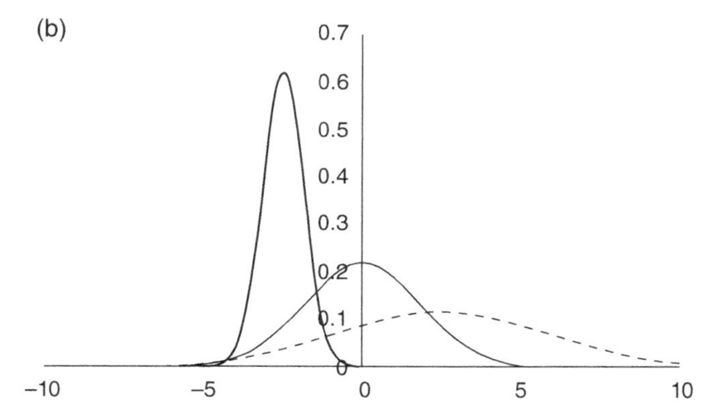

Figure 12.1 The probability distribution of (a) speed and (b) location for $x_0 = -5$, $v_0 = 0.5$, $\sigma = 0.1$ at $t = 5, 10, 15$.

Figure 12.2 shows the distributions $p(v, t)$ and $p(x, t)$ for a given set of parameters. In steady state, when t tends to infinity, the speed converges to a stationary normal distribution with

$$E(v) = 0, \quad Var(v) = \frac{\sigma^2}{2\gamma}.$$

The location converges to Brownian motion:

$$E(x) \to x_0 + \frac{v_0}{\gamma}, \quad Var(x) \to \frac{\sigma^2}{\gamma^2} t. \qquad \blacksquare$$

Examples 12.1 and 12.2 show that the presence of the damping force restricts the motion in such a way that the dependence of the location variance on time drops down from a power of three to linearly proportional, and the speed variance loses its dependence on time at all.

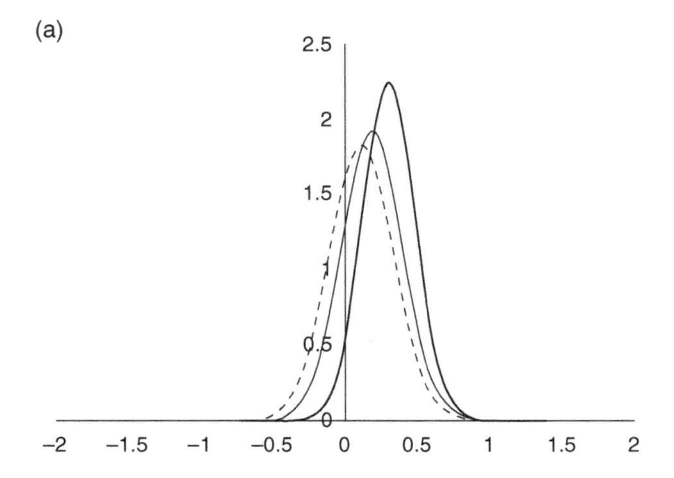

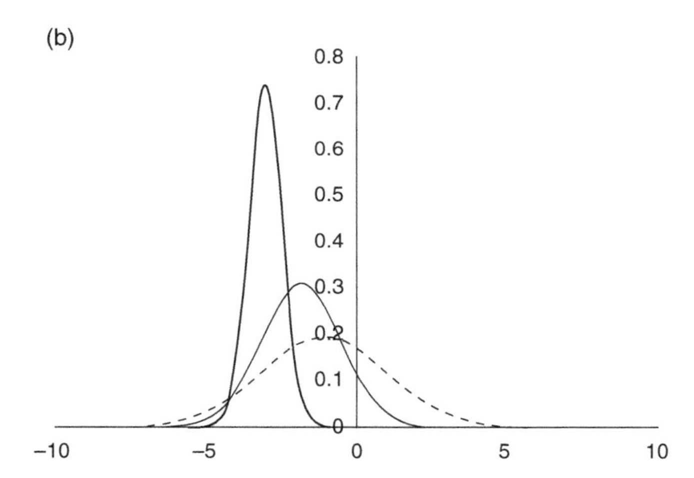

Figure 12.2 The probability distribution of speed (a) and location (b) for $x_0 = -5$, $v_0 = 0.5$, $\sigma = 0.1$, $\gamma = 0.1$ at $t = 5, 10, 15$.

Example 12.3 Suppose that a particle moves in a stationary and uniform external field, and g denotes the constant force associated with the field. The Fokker-Plank equation is

$$\frac{\partial}{\partial t}p(x,v,t) = \left(-v\frac{\partial}{\partial x} + \gamma + (\gamma v + g)\frac{\partial}{\partial v} + \frac{\sigma^2}{2}\frac{\partial^2}{\partial v^2}\right)p(x,v,t). \tag{12.14}$$

The presence of the field brings the speed to converge to a distribution with a nonzero mean. This results in a constant drift in the location distribution. Figure 12.3 illustrates numerically calculated approximations of the solution of (12.14). ■

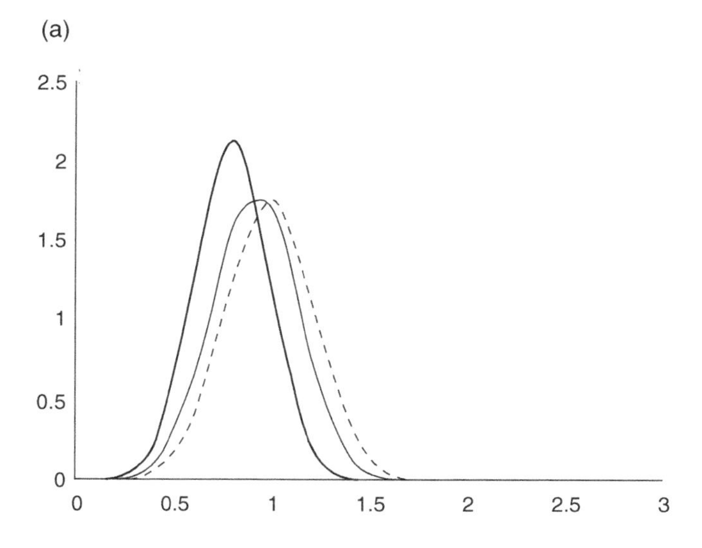

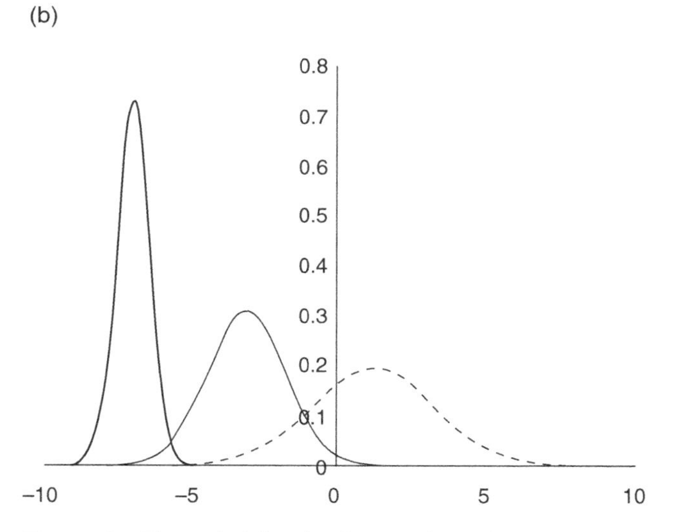

Figure 12.3 The probability distribution of speed (a) and location (b) for $x_0 = -10$, $v_0 = 0.5$, $\sigma = 0.1$, $\gamma = 0.1$, $g = 0.1$ at $t = 5, 10, 15$.

Example 12.4 Two particles, $j = 1, 2$. The particles generate identical fields, which magnitude at distance x is

$$\mathcal{U}(x) = \frac{d - |x|}{ax^2},$$

where $d > 0$ and $a > 0$ are given parameters. The force at distance x is

$$F(x) = -\frac{d}{dx}\mathcal{U}(x) = \frac{2d-|x|}{ax^3}.$$

Figure 12.4 plots the field, $\mathcal{U}(x)$, and the force, $F(x)$.

When at a high distance, the particles attract one another. At the distance $x = 2d$ the force changes the sign, and the particles repulse, when holding at a distance smaller than $2d$.

Figure 12.5 plots the speed and location distributions of the two particles for the parameters $x_0 = \pm 7$, $v_0 = 0$, $\sigma = 0.1$, $\gamma = 0.1$, $d = 1$, $a = 0.1$ at $t = 5, 10, 15, 20$. Figure 12.6 presents the impact of the field parameters, a and d, on the expected distance between the two particles. We observe that the smaller the values of a and d, the closer the particles approach each other. ■

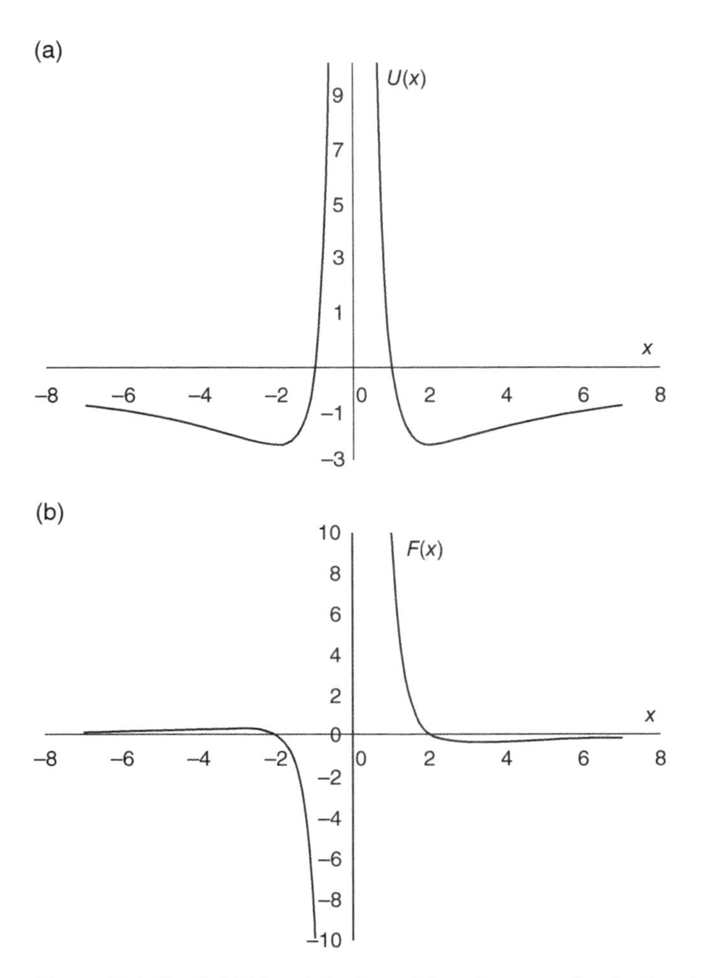

Figure 12.4 The field (a) and the force (b) at distance x for $d = 1$ and $a = 0.1$.

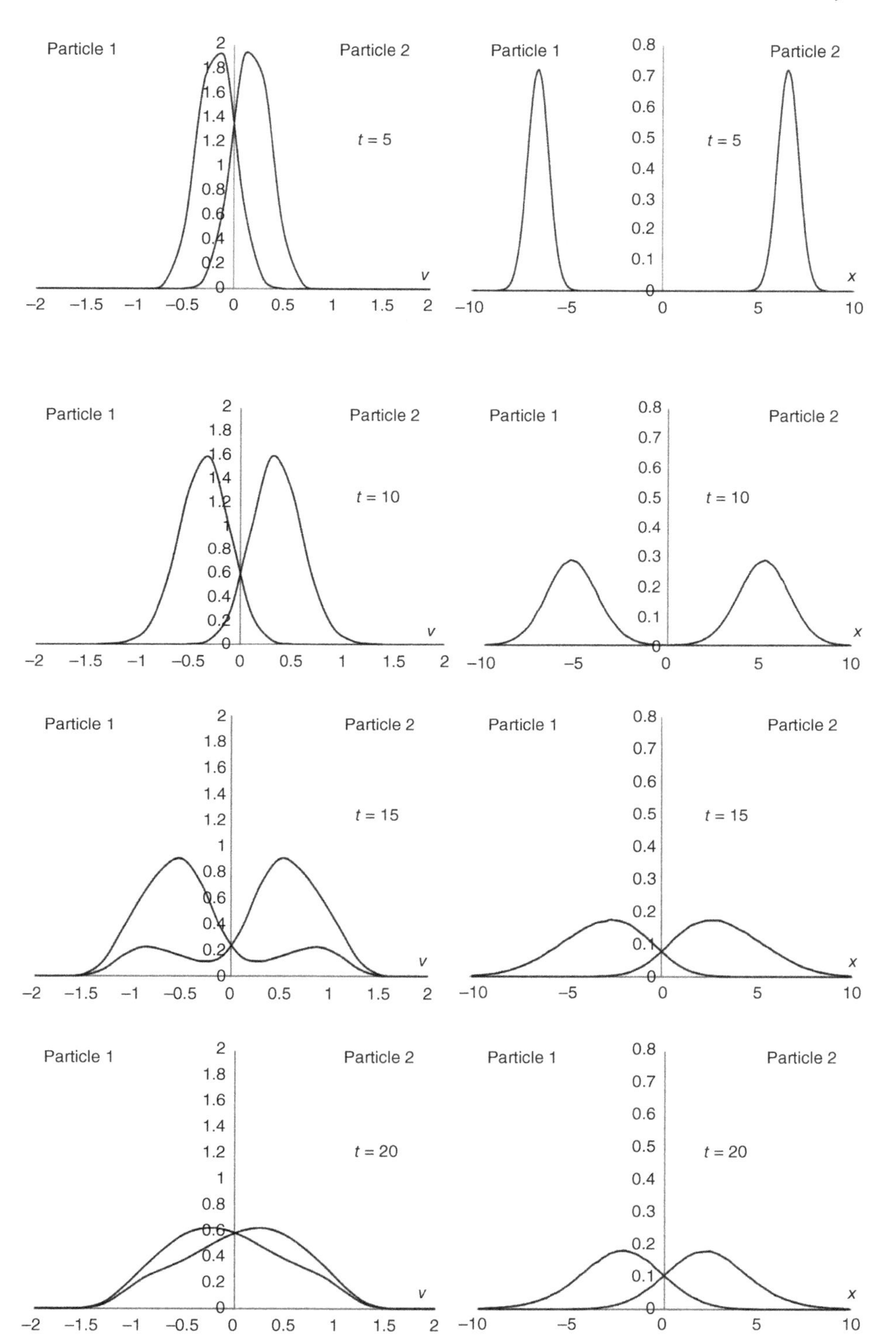

Figure 12.5 The probability distribution of speed (left) and location (right) of the two particles at $t = 5$, 10, 15, 20.

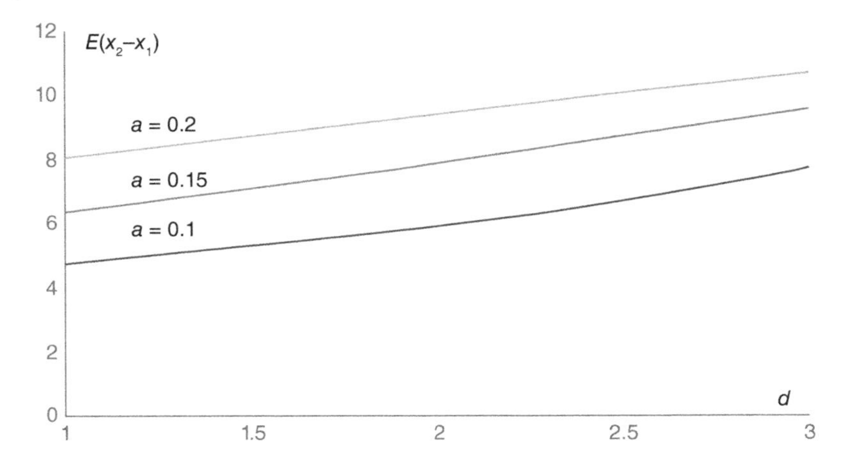

Figure 12.6 The approximate mean of the distribution of the distance between the two particles at t = 20.

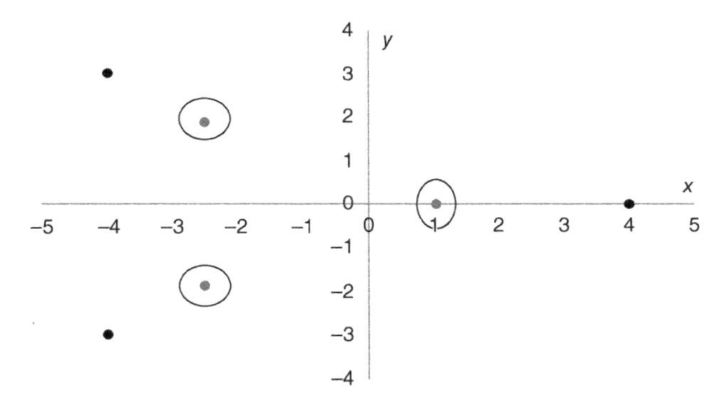

Figure 12.7 The initial and finial positions of the particles.

Consider now a group of particles, with each pair of particles interacting according to the mechanism just described. The next example shows the stable position of the particles that start moving from a given initial position.

Example 12.5 Figure 12.7 shows the initial position of the three particles (black dots); the initial speeds are set at zero. The final position fluctuates around the blue dots with the standard deviation schematically shown by the blue ovals. The values of the initial position as well as the values of the expected final position and its standard deviation are given in Table 12.1. This example used the following parameters: σ = 0.01, γ = 0.1, d = 2, a = 0.1. ■

Table 12.1 The initial and final positions of the particles.

Particle	Initial position (x,y)	Final position (x,y)	Std (x,y)
1	(−4, −3)	(−2.52, −1.86)	(0.47, 0.39)
2	(−4, 3)	(−2.52, 1.86)	(0.47, 0.39)
3	(4, 0)	(1.02, 0.00)	(0.35, 0.50)

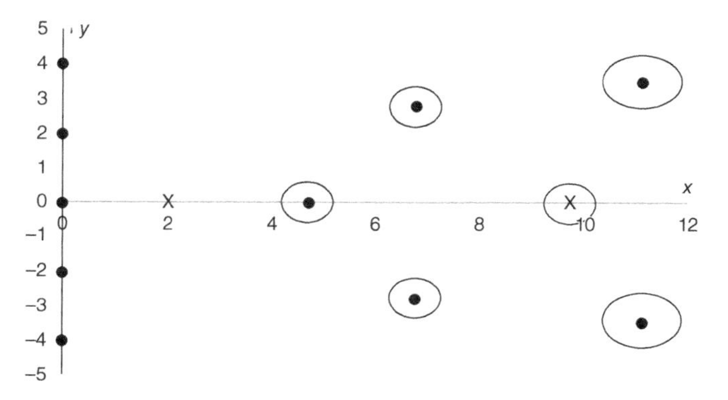

Figure 12.8 The initial and terminal positions of the particles.

Table 12.2 The initial and final positions of the particles.

Particle	Initial position (x,y)	Position (x,y) at T = 25	Std (x,y) at T = 25
1	(0, −4)	(11.13, −3.47)	(1.24, 0.96)
2	(0, −2)	(6.77, −2.79)	(0.58, 0.52)
3	(0, 0)	(4.73, 0.00)	(0.29, 0.35)
4	(0, 2)	(6.77, 2.79)	(0.58, 0.52)
5	(0, 4)	(11.13, 3.47)	(1.24, 0.96)
6 (leader)	(2, 0)	(9.72, 0.00)	(0.44, 0.22)

Example 12.6 Suppose one of the particles is a leader and the other particles are followers. The leader generates a constant force, g, that moves it along the x-axis similar to that in Example 12.3. This force has no impact on the other particles. Additionally, the leader and the followers create the fields described in Examples 12.4 and 12.5. However, the leader's field is set stronger (i.e., its parameter a is smaller) than the field of the followers to allow the latter not to stay too far behind the former.

Figure 12.8 shows the initial position of the leader (blue dot) and the five followers (black dots); the initial speeds are set at zero. The position of the particles after $T = 25$ time units is shown in Figure 12.8. The standard deviation is shown schematically by the ovals. The values of the initial position as well as the values of the expected final position and its standard deviation are given in Table 12.2. This example used the parameters $\sigma = 0.01$, $\gamma = 0.1$, $g = 0.1$, $d = 2$, $a = 0.1$ for the followers and $a = 0.02$ for the leader. ■

12.3 Summary

In this short chapter, we considered the model of the swarm dynamics by the use of Langevin and Fokker-Plank equations. These equations allow the description the swarm motion using the methods of statistical physics. The presented model cannot be considered as final or well-developed techniques; it only introduces the basic concepts and methods that require further studies, especially in the direction of the active Brownian motion.

References

Schimansky-Geier, L., Ebeling, W., and Erdmann, U. (2005). Stationary distribution densities of active Brownian particles. *Acta Physica Polonica B* 36 (5): 1757–1769.

Sjögren, L. (2015). *Lecture notes on Stochastic Processes*. Sweden: University of Gothenburg.

Tanski, I. A. (2004). *Fundamental solution of Fokker-Planck equation*. Retrieved from arXiv: http://arxiv.org/pdf/nlin/0407007.pdf.

13

Conclusions

Nir Shvalb, Eugene Kagan, and Irad Ben-Gal

In recent years, autonomous mobile robots have raised much interest in the research community as well as in industry, as they are the basis for various autonomous transportation vehicles. Compared with their human controlled counterparts, they offer advantages in regard to safety, maneuverability, and long-period endurances. Moreover, autonomous robots are likely to produce significant cost savings for commercial applications, such as inspection of infrastructures, autonomous delivery systems and autonomous security systems. These implementations are made possible due to significant advances:

1) The accessibility of various sensory systems (presented in Chapter 2 of this book) and a free-to-use global positioning system (Chapter 5) alongside with the availability of estimation and prediction methods, and new methods for environment mapping (see Chapters 3, 6, and 10).
2) The emergence of mechanical systems such as omni-wheeled mobile robots and aerial quadrotors (Chapter 3), which opens up new possibilities for in-person delivery and novel transportation opportunities.
3) The relative maturity of the motion planning research field (discussed in Chapters 1, 4, and 6) with advances in probabilistic robotics (Chapters 4, 6, and 7) that relate to motion planning where significant uncertainties in the environment, sensory systems, actuators, etc. are expected.
4) New energy sources and energy consumption reduction methods (Chapter 8).

The authors aimed to cover these main topics when considering all levels of design of such systems. The book also addressed issues pertaining to multiple agents. The benefits of this approach when compared to the use of a single agent are:

1) Tolerance for failures, implying an increase in the system's robustness.
2) Reduction in the time required for mapping.
3) Reduction of uncertainties as information is provided from several observation points.

Autonomous Mobile Robots and Multi-Robot Systems: Motion-Planning, Communication, and Swarming, First Edition. Edited by Eugene Kagan, Nir Shvalb and Irad Ben-Gal.

Companion website: www.wiley.com/go/kagan/robotsystems

4) Better coverage of a given region of interest as one minimizes the time required for event detection.
5) Design of "specialized" agents to enable complex missions.
6) A recent trend addresses the realization of multi-agent systems that could manipulate the environment as transporting a single object by several agents.

Much technological effort has been applied to cooperative swarms. Chapters 9–12 are devoted to the coordination of multiple agents. It is widely agreed that technical problems such as landing and taking off in different terrains in the aerial case, and stability issues in the planar case, should be handled individually by every agent. Nevertheless, the following hardware issues need to be resolved as well:

1) New platforms are required for various applications. An important example is the necessity for small-scale aerial agents for indoor operations, but notice that their expected payloads would be very limited.
2) Dedicated miniaturized sensors should be integrated, while the information provided by them should be fused with that of other available sensors. Also note that recent research advances show that indoor positioning will soon be available. Thus, in this regard, traditional estimation and analysis methods should be updated as well to utilize relevant sensory data.

Thus, the existing technology restricts the number of onboard sensors and the communication range.

The authors therefore believe that the decentralized motion-planning schemes such as those introduced in this book are a must. Moreover, these will enable swarm scalability and provide better reliability for the system, as opposed to systems that require a task allocation stage or full communication between the agents or systems that are controlled via a leader, a human control center, or a central processing system.

Index

Autonomous Mobile Robots and Multi-Robot Systems: Motion-Planning, Communication, and Swarming, First Edition. Edited by Eugene Kagan, Nir Shvalb and Irad Ben-Gal.

Companion website: www.wiley.com/go/kagan/robotsystems